AF685158

FIELD MANUAL:

FEDERATED SUNS™

FASA CORP.

2000

CONTENTS

CREDITS

Author
Christoffer "Bones" Trossen
Chris Hartford

Project Development
Randall N. Bills
Development Assistance
Bryan Nystul

Project Editing
Wyn Hilty
Davidson Cole

BattleTech Line Developer
Bryan Nystul

Editorial Staff
Editorial Director
Donna Ippolito
Managing Editor
Sharon Turner Mulvihill
Associate Editor
Robert Boyle
Assistant Editor
Davidson Cole

Art Staff
Art Director
Fred Hooper
Assistant Art Director
John Bridegroom
Project Manager
Brian Maslouski
Cover Art
Doug Chaffee
Cover Design
Brian Maslouski
Interior Color Section
Peter Bergting
Franx Vohwinkel
Illustrations
Loston Wallace
Christina Wald
Matthew Plog
Brad McDevitt
Layout
Brian Maslouski
Production
Matt Green
Andrea Fronc

Special Thanks To:

Bryan Nystul, for being foolish enough to give me this book; Chris Hartford, for putting in the extra effort and for creating the templates for both of us to use; both Bryan Nystul and Randall Bills, for not carrying through on their threats as the deadlines passed; Loren Coleman, for making me incredibly late (just kidding, buddy!) but at the same time giving me all those wonderful excuses to use; both Chris and Loren for setting the standard and helping me to (hopefully) reach it; and finally, all my family and friends, especially the members of the Milwaukee Emergency Services Squadron, for letting me neglect you while I spent (nearly) my every free waking hour working on this project.

Acknowledgements:

Both Chris Trossen and Chris Hartford would like to thank Bryan Nystul, Randall Bills and Dan "Flake" Grendell for spending all those late nights creating the spreadsheet, and the art department for making everything fit.

The rotary autocannon is based on an original weapon design by Jos Williams.

Much of the material found in this book is based on the work of the following authors, who have successfully shaped a living and breathing universe over the course of the past fifteen years: Adventure Architects, Randall N. Bills, Jim Brunk, Alex Bund, Loren L. Coleman, Robert Cruz, Tara Gallagher, C. R. Green, Dan "Flake" Grendell, Thomas S. Gressman, Chris Hartford, Karl Hiesterman, Todd Huettel, Christopher Hussey, Donna Ippolito, Evan Jameson, J. Andrew Keith, William H. Keith, Jr., Dale L. Kemper, Jeffrey Layton, Michael Lee, Sam Lewis, Sharon Turner Mulvihill, Jim Musser, Bryan Nystul, Mike Nystul, Blaine Lee Pardoe, Geoff Pass, Tom Peters, Boy F. Petersen, Jr., Diane Piron-Gelman, Rick David Stuart, John Thiesen, Robt. Wells, and last, but certainly not least, L. Ross Babcock, Jordan Weisman and Michael A. Stackpole. (Whew!)

HOUSE OF CARDS

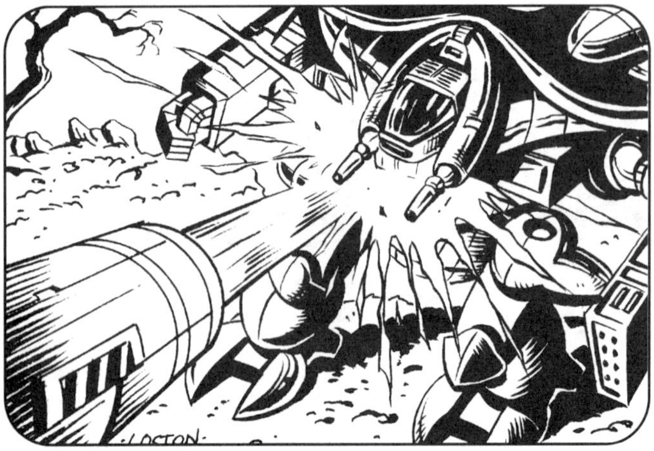

Radcliffe
Kathil, Capellan March
Federated Commonwealth
0248 Local Time, 1820 TST
16 November 3062

"CAPFlight 4825, this is Badger 320. We have concentrated Redfox activity eight-five-zero mikes due north of our position, grid zed-three-seven. Request special inspection."

"Roger three-two-zero, stand by for inspection. Two vipers inbound. ETA: forty-five seconds."

Leftenant Colonel Damien Zibler could barely make out the acknowledgement of his request over the din of the battle. Off to his right, a lance of *JagerMech*s churned out a steady hail of projectiles into the midst of an enemy formation, their tracers eerily lighting the pall of smoke hanging over the battlefield. A lightning bolt suddenly lashed out from one of his *JagerMech*s, briefly illuminating the scene of destruction as it slammed into a *Zeus*, showering the enemy group with sparks. Zibler watched flames erupt from the massive BattleMech's torso, soon engulfing the entire machine as its ammunition bins cooked off.

Behind him, he heard a succession of sonic booms, followed by the unmistakable spark-crunch of the hypersonic nickel slugs fired by his *Alacorn*s slamming into the already beleaguered group. *They won't last long against that fusillade.* He quickly turned his attention back toward the hills ahead of him and the two companies of fresh BattleMechs approaching his battered battalion. *If those fast-movers don't do something soon, we're history!*

He made the mistake of glancing at a secondary monitor, which showed him just how bad off his own *Victor* was. Armor had been breached in almost every location, and he didn't need his computers to tell him his reactor shielding was barely maintaining containment. Unfortunately, he knew the rest of his battalion was faring little better. *How the hell did we get ourselves into this mess? Katherine Steiner, that's how!*

Already the two companies belonging to the Eighth FedCom RCT—codenamed Redfox Group—were approaching attack range, and Zibler's air support was still some thirty long seconds away. "Badger Battalion, fall back to position Echo."

As the remainder of his battalion began to retreat, weapons fire lit up the advancing line. Zibler unconsciously braced in his command couch, dropping his cross hairs over the silhouette of an approaching *Salamander*. *Jim ... Jim Wendt. He was a royal ass at the academy, but ... Damn it! How did this ever happen?* He squeezed the firing stud before he even realized what he was doing, the thunder of his Gauss rifle further adding to the cacophony of the battlefield an instant before his own 'Mech was engulfed in a ball of fire. He felt more than a dozen missile impacts send chunks of armor and myomer flying, while four times that many exploded around him.

He's trying to take this out on me! Is that entire unit full of lunatics? Ordered to relocate to Halloran V, the Eighth RCT had instead launched a surprise attack on the Kathil CMM, claiming to be enforcing Archon Katherine's rightful rule over Kathil. Zibler's was one of the first Militia forces hot-scrambled to intercept the Eighth's advance toward Radcliffe.

Four of his 'Mechs were on the ground, unmoving. Two others and a tank rapidly were being consumed in a hail of missile explosions. A few more had simply broken formation and fled. *I don't blame 'em. In a stand-up battle, we're a match for the Eighth. Ambushed like this, though ...*

Swiftly closing the distance, the Eighth's 'Mechs began scoring more hits upon Zibler's rapidly thinning ranks. He punched down on his firing stud again. And again. And yet again, loosing a hypersonic slug each time into the enemy advance. *The enemy. I graduated with him.* He unconsciously stopped his own retreat, standing his ground astride a demolished mansion. *How did it come to this? No more!*

Beside him, one of his *JagerMech*s erupted into a ball of fire. He stabbed down one more time on the firing stud, smashing shards of armor from the torso of Wendt's *Salamander* before the paired lasers of an Eighth F-C *Penetrator* severed the mangled remains of his own *Victor*'s right leg, sending him crashing to the ground.

Barely aware of the alarms warning him of his reactor's rapidly degenerating magnetic bottle, Leftenant Colonel Zibler watched as the entire enemy advance burst into the telltale wall of flame that only an incendiary bomb could produce. The two Militia *Corsairs* climbed away on their own tongues of blue flame. The center of the advance, once occupied by Wendt's *Salamander*, spit forth only a staccato of debris.

In that moment, the heat within the cockpit became so searing that Leftenant Colonel Damien Zibler no longer felt a thing.

No more!

INTRODUCTION

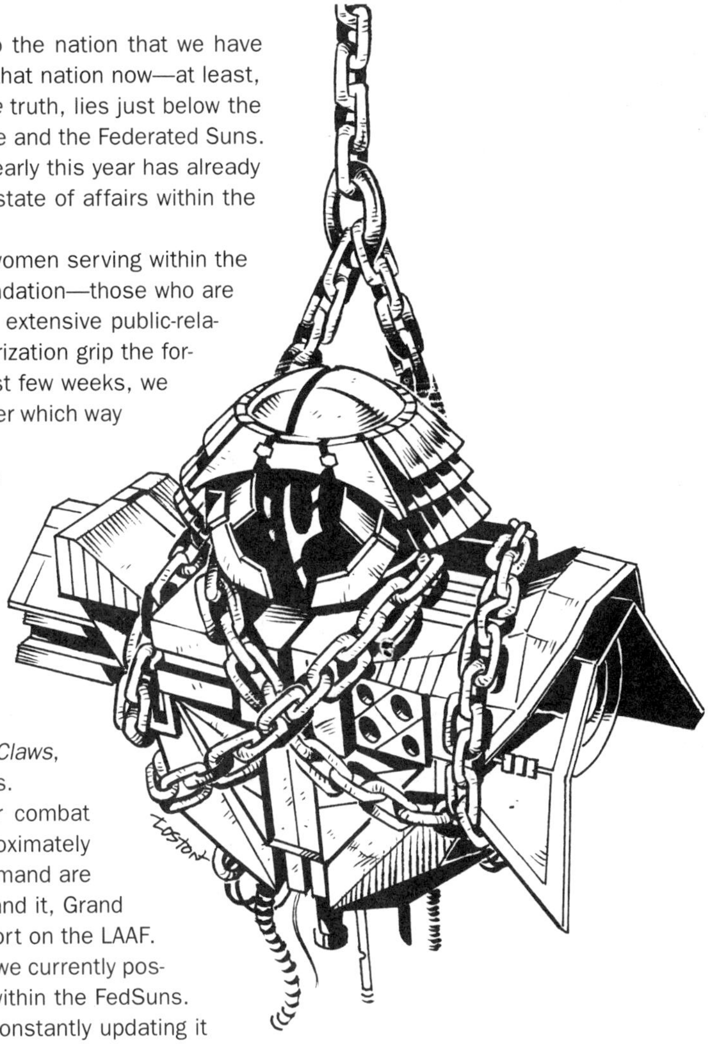

Victor,

These past several years have not been kind to us or to the nation that we have served for so much of our lives. We are virtual pariahs within that nation now—at least, so your sister would have everyone think. The truth, the *whole* truth, lies just below the layer of deceits your sister has built up within the Lyran Alliance and the Federated Suns. I certainly do not need to tell you that—Jerrard's report from early this year has already provided you with a clear and concise briefing on the current state of affairs within the former Federated Commonwealth.

What you do need to know is that many of the men and women serving within the AFFC and the LAAF are beginning to see the cracks in the foundation—those who are willing to look beyond the rosy picture painted by Katherine's extensive public-relations staff, that is. During the past year, we have seen a polarization grip the former Federated Commonwealth. And with the events of the past few weeks, we have reached a turning point, one with a bleak outlook no matter which way we head.

The information contained within this document will help us forge ahead over the next several years. I strove to paint as nonpartisan a picture of the current state of the Armed Forces of the Federated Commonwealth as I could, giving all those concerned a true glimpse of the strengths and weaknesses gripping our nation's military.

The first section of this document, entitled *Evolution of the Fox*, details the history of the Armed Forces of the Federated Suns. The second, *The Fox's Den*, briefly analyzes the structure of the current AFFC, while the third, *Sharpening the Claws*, provides a look at the training and education provided our troops.

The fourth and final major section reviews every major combat command traditionally associated with the AFFS. Only approximately half of the units serving within the Federated Suns State Command are actually covered within the pages of this report. As I understand it, Grand Duke Kell's people have taken a similar approach to their report on the LAAF. Per your request, I have forwarded Morgan all the information we currently possess on the LAAF and the traditionally Lyran units operating within the FedSuns.

As you know, I have delayed this report for some time, constantly updating it with the most current information available from our sources. But I am afraid that the time to disseminate this information has finally come.

Please do not misunderstand. I love the Federated Suns. I stood by your parents through some of the worst fighting this galaxy has ever seen because I understood it was for the best, even though in my heart I only wanted to see the bloodshed end. I did not take action against Katherine because to do so would only have plunged the FedCom into civil war. It has now gone too far, however. Since the Solaris incident, many troops loyal to you have begun a passive rebellion against Katherine by reverting to the use of the traditional Federated Suns ranks.

I believe we can no longer stand by and watch our homeland be destroyed from the inside. I do not want to see even one innocent lose his or her life because of your sister. Unfortunately, the truth of the matter is that if we do not take action now, more will suffer in the long run.

It is the decision of the damned, and the man who makes that decision is the chief damned soul. God help us. God help us all.

Ardan Sortek, Major General
Director of Operations, SLDF
6 December 3062

EVOLUTION OF THE FOX

The history of the Armed Forces of the Federated Suns is a long and distinguished one. It is a history full of heroes and glorious victories, conquerors and successful campaigns, crippling losses and earth-shattering defeats. Above all, it is a history that is tied inextricably to that of the Prince of the Federated Suns, for in all of these extreme successes and failures, the Prince has always been at the forefront, ultimately accepting the praise or blame for all the AFFS' actions.

Throughout the years, the AFFS has evolved and matured. Once a weak defense force, it has grown into the largest and most powerful military in the Inner Sphere. As the Prince's protector and the source of his wide-reaching power, the AFFS' influence and prestige increased alongside that of the Prince. Even in times of terrible crisis, the men and women of the AFFS have given their Prince their wholehearted support, knowing that he or she would return that trust, protecting them with his or her life, if necessary.

The destiny of the AFFS once seemed to be that of conqueror of the Inner Sphere, with First Lord Hanse Davion at its head. With the marriage of the Federated Suns and the Lyran Commonwealth, and the resulting union of their armed forces, more than half of the Human Sphere came under Prince Davion's influence. The combined Armed Forces of the Federated Commonwealth was the largest single military since the demise of the Star League. Unfortunately, circumstances of history, or perhaps the Powers That Be, prevented the AFFC from fulfilling Prince Davion's dream.

Stymied first by the audacity of the Dragon and then by a surprise invasion from afar, the AFFC was ultimately blocked by familial greed. Today the AFFC, or at least the Armed Forces of what was once the Federated Suns, is a fractured and in some respects broken organization. In this era of rampant nationalism, forces loyal to the Federated Suns are scattered across the Inner Sphere, many of them cut off from support and supply by the Archon who would be Princess. Though storm clouds are unmistakably looming on the horizon, the men and women of the Federated Suns' military will survive to come together again, and in doing so will prove to all the continued strength of the Federated Suns and its true Prince.

To understand the import of these words, readers must know what it is to be a son or daughter of the Federated Suns. Only then can they begin to realize the truth. This is not an exhaustive and detailed history of the Federated Suns; many hundreds of volumes have already been compiled on that subject. It is meant solely to highlight those events that have been of significant import to this nation and its armed forces.

BIRTH OF A SUN

As with the other Successor States, the genesis of the Federated Suns lies within the Terran Alliance's Democratic Proclamation and the subsequent formation of the militant Terran Hegemony. At the heart of a region of space known as the Crucis Reach lay the simple agrarian world of New Avalon. Rich with abundant natural resources, New Avalon quickly became the region's bright star. The looming collapse of the Terran Alliance brought on a rebellion that would forever change the shape of the region, and ultimately the Human Sphere.

Though New Avalon had won its independence following the Grain Rebellion, the planet would not experience peace for some time. A civil war gripped New Avalon as several powerful families battled the will of the people for dominance of the important world. Ultimately, the citizens took back control for themselves. Through it all, the members of the Davion family had fought long and hard for the freedom of New Avalon's people. Consequently, it was no surprise when Lucien Davion rose to the position of New Avalon's Prime Minister. It is for what he accomplished afterward, however, that history remembers him.

PACT AGAINST THE DEVIL

The long, downward spiral of the Terran Alliance ended abruptly in 2314, when Admiral James McKenna stepped in to end the Alliance Civil War. Under his leadership, a resurgent Terra quickly put its troubles behind it and looked back toward the stars.

Realizing that this new Terran Hegemony could reconquer the entirety of the Human Sphere if Terra's former colonies did not stand together, Prime Minister Lucien Davion committed himself to forging just such an alliance. For the next three years, he traveled extensively throughout the Crucis Reach, visiting worlds and paving the way for what was to come. By 2317, he was finally ready to make his grand goal a reality. The leaders of twenty worlds signed the Crucis Pact, forming the Federated Suns. Lucien Davion was overwhelmingly selected as the Federation's first president.

At first, the Federation existed only as a loose trade and mutual defense alliance among the Pact signators. Davion was not satisfied with that, however, and over the course of the next decade and a half, he worked to tighten the bonds between member worlds and to bring more worlds into his new nation, one that could stand up against the might of the Terran Hegemony.

FATEFUL FIRST BLOOD

Ironically, the Suns' first threat came not from Terra but from the much closer world of Chesterton. What should have been a minor border dispute turned into a disastrous campaign that cost the Federated Suns much of its prestige and Lucien Davion his position. His younger brother Charles, however, proved himself worthy of the Davion name and the Suns' presidency.

One of Charles' first acts was the formation of the Federated

Peacekeeping Force, giving the president near-autonomy over a military force loyal to the state, not to individual worlds. In the years that followed, both the Federated Suns and the FPF grew in strength, and with them the power of the Davion family.

It was the Federated Suns' third president, the son of Lucien Davion, who firmly cemented the destiny of the nascent nation and its military, however. Though Charles Davion had begun to build a military capable of dealing with the ongoing war along the Capellan Frontier, he did not survive to see his plan come to fruition.

In 2344, new President Reynard Davion left New Avalon to take command of the FPF units along the Capellan Front, marking the first, but certainly not the last, time the nation's leader would personally lead his troops into battle. Reynard's bold assault succeeded, not only in ending the war on favorable terms, but also in firmly establishing the might of the Federated Suns' military. Unfortunately, the Suns' lengthy feud with the Capellan Confederation also stems from these early events. Following Reynard's death in 2371, his son Etien took up the reins of leadership without opposition, thus establishing the Davion family's unquestioned claim to lead the Federated Suns.

DYNASTY DEFICIENT

While Etien Davion's elevation to the presidency secured his family's dynasty, it also proved that not every member of the family was qualified. During the next four decades, the Federated Suns was virtually crippled by a succession of Davion presidents who either could not or would not handle the reins of leadership. Also during these years, the entire Human Sphere plunged into the Age of War, further straining the already troubled nation.

While two of the five presidents who succeeded Reynard proved themselves worthy of the Davion name, the other three nearly brought their nation down with their insanities and tyranny. Edward Davion was the worst of Reynard's immediate successors. While he served less than three years as president, he had held a tight grip on the federation for many years while his twin brother ruled. Together, these brothers nearly tore the nation apart with their ludicrous decrees. Indeed, the people of the Federated Suns were on the brink of civil war when Simon Davion, cousin of the Twin Tyrants, stepped in.

PRINCE OF THE PEOPLE

When he ascended to the presidency following his twin brother's death, Edward Davion moved to make sure no other Davion could challenge his position. His purges missed his cousin Simon, however.

In service with the FPF on the Capellan Frontier, Simon Davion was an able and well-liked officer; he had represented the Federated Suns during the historic summit that led to the signing of the Ares Conventions in 2412. Recognizing the growing conspiracy against his cousin, or perhaps even a member of it, Simon returned to New Avalon in 2417 and removed his cousin from power, assassinating him on the floor of the High Council. Simon soon began to make the changes necessary to keep the Federated Suns together, with far more radical results than anyone had imagined.

Succeeding the man he had murdered as president, Simon Davion dispensed with the excessive bureaucracy that had never served the Federated Suns well. In its place he set up a more feudalistic order, with a Prince controlling each of five different Marches, and himself as their First Prince. Though the Davions had ruled as a virtual autocracy, these reforms officially ceded this power to the First Prince, and by extension their dynasty. Prince Simon's brilliant career was marred, however, by the resumption of hostilities with the Capellan Confederation and an invasion by the Terran Hegemony, spearheaded by their newly designed BattleMechs.

CIVIL WAR

The first cracks in this new order appeared half a century after Prince Simon came to power, when an assassin's bullet cut short his son's reign. With no clear successor in place, Richard Varnay, an important member of the Federated Suns government and descendant of one of the key figures in the New Avalon Grain Rebellion, demanded the Principality be handed to him. His audacity worked against him, however, and Ellen Davion was elevated to First Princess in his stead. Though the new Princess stripped him of much of his power, it would not be the last time the Varnay name would haunt the Davions.

Less than half a century later, Prince William Davion succumbed to a virulent disease, leaving his five-year-old son, Alexander, as his only heir. On his deathbed, the Prince named his two sisters as regents until young Alexander came of age. Unfortunately, the Prince's sister Cassandra was married to David Varnay, the heir to Richard Varnay's title—and his aspirations.

Varnay easily persuaded his wife to appoint him to the Council of Regents the two sisters were forming. To counter the Varnays, Laura Davion appointed General Nikolai Rostov, a respected and admired military commander. The fifth and final member was the ill widow of Prince William, Carmen Estevez Davion.

While the Council of Regents was in theory tasked with governing the Federated Suns until Prince Alexander came of age, David Varnay quickly used his and Cassandra's positions on the Council to advance his political aspirations. Laura blocked most of Varnay's maneuvers while Prince William's widow was alive. When Carmen's health finally failed her, however, Laura was forced to struggle against Varnay's machinations alone. She and Varnay engaged in an active political battle for Rostov's support that lasted for years.

Time was against Varnay, however. As the years rolled on, Alexander began to approach his majority, and Varnay's chances of seizing power lessened. As a measure of security, Varnay held the young Prince on New Syrtis as a virtual hostage and betrothed the youth to his niece.

The war raged on across more than a hundred worlds for some twelve years, leaving millions dead in its wake. With the support of Dmitri Rostov, son of General Nikolai, Prince Alexander Davion first smashed Laura Davion's forces in her Draconis March before turning his attention to his other aunt. This campaign was far more difficult and was further exacerbated by Rostov's refusal to commit all of his available forces to Alexander's cause.

It was Cassandra's attack on her nephew that ultimately brought Rostov back into the war. Reinforced by fresh Capellan mercenary troops and perhaps quietly supported by Confederation interests, Cassandra's offensive penetrated deep into Prince Alexander's Crucis March. Sensing an opportunity to advance his own interests within the Terran March, Rostov spearheaded a counterattack along Cassandra's left flank while Prince Alexander launched his own assault along her right flank. Cassandra's rebellion disintegrated soon thereafter, and she fled to the Capellan Confederation.

Though finally victorious in his long struggle against his aunts, Prince Alexander had one more opponent to deal with. Three years after delivering the death blow to Cassandra Varnay's forces, Dmitri Rostov launched his own coup attempt. In doing so, he sorely misjudged both his own popularity and the young Prince's resolve. When he learned that his wife had been killed by Rostov's forces, Prince Alexander mounted a furious assault, quickly ending the abortive coup when he personally dispatched the general on the battlefield.

REVOLUTION

Unfortunately for Varnay, his niece recognized his plan and betrayed him, allowing Alexander to escape. At the same time, Laura Davion drew the lines of battle against Varnay: she declared him an enemy of the state and set the Federated Suns' armed forces against him. At the same time, Varnay arrayed his own forces against Laura Davion's "illegal occupation force" on New Avalon. Caught in the middle, of course, were the people of the Federated Suns.

The Davion Civil War quickly became one of the bloodiest and hardest-fought wars on record, approaching the same level of bloodshed as the Succession Wars. Both General Rostov and David Varnay were casualties of the fighting, as were any last shreds of Laura Davion's loyalty to her nephew. The war soon degenerated into a three-way battle, with Prince Alexander struggling against his two aunts for control of the Federated Suns.

REFORMATION

Throughout the long years of warfare, Prince Alexander had one constant: the loyalty of the Federated Suns military. While it was the Davion name that gained him his first support within the military, it was the Prince's dedication to his men that ultimately allowed him to prevail. During his long struggles against his aunts, and later against Dmitri Rostov, the Prince continually won converts to his cause from forces supposedly loyal to his opposition.

Following Rostov's defeat, Prince Alexander instituted a number of sweeping reforms throughout the Federated Suns, the most radical of which were the abolition of Simon Davion's problematic system of five regional Princes and the elimination of the Terran and Outer Marches. Now there was only one Prince, wielding complete power over the entire Federated Suns and its three remaining Marches. Prince Alexander also assumed the title of Marshal of the Armed Forces of the Federated Suns, giving him absolute control of his nation's military. At the same time, he created a select group of military advisors and mandated that each potential successor would be required to serve at least five years in the AFFS before he or she could assume the position of First Prince. These reformations won the complete loyalty of the AFFS for generations to come.

Commonly considered the Federated Suns' first modern Prince, Alexander Davion ruled for the next six decades without major incident. He saw the Federated Suns through the end of the Age of War and into full membership within the Star League, quickly rebuilding his nation from the ashes of the Civil War.

GOLDEN YEARS

As a part of the Star League, the Federated Suns experienced nothing but growth. Though the Star League Defense Forces drew many of the Suns' young men and women into its ranks, the AFFS gained countless technological wonders in return. Unfortunately, the Federated Suns and its Princes also acquired a host of "peacetime" problems for which centuries of war had not prepared them.

IN BED WITH A DRAGON

In the years of detente among the member nations of the Star League, the unthinkable happened—Mary Davion, the firstborn daughter of Prince Roger Davion, fell in love with and married Soto Kurita, the youngest son of Draconis Combine Coordinator Urizen II. Rather than see the leadership of his nation potentially fall into the hands of the Kuritas, Prince Roger issued a law of succession that prevented his daughter or any of her offspring from claiming the Principality. Unfortunately, the Combine did not leave the matter alone.

After the deaths of Prince Roger Davion and Mary Davion-Kurita, Coordinator Takiro Kurita, son of Urizen II, set his plot into motion. The Coordinator objected to the proclamation of Mary's nephew as heir to the Principality in the Star League Court, demanding that Mary's son Vincent be named heir.

Unwilling to take an active role in what they deemed an internal dispute, the Lords of the Star League instead let the Combine and the Suns settle the matter themselves. This touched off the first of the so-called Council Wars, as the Combine launched an invasion in 2725, intent on taking the Federated Suns one way or another.

Coordinator Takiro's swift attack did more damage to the Combine than good, however, as it turned the Star League Council against his nation. The SLDF stepped in and quickly ended the conflict, forcing the Combine to pull back to its pre-war borders.

While the Star League officially dismissed any Kurita claim to the Federated Suns throne, this brief war had another significant consequence for the AFFS. Prince Richard Davion secretly and unofficially began to build up his military, using legal loopholes to bypass the Cameron Edict of 2650, which limited the size of the militaries each Star League member state could field. Though against the spirit and intent of the Star League, this action also kept the Federated Suns alive during the Succession Wars.

THE DEVIL AND HIS DUE

On 27 December 2766, Stephan Amaris staged a bloody coup, executing First Lord Richard Cameron and taking control of the Terran Hegemony. SLDF General Aleksandr Kerensky immediately launched a campaign that first destroyed Amaris' own Rim Worlds Republic and then took back the Hegemony and its capital world. Though many AFFS personnel, including some entire units, left their posts to join Kerensky on his long crusade, none of the Star League Council Lords, including Prince John Davion, lent any official military support to the general. Instead, they each plotted their own overthrow of the Star League.

Victorious over the usurper Amaris but stripped of his title Protector of the Realm by the Council Lords, General Kerensky led the majority of the SLDF away from the Inner Sphere rather than allow them to be drawn into the wars he could see looming on the horizon. Without Kerensky and his SLDF to stand in their way, each of the Council Lords in turn claimed the Lordship of the Star League and launched devastating military campaigns against his neighbors. Thus began the first of the Succession Wars.

TOWNE AND KENTARES

The first priority of every Successor State military was to capture the worlds of the former Terran Hegemony. Though General Kerensky had spirited away a majority of the SLDF, he left behind a vast network of warehouses and manufacturing facilities that would be invaluable to any military force. Prince John Davion gave his AFFS the mission of securing the important world of Towne for the Federated Suns.

Unfortunately for the Prince, the years of peace under the Star League had resulted in massive military cutbacks. By the time the First Succession War reared its head, the AFFS was a mere shell of its former self. The AFFS valiantly fought off the Confederation's and the Combine's first assaults but fell before the Combine's second invasion force. This debacle signaled Prince John that his military was no longer the force it once was. He immediately began a crash program to rebuild his army.

At the same time, Prince John planned a massive assault upon the Capellan Confederation to make his military seem stronger than it was. As this assault began, the Combine launched its own offensive along its entire border with the Federated Suns. Taken aback, the AFFS eventually regrouped and mounted its own counterattacks. While the first two faltered and the Combine pushed even farther into the Suns, the third counteroffensive was seemingly blessed, first destroying a major Combine supply depot on New Rhodes and then killing Coordinator Minoru Kurita on Kentares. The Combine's offensive faltered as the nation reeled at the loss of its leader.

In response to his father's death, Coordinator Jinjiro Kurita ordered the extermination of Kentares' entire population. Though some DCMS soldiers refused to follow these orders, the majority did not, and more than 52 million innocents were murdered. This despicable massacre united and invigorated the AFFS, whose new fury finally halted both the Combine and Capellan incursions.

During this brief respite, Prince Paul Davion, who had ascended to the throne after an assassin took his father's life, continued the reformations within the Federated Suns military. Assuming many of the powers formerly granted to the Star League's First Lord, the Prince soon forged the Federated Suns and its military into a united body that could stand against its enemies.

WAR THROUGHOUT THE AGES

The Succession Wars continued for the next two and a quarter centuries, ruining hundreds of worlds, bankrupting the economies of the Successor States and reducing House militaries to a mere shell of their former selves. The Armed Forces of the Federated Suns proved themselves the most capable military in the Inner Sphere, regaining all of their lost territories and more.

During the height of the Third Succession War, Princess Melissa Davion made several radical changes within the AFFS, the most significant of which were the new focus on combined-arms warfare and a streamlining of its bureaucracy. The AFFS emerged from these reforms stronger than ever. The Federated Suns nevertheless experienced a number of internal difficulties throughout the remainder of the Third Succession War, from the near rebellion Prince Joseph experienced when he attempted to denude the MechWarrior Brotherhoods of some of their power—and the one his son Peter faced from the Warrior's Cabal, when he strove to finish what his father had begun—to Prince Andrew's action against mutineers in the Draconis March.

SUNS DOMINANT

The death of Prince Andrew Davion and the ascension of Ian Davion to the throne ushered in a new era for the Federated Suns. The Prince, commonly known as "The Hound," and his brother Hanse, known as "The Fox," commanded their nation's military from the front, which breathed renewed life into the AFFS' demoralized troops. The Suns won a number of major victories because of the devotion and skill of these brothers, and even when Prince Ian died on Mallory's World fighting to protect his men, the AFFS did not miss a beat.

THE FOX ASCENDANT

Hanse Davion became First Prince of the Federated Suns in 3013 after Ian's death, following an attempt on his life by radicals within the Capellan March. One of his first actions as Prince was the creation of the New Avalon Institute of Science, a university devoted to the recovery of technologies lost during the long centuries of the Succession Wars. Though constantly targeted by dissident elements within Duke Michael Hasek-Davion's Capellan March, the Prince's attention was focused almost entirely upon the fate of his nation. As a result, he only narrowly avoided an elaborate Capellan plot to replace him with a double in the last days of the Third Succession War, which would undoubtedly have delivered the Federated Suns into Chancellor Maximillian Liao's lap.

Of course, Prince Hanse's greatest accomplishment was the alliance between his Federated Suns and Archon Katrina Steiner's Lyran Commonwealth. Signed in secret on Terra in 3022, the Federated-Commonwealth Alliance Document truly signaled the beginning of a new era. The leaders of the Inner Sphere's other nations scrambled to enact their own mutual defense pact in an effort to counter this new alliance.

CONQUERING THE STARS

Prince Hanse Davion and Archon-Designate Melissa Steiner were wed on 20 August 3028, signaling to all the official alliance between the Federated Suns and the Lyran Commonwealth. Simultaneously, Prince Hanse Davion set his AFFS on a massive military operation aimed at winning him the Capellan Confederation. This Fourth Succession War was the shortest of the Succession Wars but also the most fiercely fought. Where limited long-term operations had become the norm in the Inner Sphere, Prince Hanse's Operation Rat broke the mold by dropping dozens of combined-arms regiments onto each world, quickly overwhelming any opposition.

By the end of this war, in which the armed forces of each Successor State became involved, the Federated Suns had crushed the Capellan military and taken half of the Confederation's worlds. Though not as successful, the Lyran Commonwealth, too, won victories against the Free Worlds League and, more importantly, the Draconis Combine.

Nine years later, Prince Hanse attempted to repeat his triumph, this time against the Draconis Combine. The Prince was not as successful this time. A surprise Combine counterattack lead by then-Deputy of Military Affairs Theodore Kurita stymied the AFFS advance and ultimately persuaded Prince Hanse to halt his invasion.

A NEW AGE OF WAR

Though the Federated Commonwealth would not officially come into being until the firstborn son of Hanse and Melissa Davion came of age and took up the leadership of both nations, the armed forces of the two realms merged in 3042. Under a united High Command of the Armed Forces of the Federated Commonwealth, members of the former AFFS trained extensively with their Lyran counterparts, steadily bringing the AFFC's level of training up to that of the AFFS. At the same time, procurement officers scrambled to supply the AFFC with the technologies being unlocked by Prince Davion's NAIS.

Even with these almost-prescient preparations, the AFFC

could not stand against the Clans when they invaded in 3049. The AFFC fell back in the face of the Clans' devastating assaults as Prince Hanse scrambled to strip garrisons from the Federated Suns State and reposition them within the Lyran State. These reinforcements helped slow the Clan advance and even allowed the AFFC to win several victories against the invaders, but it was only through luck and the audacity of ComStar that the invasion was ultimately halted. Because of their sacrifice, the Inner Sphere gained a fifteen-year respite from the full wrath of General Kerensky's children.

The Federated Commonwealth felt even greater pain with the end of the Clan invasion. Prince Hanse Davion fell to a heart attack just after the Truce of Tukayyid, while a bomb claimed the life of Archon Melissa Steiner-Davion just a few years later. Thus was the Federated Commonwealth born, not amidst celebration but mourning.

END OF INNOCENCE

The Clan invasion had far greater implications for the Inner Sphere than just the loss of worlds. It forced the leaders of the Successor States to work together for the common good, a level of cooperation not seen since the Star League era. The Federated Commonwealth and their long-time enemy the Draconis Combine fought side by side against the Clan horde while the manufacturing might of the Free Worlds League kept both armies supplied with the equipment they needed to continue the fight.

In order for League Captain-General Thomas Marik to agree to help the Commonwealth and the Combine, Prince Hanse had to offer the services of his NAIS to help Marik's young son, Joshua, who was suffering from leukemia. Even after the Truce of Tukayyid, the Captain-General allowed the shipments to the Federated Commonwealth to continue. In 3057, however, Joshua Marik's health turned worse, and leukemia quickly claimed the child's life.

Rather than lose the manufacturing power of the Free Worlds League and risking his nation falling to the Clans, Archon-Prince Victor Steiner-Davion put a double in Joshua's place, hoping to convince Captain-General Marik his son still lived. Unfortunately, Marik discovered the substitution and, mistakenly believing that the Archon-Prince was attempting to place a pawn in the League's line of succession, launched a punitive war against the Federated Commonwealth. Hoping to capitalize on Marik's operation, Capellan Chancellor Sun-Tzu Liao set his own nation on the Federated Commonwealth as well. No one could have anticipated the results of this war.

This coordinated assault targeted the Federated Commonwealth's Sarna March, the region created from the Commonwealth's gains during the Fourth Succession War. With a majority of the AFFC reserves still stationed along the Clan front, the Sarna March quickly degenerated into the aptly named Chaos March. While each of the nations involved had claimed a portion of this region by the end of the brief war, none had the military forces it needed to take the rest. The remaining worlds had gained their independence, whether they wanted it or not.

At the same time, with confusion gripping the entire Federated Commonwealth, Katherine Steiner-Davion seized power on Tharkad and seceded the Lyran State. Soldiers in the Federated Suns and Lyran States were suddenly unsure of who their commanders were or where to place their loyalties.

The unrest simmering in the former Federated Commonwealth could well have escalated into civil war had Prince Victor Steiner-Davion not stepped in to quash the rebellion before it began. While countless AFFC soldiers stationed in the Lyran State professed a desire to return to the Federated Suns, the Prince ordered all AFFC personnel to remain at their current stations and focused his energies instead on the still-looming threat of the Clans.

SABRES RATTLING

In early 3058, Clan Jade Falcon attacked the Lyran Alliance in force, penetrating deep into Archon Katherine's new realm. With their capital world in jeopardy, Archon Katherine's advisors gathered a large force on Coventry to stop the Falcon advance. Prince Victor, already on his way to Tukayyid for a series of wargames with a multinational task force, diverted the entire force to Coventry. By the time the Prince arrived, the Lyran defenders and the Jade Falcons had been locked in heavy combat for months. Rather than continue the fight, the Falcons retreated to their Occupation Zone.

Following this effort, Archon Katherine invited the leaders of the Inner Sphere together, an event that ultimately, and quite surprisingly, led to the re-formation of the Star League. Devoted solely to removing the threat of the Clans forever, this new Star League became a military association rather than a united government, and it soon proved its potency.

The Star League Council, realizing that only a showing of massive force would persuade the Clans to cease their invasion forever, chose to eradicate an entire Clan. The target of this massive military campaign: Clan Smoke Jaguar.

The revived Star League Defense Forces gathered more than a hundred BattleMech regiments from the Inner Sphere's nations, and even from two Clans. Under Prince Victor's leadership, the main body of this army, including nine of the AFFC's most elite units, descended upon the Jaguars' Occupation Zone. At the same time a second task force, commanded by AFFC Marshal of the Armies Morgan Hasek-Davion, traveled in secret to the Jaguars' homeworld of Huntress, located some one thousand light-years outside the Inner Sphere. Both sides fought bitterly in this bloody campaign. In the end, though, Prince Victor presided over the death of the Smoke Jaguars. The SLDF's victory in the Great Refusal, fought later on the Clans' homeworld of Strana Mechty, put an end to the Clan invasion once and for all.

When Prince Victor returned to the Inner Sphere, he was universally hailed as the Conqueror of the Clans. He also returned a man without title or position.

FAMILIAL AVARICE

When he accepted the leadership of the SLDF, Prince Victor placed his sister Yvonne Steiner-Davion as regent on New Avalon. While the Prince concentrated on putting an end to the Clan threat, Archon Katherine quietly took her sister's place as regent after manipulating her sister and the people of the Federated Commonwealth into accepting her with open arms. While many in the AFFC opposed the Archon's moves, there was little they could do. Archon Katherine quickly consolidated her position on New Avalon, illegally taking on the title of First Princess and in name, if not in actuality, reuniting the Federated Commonwealth.

At the same time, the Capellan Confederation embarked on a campaign aimed at reclaiming the St. Ives Compact. At first making use of troops under Star League colors, Capellan Chancellor and Star League First Lord Sun-Tzu Liao soon replaced the SLDF units with his own, entangling the entire Compact in a desperate struggle for survival. With the Archon apparently unconcerned with the outcome of this war, sympathetic AFFC officers gave the Compact as much support as they could. Field Marshal Duke George Hasek went so far as to assign AFFC troops to the defense of the beleaguered nation.

Though effectively stripped of his title, Prince Victor had not been completely removed from power. When he returned from Clan space, he assumed the position of Precentor Martial, the commander of ComStar's entire military, and after Coordinator Theodore Kurita was chosen to replace Chancellor Liao as First Lord, Prince Victor was named the SLDF's Commanding General. The Prince immediately poured as much ComStar and SLDF support as he could, without direct Council approval, into the St. Ives Compact.

TO STAVE OFF THE INEVITABLE

The coming months and years could very well prove to be the most difficult the Federated Suns have ever faced. Elements of the AFFC continue to defend the St. Ives Compact. Many of its most senior officers have left its service, mustered out by Archon Katherine's silent purge. Though some of these veterans have found places in the revived SLDF, many more have simply walked away in disgust.

The men and women of the AFFC have retained loyalties to their home nations since the birth of the united military; that was expected and even unofficially encouraged. Unfortunately, those feelings have grown even stronger since the secession of the Lyran Alliance, and have been further fed by the machinations of Archon Katherine. So strong are these feelings that during the first salvos of the war in St. Ives, an AFFC unit in service to the SLDF "peace-keeping" mission fired upon another AFFC unit assigned to the defense of the Compact.

This phenomenon is by no means unique to the Federated Commonwealth. Indeed, generals in the Federated Suns and the Lyran Alliance have recently begun rating their component units not only on traditional classifications but also on their loyalty to Archon-Princess Katherine, Prince Victor or their own nation.

With the Federated Suns and the Lyran Alliance under the effective control of Archon-Princess Katherine, traditional AFFS units, and those outwardly professing a loyalty to Prince Victor, have been experiencing consistent supply and funding problems. The Archon is apparently trying to starve out any loyalty to her brother, a process unfortunately helped by the heavy debt the AFFC incurred during the action against Clan Smoke Jaguar.

While a number of potential problems simmer within its ranks, the AFFC continues to defend its nation and uphold its commitments. It has a long and proud history and has suffered through far more trying times than those faced today. No matter its difficulties, the Federated Suns military has always emerged stronger, more focused on its mission and loyal to its Prince. The AFFC will continue to do so.

THE FOX'S DEN

The Armed Forces of the Federated Commonwealth is a military in turmoil. Of course, that was not always the case. The AFFC organization was officially formed in early 3041 to oversee the combined militaries of the Federated Suns and the Lyran Commonwealth. At its height, the AFFC successfully directed the actions of several thousand regiments of men and 'Mechs (though only a fraction of that total were 'Mech-only regiments) against the greatest threat the Inner Sphere has known—the Clans. During the Clan war, the AFFC kept its units supplied and combat-ready across the Federated Commonwealth's expanse of more than a thousand light-years, despite transport shortages and the brutal losses experienced on the Clan front.

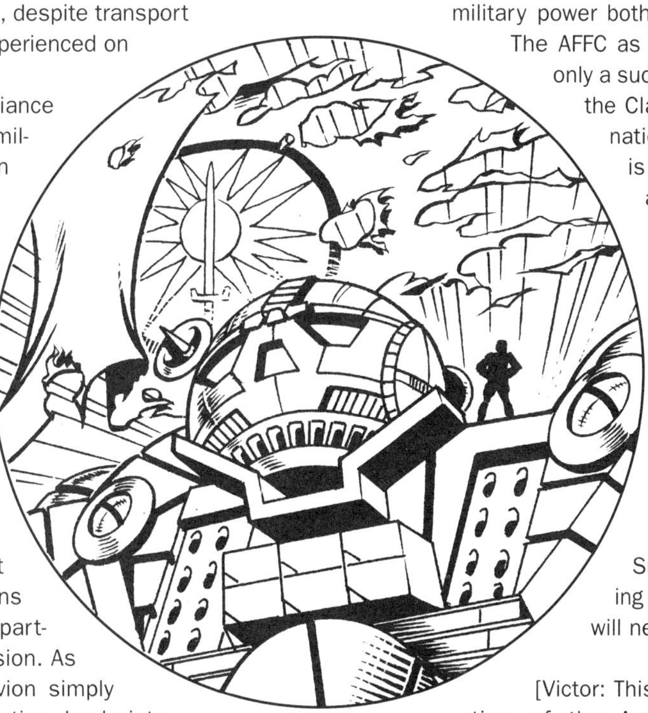

The secession of the Lyran Alliance and the accompanying loss of the military forces assigned to the Lyran State Command, especially the majority of its JumpShip fleet, threw the AFFC into a state of confusion made worse by the rapidly deteriorating Sarna March. With the loss of half of its combat units and personnel, as well as many of its senior officers and directors, the AFFC's huge bureaucracy was suddenly rendered moot. At the same time, the Federated Suns' State Command was left with a number of vacant positions because a number of personnel departed in the wake of the Lyran secession. As a result, Prince Victor Steiner-Davion simply folded the AFFC's grand organization back into what was once the Armed Forces of the Federated Suns.

Though pundits have argued for the AFFC's grand organization to assume its former role, in light of Archon Katherine Steiner-Davion's assumption of the throne on New Avalon, the practical reality is that both realms are still operating as independent nations. With the deterioration of relations between the Federated Suns and the Lyran Alliance, there seems little chance for reconciliation. And the Federated Suns' de facto Prince, Archon Katherine, has not helped matters. Her preoccupation with the Lyran Alliance has shown many Federated Suns citizens where her true loyalties lie, further widening the gap between the two supposed allies.

The Archon's lack of attention is not the only trouble plaguing the AFFC, however. With no other nation willing or capable, the Federated Suns ended up bearing a majority of the financial burden for Operations Bulldog and Serpent. At the same time, a widespread conspiracy of graft and corruption was embezzling hundreds of millions of C-bills from the Federated Suns government. Both of these situations sapped much-needed funding away from the military, reducing spending to the absolute minimum. New equipment purchases and R&D programs were severely curtailed as a result. The AFFC functioned on autopilot during these years, simply subsisting instead of thriving.

Despite the troubles the AFFC has experienced over these past few years, there is no doubt that it is still the preeminent military power both in and out of the Inner Sphere. The AFFC as a whole is riding the high of not only a successful military campaign against the Clans but also a wave of resurgent nationalism. Unfortunately, that wave is working both ways. Many units are professing loyalties to their former Prince, Victor Steiner-Davion. Others are rallying behind Archon Katherine Steiner-Davion. The eventual outcome could potentially rip both nations asunder.

The bright side to this gloom is that the AFFC is on track to regain its former glory. Under the leadership of its most senior officers, the Federated Suns' military continues on, readying itself for the storm that all hope will never come.

[Victor: This section deals with the organization of the Armed Forces of the Federated Commonwealth as it exists today. Regardless of the Lyran secession and your sister's attempted "reforms," the bulk of the former Armed Forces of the Federated Suns continues to operate under the same principles and traditions as it did during the years of the united Federated Commonwealth. Those are clearly spelled out in the pages that follow. —AS]

COMMAND STRUCTURE

The structure of today's AFFC differs very little from that which Prince Alexander Davion put into place during the mid-twenty-sixth century. To be sure, Prince Hanse Davion made quite a few alterations to the AFFS in order to pave the way for the united Armed Forces of the Federated Commonwealth. Most of these changes were purely administrative, however, and had little impact on the day-to-day operations of the Federated Suns military.

Many historians throughout the Inner Sphere consider the AFFC the most professional military organization in known space, even while condemning the Federated Suns for being the most aggressive in the Inner Sphere. The reasons for these assertions are clear: the Federated Suns were forced by their neighbors into a number of military campaigns early in their history, demanding they build a strong and capable military, lest their nation fall. That strong military tradition continues today.

FIRST PRINCE

The Prince of the Federated Suns has always been heavily involved in the matters of his or her military. It wasn't until the days following the disastrous Civil War that the Prince took direct command, however. Since the days of Alexander Davion, the Prince has maintained complete and absolute control of the military as Supreme Marshal.

The Prince's role changed somewhat with the creation of the AFFC. In the years leading up to the Federated Commonwealth, Prince Hanse Davion shared responsibility as the Supreme Marshal of the AFFC with his wife, Archon Melissa. That responsibility reverted to Archon Prince Victor Steiner-Davion when the Federated Commonwealth nation officially came into being upon the passing of both his parents.

By law, the Prince must serve in the military for at least five years before he or she can assume the title of First Prince and Supreme Marshal. While she is legally unable to assume the title of First Prince because of that proviso, Archon Katherine has nevertheless taken on every aspect of the position.

The Prince's Champion

Although it was somewhat superseded in the reorganizations that paved the way for the Federated Commonwealth and its military, the Prince's Champion is still a powerful position. In effect the Prince's right hand in both military and civilian matters, this title was traditionally bestowed on the Federated Suns' most senior field marshal. This officer would take command of the entire Federated Suns in the event the Prince was unable to do so, and more commonly would preside over the High Command when the Prince was occupied by other matters. Prince Hanse Davion appointed then–Field Marshal Ardan Sortek as Champion following the death of Field Marshal Yvonne Davion in 3035.

The creation of the position of Marshal of the Armies, in addition to the dual responsibility the Prince and the Archon shared for their nations, freed the Champion from many of these duties. Though legally the Prince's second, the Champion freely delegated these powers in order to concentrate his energies on his other responsibilities within the AFFC. However, the position of Champion remained not only as the highest honor the Prince could bestow upon an officer, but also as a potential regent in an emergency.

Since the breakup of the Federated Commonwealth and Archon Katherine's assumption of control on New Avalon, the position has returned to the forefront. Though the Champion can only be removed by the First Prince, Archon Katherine, as regent of the Federated Suns, dismissed Field Marshal Sortek and replaced him with Field Marshal Simon Gallagher. Because of this, and because the Archon is using Field Marshal Gallagher to command the Federated Suns military, effectively bypassing the Marshal of the Armies, many of the AFFC's most senior officers are quietly working against her.

MARSHAL OF THE ARMIES

The Marshal of the Armies was created following the unification of the AFFS and the LCAF. As the supreme commander of the Armed Forces of the Federated Commonwealth, the Marshal of the Armies is directly responsible for the entire AFFC and is answerable only to the First Prince (or, before the Lyran secession, the Archon Prince). Practically, if not officially, this officer took on many of the responsibilities of the Champion, including supervising the High Command.

To date, only three officers have served as Marshal of the Armies. Jackson Davion has occupied this post since the death of Morgan Hasek-Davion, who in turn was elevated to the position upon the retirement of Ran Felsner. Though his powers have eroded somewhat since Archon Katherine assumed the leadership of the Federated Suns, the Marshal of the Armies still retains significant authority and prestige within the AFFC.

AFFC HIGH COMMAND

Composed of the commanders of each major military department, the commanders of the three Regional Commands, the Marshal of the Armies and the First Prince (and/or his Champion), the High Command is the chief governing and advisory body of the AFFC. While the Prince retains absolute command of the AFFC, he would be foolish not to listen to the advice of his most senior military commanders. As a result, the High Command often sets the tone for the rest of the AFFC, enacting policies and determining its long-term goals. The High Command meets, as it has for centuries, in the Strategy Room—otherwise known as the Fox's Den—in the AFFC's headquarters on New Avalon. Surrounded by the most advanced computer systems and holo-tables, the High Command has immediate access to the Federated Suns complete information networks.

In addition to their military duties, the members of the High Command sit on the Privy Council. Together with the leaders of each of the Federated Suns Ministries, they advise the Prince on matters of interstellar relations and civil policy, effectively forming a senate of sorts.

The High Command consists of twelve permanent members, not counting the Marshal of the Armies, the Prince or his Champion. Prior to the Lyran secession, the commanders of the Federated Suns and the Lyran State Commands, as well as the six Lyran March commanders, were also members of the High Command. These two field marshals were directly subordinate

to the Marshal of the Armies; each was in charge of the military forces assigned to each state. As with the rest of the excess AFFC organization, the Federated Suns State Command was absorbed into the staff of the Marshal of the Armies.

MILITARY DEPARTMENTS

The commanders of the AFFC's nine major military departments report directly to the Marshal of the Armies. These departments are mostly administrative and support organizations; while the AFFC's combat forces are directed through a separate chain of command, these departments ensure the AFFC is combat-ready.

Department of the Army and the Navy

This department handles all personnel matters within the Regular Army, including filling vacant billets, processing transfers and dealing with all complaints from the AFFC's combat forces. Members of this department coordinate as necessary with members of the Departments of Military Intelligence and Military Justice. Field Marshal Gregory Sykes commands the Department of the Army and the Navy; he is assisted by Marshal Todd Revons, who commands the AFFC's Regular Army, and Fleet Admiral Alistair Buchwald, who commands the AFFC's Navy.

Department of Mercenary Relations

Because of the sheer size of the Federated Suns, the AFFC is one of the Inner Sphere's largest employers of mercenary forces. The Department of Mercenary Relations handles all facets of the AFFC's dealings with mercenary units, from contract negotiations to payment and supply. Liaison officers from this department are assigned to each mercenary unit in the AFFC's employ to assist these units in administrative functions, such as supply requests, and to pass on AFFC orders. Field Marshal Carlos Post commands this department.

Department of Military Administration

The Department of Military Administration is responsible for most of the tasks that do not particularly fit into the missions of the AFFC's other departments. This department is responsible for such diverse assignments as payroll and record-keeping services for the AFFC, the Medical Corps and the Chaplain Service. The Department of Military Administration is commanded by Field Marshal Andre Mossbauer.

Department of Military Communications and Research

Organized at the same time as the AFFC, the Department of Military Communications and Research is perhaps the most secretive of the AFFC's many branches. This department coordinates all research and development programs administered outside the AFFC's direct control, such as with manufacturers and educational institutions. This department also oversees a huge staff that conducts their own highly classified R&D programs. Commonly known projects include HPG research, including development of the "Fax Machine" used during the Fourth Succession War, and the reverse-engineering of captured Clan equipment. Field Marshal Sidney Zweibel, a renowned scientist and researcher in his own right, commands this department.

Department of Military Education

Responsible for the recruiting, training and education of the AFFC's entire complement, the Department of Military

Education is perhaps the AFFC's most important division. This department runs each of the Federated Suns academies and training centers and sets the standard of training for all non-AFFC educational institutions. Additionally, it oversees the AFFC's library and combined data bank, including copies of the Star League memory cores as well as its navy's astro-navigation charts. This department is run by Field Marshal Angela Kouranth.

Department of Military Intelligence

The military counterpart of the Ministry of Intelligence, Investigations and Operations (MIIO), the DMI is primarily concerned with gathering intelligence and information on all enemies of the Federated Suns, both current and potential. Secondary missions include active operations against the Federated Suns' enemies, performing background checks on AFFC personnel, rooting out enemy agents operating within the AFFC and the Federated Suns, and maintaining the security of classified information and projects. As part of these missions, the DMI coordinates with each of the AFFC's other major departments as well as with other Federated Suns' agencies, such as the civilian MIIO, and even agencies within other nations. The DMI is the only military department that is wholly responsible for its own training.

The Department of Military Intelligence is divided into a number of subdivisions, each tasked with a different aspect of its mission.

MI1 – Command: MI1 is the command division of the DMI. The directors of each of the DMI's six other divisions are members of the MI1 Command Council, as is the commander of the DMI, Field Marshal Stephanie Day.

MI2 – Analysis and Speculation: The personnel of MI2 take all of the raw intelligence and information gathered by the DMI and all other associated agencies and turn that into usable information and predictions. MI2 personnel are usually assigned to each major operation to analyze immediately any relevant intelligence. Marshal Daniel Darwith directs MI2's efforts.

MI3 – Electronic Information Gathering: Information is gathered in one of two ways, either through human interaction or through electronic measures. MI3 concerns itself with the latter, known as ELINT. MI3 operates listening posts and surveillance satellites throughout the Inner Sphere and usually has communications specialists involved in each military operation. MI3's director is Marshal Corbett Draven.

MI4 – Covert Operations: Commonly referred to as the "Stealthy Foxes," MI4's mission is HUMINT, intelligence gathering through human interaction. At one point or another in its long history, MI4 has run agents and operations in every nation in the Inner Sphere. Marshal Jemaal Rehm is the Stealthy Foxes' director.

MI5 – Counterinsurgency: Where MI4 deals primarily with operations outside the Federated Suns, MI5's mission is to ferret out enemy agents and spies operating within the Federated Suns and the AFFC. Marshal Melora Li-Vehn is the director of this division, sometimes known as the "Relentless Hounds."

MI6 – Special Forces: The men and women of MI6 are some of the best-trained and most dedicated personnel in the AFFC. While MI6 is probably the most recognized branch of the DMI among the Federated Suns' people, the actual operations and personnel of MI6 are among the most guarded secrets. The reason, very simply, is that MI6, known as the "Rabid Foxes," is tasked with special military actions such as hostage rescues, extractions in foreign territories, sabotage of enemy facilities and the like. Because of the nature of their missions, the operators of MI6 have even been called "Ninja Brain-Surgeons" in the press, a term many have come to enjoy. The Rabid Foxes' commanding officer is Marshal Christian Robert McCarthy.

MI7 – MIIO Liaison: Because the civilian MIIO is tasked with many of the same responsibilities, the operations of each agency could potentially interfere with its sister. As a result each has set up a liaison division, to coordinate the operations of each and provide a conduit to pass information and raw intelligence back and forth. Marshal Daniel Robards oversees the liaison division.

Department of Military Justice

This department deals with all matters of justice within the AFFC, from investigation and prosecution to punishment and corrections. The Department of Military Justice employs its own host of lawyers and magistrates and oversees the AFFC's Military Police corps. The AFFC's chief justice, and commander of this department, is Field Marshal Kalena Sakharov.

Department of Strategy and Tactics

Another department that came into being when the AFFC was formed, the Department of Strategy and Tactics was once a minor subdirectorate of the High Command, one that has grown steadily in size and responsibility since the beginning of the Fourth Succession War. The Department analyzes each current or potential military campaign and formulates the grand strategies the AFFC will employ. It also coordinates with the Department of Military Education to develop training curricula that include the latest military campaigns. Field Marshal PAul Gevin commands the Department of Strategy and Tactics.

Department of the Quartermaster

The Department of the Quartermaster is responsible for keeping the AFFC supplied. One subdivision is responsible solely for coordinating with the AFFC's many departments and military industries throughout the Inner Sphere for purchasing new and replacement equipment. Other divisions disburse the supplies and keep the AFFC's equipment operating. The department has set up depots within each Combat Theater to keep the AFFC operating. In major military operations, the department sets up forward supply depots to keep the supply lines short. While the department set up a forward depot on Kittery to support the Compact's defense against the Capellan Confederation, persistent rumors seem to indicate that other, unauthorized depots are being built throughout the Federated Suns. Field Marshal Melvin Israel Aleixandre commands this always-important department.

REGIONAL COMMANDS

With only a few notable exceptions, all of the AFFC's combat forces are assigned to its Regional Commands. Where the different departments keep the AFFC functioning and support its combat forces, the Regional Commands actually prosecute the wars and defend the Federated Suns borders.

The Federated Suns is broken into three major divisions, called Marches. Each of these Marches serves as an independent political and military territory, allowing the Prince and his military leaders to exert better control over the massive Federated Suns and its military. These Regional Commands are each commanded by an AFFC field marshal who also holds a seat on the High Command.

In a tradition that dates to Prince Alexander, the commanders of the border regions, the Capellan March and Draconis March Regional Commands, commonly hold dual positions as the Duke of the March, the political leader of the region. This long-standing tradition has proved its worth many times during the Federated Suns long history, allowing these regional commanders to marshal all of the resources of their March behind an operation without having to lobby civilian leaders for support.

In contrast, because the First Prince is the titular leader of the Crucis March, the commander of this region is responsible solely for the military forces under his command. As the Crucis March sees very little combat within its borders, it is often called upon to reinforce and support either the Draconis or Capellan Marches during war. The officer appointed to this position is traditionally one of the First Prince's most trusted advisors and, since Prince Hanse gave up direct military command of this March, has also been the Prince's Champion.

The commanders of these regions are often given the latitude to prosecute any military actions within their regions as they see fit, with the High Command and the AFFC's bureaucracy giving them the support they need. Many times in the past, though, the commanders of the two border regions have taken it upon themselves to launch their own operations without the consent or even knowledge of the First Prince. Few of these initiatives have ever succeeded, however, as without the endorsement of the Prince and the Command Council, their forces lack the support they need from the Department of the Quartermaster.

Field Marshal Duke George Hasek commands the Capellan March Regional Command, Field Marshal Duke James Sandoval leads the Draconis March Regional Command, and Field Marshal Simon Gallagher commands the Crucis March Regional Command.

Combat Theater

Each Regional Command is further divided into two or three Combat Theaters, preventing the supply and command lines from stretching too thin. These theaters, formerly known as Operation Areas, are purely military structures, serving no civilian governing purposes. The Combat Theaters are given either coreward or edgeward designations, depending on their relative distance from Terra. Field marshals command Combat Theaters.

Polymorphous Defense Zone/Combat Region

The smallest astrographic subdivision within the Federated Suns, the Polymorphous Defense Zone (PDZ) encompasses a region that includes anywhere from ten to more than three dozen worlds. Each PDZ takes its name from its command and logistics center. Directly responsible for the defense of its worlds, each PDZ is permanently assigned a March Militia unit, plus additional conventional and/or BattleMech Regular Army forces.

The AFFC uses several additional terms to describe its PDZs. The ten subdivisions within the Crucis March are still officially referred to by their historical name of Combat Regions. During the years of the united Federated Commonwealth, these areas were simply known as Commands. PDZs are normally commanded by AFFC marshals, though those officers are given a temporary promotion to field marshal if an officer of that rank is assigned to the PDZ as a subordinate.

TEMPORARY COMBAT COMMANDS

The permanent structure of the Regional Commands is more than sufficient for taking care of the Federated Suns defensive needs and prosecuting the type of warfare common during the Succession Wars. What it could not hope to support, however, were the type of massive operations in the Fourth Succession War and, more recently, Operations Bulldog and Serpent.

In cases such as these, the Prince or the Marshal of the Armies will appoint a senior officer, often a field marshal, to command an entire task force. Though nominally subordinate to a March commander, the leaders of these task forces are frequently given free rein to conduct the operation as they see fit. In addition to the sometimes scores of regiments assigned to a task force, the commander is often given logistics support equal to an entire Combat Theater's. Task forces are not permanent organizations and normally last only as long as the campaign.

COMBAT ORGANIZATIONS

Many of the AFFC's longest-serving units are part of a larger organization, such as the Davion Brigade of Guards. Though historically these organizations effectively functioned as an independent military, with their own regulations and supply lines, they were all eventually absorbed into the Armed Forces of the Federated Suns' main chain of command.

The commanders of these organizations still retain much of their authority, coordinating reassignments with the High Command and ensuring that the units under their command are

AFFC STANDARD REGIMENTAL COMBAT TEAM ORGANIZATION

Element	Combat Strength
1 BattleMech regiment	132 BattleMechs
3 vehicle regiments	324 armored vehicles
5 infantry regiments	3,591 infantry troopers
1 artillery battalion	36 artillery pieces
2 aerospace wings	40 aerospace fighters

taken care of. Unconfirmed reports even indicate the commanders of some of these organizations have been secretly funneling supplies to their subordinate units within the Lyran Alliance. In many instances, the Prince has used these officers to bypass the High Command and issued orders directly to their units.

FORCE STRUCTURE

The AFFC groups its forces into five broad categories, described below.

REGULAR ARMY UNITS

The Regular Army contains the best-trained and best-equipped soldiers in the AFFC. The Regular Army fields thousands of regiments, including not only the high-profile BattleMech regiments but also conventional armor and infantry units. Approximately half of the Regular Army's forces are organized as independent units, while the other half form RCTs or other large combat groups. Each Regular Army unit is assigned its own integral DropShip transport.

Many of the Regular Army units under the AFFC's authority fall under the purview of a larger military organization, such as the Avalon Hussars or the Deneb Light Cavalry. Many more, however, are considered independent. A majority of these "orphaned" units have their origins in the Lyran Commonwealth, where they belonged to a larger organization. Every ex-Lyran unit with such an affiliation has a small staff within the Department of the Army and the Navy that, at least in theory, looks after their interests. While they do not have a specific voice in the AFFC's highest levels, they are assisted by their own specialists within the department.

RCTS

At the heart of the AFFC's combat doctrine is the combined-arms Regimental Combat Team, or RCT. As with every other modern military, the BattleMech is the AFFC's primary weapon. The AFFC does not consider the BattleMech its only effective weapon, however.

On paper, the RCT consists of a single BattleMech regiment, three armored regiments (traditionally one heavy and two medium), five infantry regiments (one jump, two mechanized and two foot), an independent artillery battalion and two aerospace fighter wings that are permanently attached to the BattleMech regiment. After the Fourth Succession War, the AFFC began to convert the foot infantry regiments in its RCTs to mechanized infantry regiments, a process that was only partially completed before the appearance of the Clans. Since the end of the Clan war, many RCTs have converted one of their infantry regiments to an armored infantry battalion. In light of these changes and the heavy action the AFFC has been involved with over the past fifteen years, there are more than a few variations on this "standard."

Each RCT receives its designation from the BattleMech regiment assigned to it; each component unit retains its own name, however. RCTs are commonly commanded by marshals.

MILITIA UNITS

Militia units are mustered to defend their home territories from invasion. Though they receive neither the latest equipment nor the best recruits, these militia forces are staffed mostly with members native to the area they are defending. Combined with the fact that many of the militia's men and women are long-time AFFC veterans, these forces make up in intensity what they lack in equipment.

The AFFC fields two different types of militia units, detailed below.

March Militia

The primary mission of each AFFC March Militia unit is to defend its home region from invasion by a foreign power. Most PDZs (or Combat Regions) in the Federated Suns is permanently assigned a March Militia, similar in size and composition to an RCT. At the heart of each March Militia is a full BattleMech regiment. Supporting this 'Mech unit are usually two armored regiments, five infantry regiments, an artillery unit and anywhere from a few light squadrons of aerospace support to a full aero regiment.

These March Militias do not field the latest combat equipment and vehicles, as most of this is routed first to the Regular Army units. They are, however, on the second supply tier and are eligible to receive new equipment once the majority of the Regular Army has been supplied. Additionally, the March Lords have been known to divert supplies from other units to their favored Militia forces.

March Militias are limited to their own PDZ and any world that borders their PDZ. They may operate outside those areas only under orders of the March commander or higher. Each March commander is assigned a very limited amount of transport capability that he can parcel out among his command as needed. As a result, few March Militias have ever received permanent transport capabilities.

Planetary Militia

The planetary militias are the AFFC's final line of defense.

These militias are organized and equipped by the governments of each Federated Suns world. Training is accomplished either through AFFC training facilities or by those operated by local militias, if they have sufficient resources. Although planetary militias are technically not an element of the AFFC, each PDZ's commander, or higher, has the express authority of the First Prince to take command of and mobilize any or all of the planetary militias under his or her purview.

Planetary militias must rely upon taxes collected from the world's residents to equip their forces. As a result, planetary militia units rarely see anything but the oldest and most outdated equipment unless they have a benefactor of some sort. The AFFC often assigns older equipment that has been retired or otherwise replaced to these militias rather than scrap them.

Planetary militias, with few exceptions, field only conventional armor and infantry units. Militias commonly number anywhere from a mixed regiment on the smallest planets to dozens of active and reserve regiments on the most populated ones. Few, if any, of these militias possess their own transport capability.

TRAINING UNITS

The AFFS was the first major military to add dedicated training units to its rolls. Staffed with inexperienced students under the tutelage of veteran officers and trainers, these units give the students the real-life seasoning they require before they can move on to a Regular Army unit. These training units, devoted almost exclusively to the training of MechWarriors, are located all over the Federated Suns. Most are designated as Training Battalions, attached to a major AFFC academy or training facility. Both the NAIS College of Martial Sciences and the Albion Academy, with their massive cadet corps, also field large training units; between them, the two academies fill five full BattleMech regiments.

The quality of equipment varies widely among these training units, depending on the educational institution they are attached to. Both the CMS and the Albion Academy groups field some of the latest and best equipment, while the First Bell Training Battalion is lucky to keep its outdated equipment functioning. Each of the Training Battalions is treated as a March Militia unit as far as supply and transport are concerned. The CMS and Albion regiments, however, are treated in every way as Regular Army units, leading to many jealous mutterings from the other training units.

MERCENARY UNITS

While the Federated Suns has historically fielded the largest military in the Inner Sphere, it is also consistently the one power that employs the most mercenary units year after year (though recent reports indicate the Word of Blake is attempting to assume that title). The mercenary units the AFFC employs come from a variety of backgrounds. Well over half have worked with the AFFC—or the earlier AFFS—in the past and remain in their employ; the rest are recruited from a variety of sources, including through

the Mercenary Review and Bonding Commission on Outreach.

The Federated Suns employs mercenary commands of all sizes and specialties, though most are BattleMech units. The AFFC assigns these units to a variety of tasks. Most common are garrison assignments in the Federated Suns border regions, shoring up the AFFC's overtasked March Militias. Mercenaries are also frequently given raiding missions and, in times of war, combat support missions. In most cases, mercenary units are only employed in roles that support Regular Army forces. The AFFC does quietly employ mercenaries to prosecute campaigns it could never officially be involved in, though.

The AFFC offers a number of benefits to those mercenary commands it employs. Most are given the freedom to accomplish their assignments with only the most basic guidance from the High Command. All are given the option to purchase equipment through the AFFC's Department of the Quartermaster. Longer-serving units are given better compensation and are even extended lines of credit through the Quartermaster.

The AFFC employs scores of mercenary units throughout the Federated Suns. In addition, many worlds and even a number of government officials employ mercenary commands, each of which can be legally dragooned into AFFC service. While it is outside the scope of this work to detail each of these units, a number of the longest-serving major commands that are currently stationed within the Federated Suns are briefly described below.

1st Cunningham's Commandos

CO: Colonel James Longstreet

Cunningham's Commandos were the first mercenary unit to be permanently assigned as an opposing force (OpFor) unit for a Federated Suns Combat Training Center. During the height of the Clan war and the later invasion of the Sarna March, the unit was one of the few left to guard the Capellan border. When it was clear that the Capellan-St. Ives war would not end quickly, Duke Hasek called upon the Commandos to escort relief supplies to the Compact. Though not assigned to combat duties, the unit nevertheless was "forced" to defend itself on a number of occasions.

The Fighting Urakhai
8th Striker CO: General Bryan Holstead
DeMaestri's Sluggers CO: Colonel Ross DeMaestri

The two Fighting Urakhai regiments in service with the Federated Suns were separated from their third regiment following the Lyran secession. After suffering heavy casualties in the Chaos March, the two regiments were assigned to the Periphery border of the Crucis March to give them time to rebuild.

Hansen's Roughriders
CO: Colonel Wolfgang Hansen

After fighting long and hard along the Clan front, the Roughriders returned to the Federated Suns and took up station along the Capellan March's rimward reaches. Following the Capellan Confederation's invasion of the St. Ives Compact, Field Marshal Duke George Hasek sent the unit to Bromhead, where they have conducted a number of reconnaissance raids deep into the Confederation.

Harlock's Warriors
CO: Colonel Samantha Viele

Originally hired to help fortify the Clan front, Harlock's Warriors were rotated into the Sarna March when the Free Worlds League launched Operation Guerrero. The unit has remained there since, under orders from Prince Victor, to help prevent any more Capellan or League incursions into the Chaos March. Over the past two years, the Warriors have experienced a number of severe supply problems that have only begun to turn around since they have been able to draw supplies through ComStar channels.

Illician Lancers
59th Strike Regiment CO: General Brenda Bradley
21st Illician Rangers CO: Colonel Roy Barnard
4th Illician Rangers CO: Colonel Alicia Ramaley
9th Illician Rangers CO: Colonel Kenneth Koppell

One of the Federated Suns' longest-serving mercenary units, the Illician Lancers are also one of the largest, fielding four full BattleMech regiments. The unit was recently rotated into the Capellan March, where they became the heart of its rimward defensive line. Their postings were shuffled around somewhat when the Confederation invaded the St. Ives Compact, with the Fifty-Ninth Strike Regiment currently on "loan" to the SIMC and the other three regiments moving up and down the border, keeping the CCAF on constant watch for a Lancer incursion.

Vanguard Legion
CO: Colonel Chad Dean

Hired to replace the loss of Laurel's Legion to the Confederation, the Vanguard Legion, like many mercenary units, was formed from veterans of the Clan war. However, the commander—Chad "G" Allen—felt that his old unit, the Twenty-fourth Arcturan Guards, was left to die at the hands of the Falcons by the Lyran State Command and has turned his unit against anything "Lyran." After several years in the employ of the Draconis Combine, working alongside ComStarÕs Explorer Corps to locate the Clan homeworlds, they did not renew their contract when their commander was killed in an accident. However, the very lucrative terms of the Explorer Corps contract enabled the unit to expand to a full regiment in 3061. The new commander shares his predecessor's dislike of Lyrans, which has put the unit at odds with the AFFC's Department of Mercenary Relations. A heavy regiment, the Vanguard's philosophy is to hit hard and fast, depriving the enemy of any chance to erect a suitable defense.

Lexington Combat Group
32nd Recon CO: Major General Malcolm Feinman
Marie's Golden Hammers CO: Colonel Marie Stevens
Frederic's Gazelles CO: Colonel Frederic Zaffson

The Federated Suns have enjoyed the services of the Lexington Combat Group for more than a century. The unit's three BattleMech regiments and supporting units have seen action throughout the Inner Sphere during its service to the Federated Suns. Though technically assigned to the Capellan March's Periphery region, the entire unit has been hopping between worlds throughout the March to help shore up the Capellan border.

Screaming Eagles
CO: Colonel Tam Lafayette

Once fielding two full BattleMech regiments, the Eagles have suffered from the ravages of war. Now fielding only a single 'Mech regiment, the Eagles have been assigned to the Woodbine Combat Training Center for the past decade as its OpFor unit.

12th Vegan Rangers
Alpha Regiment CO: General Tom Stancel
Beta Regiment CO: Colonel Miller al-Nahib
Gamma Regiment CO: Colonel Christie Desantis

Though spared the ravages of the Clan war, the Twelfth Vegan became deeply embroiled in the League invasion of the former Sarna March. During the course of that bitter war, the unit suffered heavy losses and was forced to disband one of its four regiments just to keep the others at operational strength.

The unit's three regiments remain stationed within the Chaos March, awaiting the chance to repay their losses.

FORCE COMPOSITION

Even after more than two decades of alliance with the former Lyran Commonwealth Armed Forces, the AFFC is still more a new incarnation of the old AFFS than a hybrid of the two organizations. In fact, the generals overseeing the nascent AFFC used the organization and force structures of the AFFS extensively in forming this combined military. The lessons learned during the many major military actions fought in this century simply reinforced and refined the combat doctrines the AFFS had been practicing for many years. Combined-arms warfare is the standard in the AFFC, as it has demonstrated to the rest of the Inner Sphere, as well as the Clans, time and again.

AFFC STANDARD BATTLEMECH ORGANIZATION

Element	Component Units	Combat Strength
Lance	4 BattleMechs	4 BattleMechs
Company	3 lances	12 BattleMechs
Battalion	3 companies	36 BattleMechs
Reinforced Battalion	3–4 companies + 1 command lance	40–52 BattleMechs
Regiment	3 battalions	108 BattleMechs
Reinforced Regiment	3–4 battalions + 1 command company	120–132 BattleMechs

BATTLEMECH UNITS

Regardless of its reliance upon a combined-arms principle, the AFFC's combat unit of choice is the BattleMech, much as it is in every other modern military. 'Mech units form the core of each of the AFFC's heavily used RCTs, as well as a number of high-profile independent regiments. With few exceptions, BattleMech-equipped forces are the first units to see combat in any military action and, in those cases where a retreat is called, the last to leave.

AFFC BattleMechs are normally grouped into companies and battalions with elements of like capability and performance. From these, entire regiments are formed, each with different capabilities and strengths. The High Command does not try to dictate a standardized table of organization and equipment for its entire BattleMech force, instead often delegating responsibility for a particular regiment's TO&E to its parent organization and commanding officer, so long as the materiel is available.

On the average, Regular Army 'Mech regiments tend to field a slightly heavier 'Mech complement than the AFFC's various Militia and training units, though that is not the rule. Many Regular Army units, such as the Davion Light Guard RCT, field only a light 'Mech force. In most cases, the lighter 'Mech units are also assigned supporting armor elements that complement their capabilities.

While the AFFC's BattleMech corps is slightly lighter than that of the average Successor State military, it also fields the numerically largest 'Mech corps. The AFFC's needs are so great, in fact, that in time of war it is forced to rely upon the industries of other nations to keep its 'Mech units operational.

Today, after almost two decades of effort, slightly more than 90 percent of the AFFC's Regular Army is equipped with upgraded technologies. Even the March Militias field more than 70 percent upgraded equipment. Because of the sheer devastation of the Clan invasion, a majority of the Regular Army's BattleMech corps is made up of designs produced within the past decade or so, while the rest is simply older equipment refitted with new technologies. Recently debuted chassis like the *Stealth,* the *Gunslinger,* the *Cestus* and the *Nightstar* are already becoming as identified with the Federated Suns as the *Javelin* or the *Enforcer* was in the past.

Additionally, in an effort to boost the AFFC's lagging morale in the wake of the Lyran secession, Prince Victor commissioned upgrades of the *Enforcer* and the *JagerMech.* Both of these designs are quickly finding their way to Regular Army units.

The AFFC also fields several OmniMech designs of Inner Sphere origins. Though most of these are manufactured by corporations based outside the Federated Suns, the AFFC has recently debuted its own home-built OmniMech chassis. In addition, many of the Regular Army units that fought against the Clans operate some Clan-based OmniMechs that they scavenged from the battlefields. In all, however, OmniMechs of any origin account for less than 7 percent of the Regular Army's BattleMech complement.

In the March Militias, the percentage of new designs is far lower. A great majority of the BattleMechs found in these Militias are classic designs, many of which have been refitted with upgrade kits. Only a very few new designs, such as the Watchman or Sentry, have found their way into the AFFC's militias in significant numbers. This trend is likely to continue until the AFFC is able to completely equip its Regular Army with the latest 'Mech designs.

INFANTRY UNITS

The conventional infantryman forms the backbone of the AFFC, accounting for some two-thirds of the Regular Army's regiments currently in service. That percentage grows even larger when the various March and planetary militia units are taken into account.

Within the AFFC, conventional infantry units are broken

AFFC STANDARD INFANTRY ORGANIZATION

Element	Component Units	Combat Strength
Squad	7 men	7 troopers
Platoon	4 squads	28 troopers
Company	3 platoons	86 troopers
Spec. Ops Company	4 platoons	112 troopers
Battalion	3 companies	258 troopers
Spec. Ops Battalion	4 companies	448 troopers
Regiment	3–4 battalions	774–1,032 troopers
Brigade	4–8 regiments	3,096–8,256 troopers

into a number of different categories and missions. The most common infantry units are foot and motorized regiments, though these units are generally employed only in static roles due to their lack of mobility. Mechanized infantry units are equipped with armored personnel carriers and often some organic armor elements, giving them the mobility needed to keep up with other units. Jump infantry are the least common infantry unit found in the AFFC, simply because they are the most difficult to train and the most expensive to operate, and thus are rarely found outside an RCT.

Apart from their designation into foot, motorized, mechanized and jump categories, each of the AFFC's conventional infantry regiments is grouped into a few different missions. RCT Infantry units and March Militia Infantry units are assigned to RCTs or March Militias, respectively. Forward Deployment Infantry units normally follow the first waves of an invasion force, reinforcing and supporting the front lines. Primary Relief Infantry units are commonly assigned to clean-up and garrison duties, holding an invasion force's rear areas. Special Infantry units possess unique skills, such as the mountain troops, and are available for special assignments, usually through the High Command. Home Guard Infantry units man the final line of defense as part of the planetary militias.

The AFFC groups its Forward Deployment and Primary Relief Infantry regiments into semi-permanent brigades during peacetime and while at war. These brigades normally consist of anywhere from four to eight individual regiments and are often assigned one or more armored regiments to boost their capabilities.

Most of these conventional infantry regiments have their origins on a particular world or group of worlds and still recruit a majority of their members from their homeworld(s). Unlike in other services, once a member is assigned to a particular regiment, he or she can expect to remain in that unit unless he/she requests a transfer. The only exception is with the highest-profile regiments, such as those assigned to the Davion Brigade of Guards, whose members hail from worlds across the Federated Suns and for which only the most qualified and loyal personnel are recruited.

In addition to these conventional infantry units, the AFFC employs a number of highly specialized infantry forces. Most Regular Army RCTs and Forward Deployment brigades are assigned one or more engineering battalions to aid in demolition and construction missions. Special Forces units specialize in sabotage and rescue missions behind enemy lines. Marine units are used almost exclusively in spaceborne boarding actions. Each Regional Command is assigned a number of these extraordinary units, while the High Command retains responsibility for the rest and parcels them out as needed. Though these units often operate in strengths of less than a platoon, they are assigned into reinforced battalion-sized forces. Because of the nature of these units, the personnel assigned to them frequently do not share similar backgrounds.

BATTLE-ARMOR UNITS

Introduced after the Clans so ably demonstrated their devastating capabilities, powered-armor infantry units, or battle-armor units, have quickly become among the most sought-after elements throughout the AFFC. More flexible than any conventional infantry unit, and far tougher, power-armored units give commanders a unique advantage on the battlefield.

The AFFC was the first military in the Inner Sphere to field its own battle-armor designs, though unfortunately too late to make any difference on the Clan front. The specialized Sloth and Infiltrator suits quickly followed those first suits and gave AFFC commanders more capabilities. Since those debuted, almost a decade ago, the AFFC and the NAIS have worked closely together to develop a battle-armor suit that more closely resembles the capabilities of the Clan Toad suit. The Cavalier, which is the FedCom version of the standard Inner Sphere power armor, and the Infiltrator Mk II, an extensive upgrade of the original Infiltrator, are the result, and though they have just been introduced, they are already proving to be the most popular suits in the field.

The few companies in the Federated Suns that manufacture battle-armor suits have been working overtime to fill the AFFC's orders. To date, the High Command has assigned most infantry regiments attached to an RCT at least a company of battle armor, and several RCTs even field independent battalions. A number of independent Regular Army 'Mech regiments have also been assigned battle-armor support, as have the AFFC's special operations units. A great majority of these battle-armor units are further assigned their own organic APC transport elements, allowing them to keep up with the 'Mech and armor units they are assigned to support. The High Command has yet to permanently assign any battle-armor units to a March Militia.

CONVENTIONAL ARMOR UNITS

While far less predominant than infantry units, armored regiments outnumber the Regular Army's BattleMech complement by a ratio of slightly more than five to one. Just as the AFFC fields a variety of 'Mech designs, so too does it operate a myriad of armored vehicle designs, with no single class dominating the AFFC's armor corps.

The AFFC categorizes its armor regiments by broad mission, similar to the way it categorizes its infantry units. Regular Army armor regiments are assigned the majority of the combat missions and are among the first elements to land on an enemy planet or to engage in combat along a new front. The armor regiments assigned to each RCT fall into this category, as do those that operate independently or with Forward Deployment Infantry elements. Reserve armor regiments take on garrison duties as a part of March Militias or independently, either as a permanent assignment or in support of a larger military operation.

As with its 'Mech corps, the AFFC fields vehicle designs of varying technology base. The Department of the Quartermaster waited until after the Clan invasion had stalled before soliciting for new vehicle designs and field refit kits, each using lostech advancements made in the previous decades. As a result, the Department's Armored Vehicle Upgrade Program (AVUP) did not truly begin until late in 3053.

AVUP's administrators have concentrated their efforts on upgrading the AFFC's Regular Army armor units first, with special emphasis placed on units assigned to RCTs. The process has gone exceedingly well, surpassing the expectations of the plan's administrators. Each of the AFFC's RCTs, and many of its independent units, are now fielding brand-new vehicles. Bulldogs, Challengers and Alacorns are all found in concentrated numbers in these regiments. With so many armored regiments to supply, however, the process will still take many years to complete. While some 80 percent of the Regular Army armored regiments have received significant numbers of these vehicles, slightly less than 40 percent of the Regular Army's total vehicle complement mounts upgraded technology.

With the Department of the Quartermaster giving full priority to the Regular Army, the AFFC's Militia units have yet to see any of these new designs. On the other hand, they are still well-equipped with armor. After centuries of combined-arms warfare, the average TO&Es of both Regular Army and March Militia armor units look identical. Indeed, the only differences lie in the quality of training and the missions given to these units. Heavy and medium formations are the most common, and as a result, tanks such as the Patton/Rommel, the Demolisher, the Hunter and the Pegasus are all commonly fielded. Most of the larger planetary militias also field armor forces, though the majority of their regiments average in the light to medium classes.

AFFC STANDARD ARMOR ORGANIZATION

Element	Component Units	Combat Strength
Platoon	4 vehicles	4 vehicles
Company	3 platoons	12 vehicles
Battalion	3 companies	36 vehicles
Reinforced Battalion	3–4 companies + 1 command platoon	40–52 vehicles
Regiment	3 battalions	108 vehicles
Reinforced Regiment	3–4 battalions + 1 command company	120–132 vehicles

AFFC STANDARD BATTLE ARMOR ORGANIZATION

Element	Component Units	Combat Strength
Squad	4 troopers	4 troopers
Platoon	4 squads	16 troopers
Company	4 platoons	64 troopers
Battalion	4 companies	256 troopers

AEROSPACE UNITS

The AFFC's combined-arms paradigm is not limited to just its ground forces, as the enemies of the Federated Suns have often learned throughout the years. Just as with every other force type, the AFFC groups its aerospace forces into several different categories. Regimental Wings provide dedicated support for the BattleMech regiments they are permanently attached to. Naval Wings are permanently assigned to a naval task force or WarShip and provide for local defense of these valuable assets and long-range interception of enemy naval forces. Independent Wings are not permanently attached to any ground unit and instead can be tasked to support nearly any type of combat unit. Ground Support Wings consist mostly of conventional aircraft and are dedicated to close air-support missions. Air Guard units are aerospace forces assigned to a militia unit.

When assigned together, two or more aerospace wings are often grouped into an aero regiment, allowing each component wing and squadron to share combat assignments. In these cases, if one has not already been appointed, the senior wing commander takes command of the entire aerospace complement and acts as the "Air Boss," coordinating the entire regiment's operations. The wings assigned to an RCT or large ground force are most frequently organized in this way, as are the wings attached to a WarShip or other naval flotilla. Additionally, an RCT's aerospace complement is assigned a

AFFC STANDARD AEROSPACE ORGANIZATION

Element	Component Units	Combat Strength
Flight	2 fighters	2 fighters
Squadron	3 flights	6 fighters
Wing	3 squadrons + 1 command flight	20 fighters
Aero Regiment	3 wings	60 fighters

battalion of engineers who are trained to quickly construct and repair temporary forward airfields.

As with its vehicle corps, the AFFC only began the upgrade process for its aerospace forces within the past decade. Classic fighters such as the *Sparrowhawk,* the *Corsair* and the *Stuka* still dominate aero wings throughout the AFFC, though many have now been upgraded to the latest models. The Department of the Quartermaster has given both Regimental and Naval Wings priority in the assignment of new equipment, and as a result almost 70 percent of these wings now use upgraded fighters. In the rest of the AFFC's aero wings, however, the rate drops to an average of about 30 percent.

NAVAL UNITS

Unlike the navies operated by the Inner Sphere's other major powers, the Federated Suns Naval Service is devoted solely to the support of the AFFC, which in turn is primarily tasked with the defense of the people of the Federated Suns. Though on the surface this mission may not seem different from that of any of the other Successor StatesÕ naval forces, its most basic meaning pervades the entire AFFC Naval Service. To accomplish this mission, the Naval Service divides its forces into three separate categories.

The Naval Transport is the largest grouping in the Naval Service and is tasked with transporting the AFFC's component units to and from their assignments. Approximately half of the JumpShips of this division operate independently and could end up transporting just about any type of unit or cargo. The rest are grouped into squadrons composed of enough JumpShips to transport an entire RCT or brigade. Though most of the AFFC's Regular Army units possess their own organic DropShip capabilities, this division can call upon a sizable DropShip fleet to assist in transporting those units that do not. Because the AFFC is still reeling from a severe shortage of Jump-capable spacecraft, none but the highest-profile Regular Army units are permanently assigned JumpShips. This trend will likely continue, even in light of the ongoing dragooning of civilian transports, until the AFFC can recoup its once-strong JumpShip fleet—a goal it hopes to reach by 3067.

The second major category is Naval Escort. Assault DropShips form the basis for these squadrons and flotillas, which are often called upon to provide escort for important convoys as well as major invasion forces. These escort units are normally mated with one or more Naval Wings and can intercept enemy naval forces in deep space just as well as they can provide close air support to their ground-based cousins. If the AFFC were ever to be involved in a large-scale naval battle, these units would form the first line of defense for the AFFC's WarShip fleet. Naval Escort squadrons are assigned permanent JumpShip transport only if their mission warrants it.

The AFFC's final spaceborne classification is Naval Assault. At the heart of each Naval Assault unit is one of the Federated Suns few WarShips, around which a small fleet of assault ships and Naval Wings have been arrayed. As the mission of the Naval Service is effectively to support the AFFC's ground forces, Naval Assault units are geared more toward defense and maneuverability than all-out assault. As a result, the AFFC's WarShip fleet is noticeably lighter than any other in known space, with ships like the corvette-class *Fox* and the light cruiser-class *Avalon* dominating. Though they typically mount far less firepower than the average Inner Sphere WarShip, each has the maneuverability to keep up with most military DropShips, giving them a distinct advantage in the roles they are designed for. Naval Assault units have their own lithium-fusion-battery-equipped JumpShips permanently attached to transport those DropShips that the WarShips cannot carry.

AFFC WarShip Fleet

Fleet Admiral Alistair Buchwald commands the current AFFC fleet. An officer within the Federated Suns Admiralty for more than forty years, he has amassed a tremendous number of contacts, both naval officers and contractors, throughout the entire Inner Sphere. As the Admiraltys Director of Procurement, Buchwald took charge of the AFFC's WarShip design program, using his vast array of contacts to speed the process of rebuilding the nation's crippled JumpShip fleet and building a new WarShip fleet. Though not a member of the High Command, he has been very critical of the Archon's decision to cut naval funding in favor of additional social programs that amount to little more than PR campaigns, which has forced him to make some difficult cuts within the Admiralty.

Two major classes of ships, the *Fox* corvette and *Avalon* cruiser, form the backbone of the AFFC Admiralty. A third, the *Mjolnir*-class cruiser, designed prior to the breakup of the Federated Commonwealth, was redesigned as a battleship, and now serves the fleets of the Lyran Alliance. A fourth, the *Durendal*-class destroyer, exists only on the drawing board.

Though the principal design and construction were carried out within the Draconis Combine, the unfinished hull of a *Kirishima*-class cruiser was delivered to the New Syrtis Shipyards for final fittings and trials. Performance-wise, the Covenant—both the vessel and class name were changed—is identical to the Combine's *Kirishima.* Upon delivery to New Syrtis, however, the shipyard made quite a few external and internal cosmetic changes, finally delivering a ship that resembled the Combine Kirishima as little as possible. In fact, few outside the High Command and the New Syrtis yards know the Covenant's true origins, though they overlooked a number of Japanese-labeled access panels and consoles, leading more than a few crew members to wonder abou their ship's heritage.

The AFFC fleet currently consists of the *Fox*-class FCS *Fox,* FCS *Indomitable,* FCS *Intrepid,* FCS *Indefatigable,* FCS *Antrim,* FCS *Murmansk,* FCS *Brest,* FCS *New Syrtis,* FCS *Robinson,* the *Avalon*-class FCS *Melissa Davion,* FCS *Alexander Davion,* FCS *Lucien Davion,* FCS *Simon Davion* and finally the *Covenant*-class FCS *Covenant.*

The *Fox*-class FCS *Rostock* was lost on 5 March 3060 during Task Force Serpent's assault on the Smoke Jaguar capital of Huntress, and the *Fox*-class FCS *Antrim* is undergoing repairs and is expected back in service in early 3063.

The following vessels are currently under construction: the *Fox*-class FCS *Kathil* (formerly the FCS *Sarna)* [expected 3063], FCS *Bryceland* [expected 3064], FCS *Kentares* [expected 3064], FCS *Admiral Corinne Donnings* [expected 3065], FCS *Admiral Michael Saille* [expected 3065], the *Avalon*-class FCS *Robert Davion* [expected 3063], FCS *Andrew Davion* [expected 3065] and FCS *Hanse Davion* [expected 3065].

THE AFFC UNIFORM

Prior to the formation of the AFFC, the Armed Forces of the Federated Suns wore a uniform steeped in history and tradition. With the united AFFC, however, came compromise. In most operational respects, the AFFS served as the template for the new AFFC, especially in organization and combat doctrine. When it came to the uniforms for this new AFFC, though, it was the LCAF that provided most of the inspiration. Surprisingly, the men and women of the AFFS raised few objections to these changes. The combinations that changed the most were the dress uniforms, and many soldiers quite frankly felt the AFFS uniforms were shockingly bland and unstylish in comparison with those worn by other militaries.

Likewise, the AFFC introduced a new rank system and several new decorations to the uniform. As the existing decorations of each realm remained, there were no objections to the creation of a few new major awards. The most strenuous objections came with the introduction of a new rank system that combined elements from each nation's military into what many considered a "bastardized attempt at making it idiot-proof," a quotation taken directly from a memo circulated by a ranking field marshal to his staff in 3042. Others were less polite. What this system attempted to do is create a simplified rank structure. Unfortunately, both the AFFS and the LCAF had possessed far more robust rank structures, and as a result the rank and file of the nascent AFFC immediately voiced their displeasure with the new system. Despite their objections, however, the system remained in use.

The modern AFFC experienced a radical shift in its uniforms following the secession of the Lyran Alliance. Traditional Federated Suns units in both nations slowly began to wear a uniform that more closely resembled the former AFFS style. Soon many of the units and personnel stationed within the Federated Suns began to wear the new style, and the commanders of the Capellan and Draconis Marches eventually authorized the new uniforms for wear within their own regions.

These new uniforms and the accompanying classic rank structure are not used unilaterally, however, even within the regions that have officially adopted them. On the whole, units and staff departments claiming allegiance to the Federated Suns have taken to these changes, while those claiming allegiance to Archon Katherine and/or the Lyran Alliance either wear the standard AFFC or the LAAF uniform. Units that are generally neutral also continue to wear the standard AFFC uniform.

AFFC UNIFORMS

Apart from its shift in preferred dress uniforms, the current AFFC retains the same basic uniforms its soldiers have worn for the past several centuries. As with its dress combinations, though, all of these uniforms have undergone some evolution since they were introduced in 3042. The following section details the uniform combinations that are most commonly found throughout the AFFC and traditional Federated Suns units in service with the LAAF. All personnel are authorized to wear patches denoting their major unit or command on their right shoulder and a patch denoting the department or March of assignment on their left.

Dress Uniform

The dress uniform combination now commonly worn by traditional AFFS units is very similar to that introduced to the Federated Suns State Command in 3042. The uniform coat is a relatively plain design of a deep green, the same color associated with the Federated Suns traditional uniforms. The cuffs are set apart in a green, silver or gold color, depending upon whether the wearer is enlisted, an officer or a general officer. Subtle piping, in the color associated with the wearer's branch of service, also sets apart this cuff area and highlight the rest of the uniform coat. The high collar completely hides the plain shirt or blouse worn underneath. Epaulets bearing rank insignia are mounted on the coat's shoulders, and decorations are worn on the coat's left breast. Additionally, all personnel wear a green cape with this uniform to official state functions; enlisted personnel wear a half-cape, while officers wear a full cape with silver trim and general officers wear the full cape with gold trim. Officers who have graduated from a Lyran academy often choose to wear their school rag around their waist, with the end of the rag extending down their left leg.

The uniform's trousers are likewise a deep green, with a single stripe indicating branch of service running down each leg. The trousers are of the same stirrup-design as the AFFC uniform, extending over the wearer's black shoes or boots. MechWarriors often add metal stirrups.

This uniform is also commonly worn by many ranking AFFC staff and officers while not in the field. When worn on a daily basis, most officers shed the coat and simply wear their rank insignia on the shoulders of a simple shirt or blouse. This shirt is a pale shade of the wearer's service-branch color.

The colors of the service branches are as follows: MechWarriors wear red, aerospace pilots blue, infantry light green, armor tan, naval crew black, technical and service gray, administrative yellow, medical white, legal silver, and logistics and transportation brown.

Field/Combat Uniform

The AFFC's field uniform is based on a jump suit, colored or camouflaged as dictated by the local commander based on conditions. This jump suit is commonly worn by all AFFC personnel when the situation does not warrant a different uniform combination. The jump suit is constructed to allow a full range of movement by its wearer and provides ample storage space in the form of several pockets.

Combat and other field troops add a simple field jacket, made with a bullet and laser-resistant polycarbonate weave, giving them a basic level of protection against enemy fire and the elements. Infantrymen also wear form-fitting body armor and packs or other load-bearing equipment. A helmet provides head protection for the wearer and displays his or her rank. The helmets of most Regular Army soldiers are also equipped with an integral night-vision and rangefinder scope.

MechWarrior Uniform

MechWarriors wear a far more abbreviated uniform while operating their combat machines. At the very least a sleeveless and legless bodysuit is worn underneath a cooling vest. Many Regular Army MechWarriors have been issued full-body cooling suits, giving them superior heat management inside the cockpit and significant environmental protection outside.

Aerospace Pilot Uniform

Like their MechWarrior brethren, AFFC aerospace pilots wear a full-body suit, though one that incorporates a number of specialized bladders that assist the pilot in dealing with the stresses of high-G maneuvers. In addition, this aerospace suit is airtight when its helmet is sealed, allowing the pilot to survive for up to an hour in a hostile environment with just the small oxygen tank attached to the suit. Outside the cockpit, pilots commonly wear a highly stylized jacket. Rank insignia, when worn, is displayed on the helmet and the bodysuit's shoulders.

Technician Uniform

Technical crews wear the same jump suit as infantrymen, though they add a vest with a number of small pockets and hooks sewn into it to hold their most commonly used tools. Rank insignia is worn centrally on the chest of this fire-retardant and electrically insulated vest. Other protective gear is worn as necessary.

AFFC RANK STRUCTURE

As with the uniforms themselves, the rank structure of the AFFC is undergoing a radical shift. Officially, all elements of the AFFC continue to operate under its standard rank structure as defined in 3042. With the rampant rise of nationalism and the growing animosity between those claiming allegiance to the Federated Suns and the Lyran Alliance, however, it was only a matter of time before the old AFFS system came back into use, especially in light of the LAAF's return to the classic Lyran system.

Because the classic AFFS rank structure is the most pervasive within the Federated Suns, it is described below. The AFFC rank that most closely equates to each (if any) is listed in parentheses next to the title. Naval ranks are named and described within each section. The rank insignia themselves have been given a more modern look.

Enlisted Ranks

The insignia for each of the enlisted ranks is based on a system of stripes, known as "Scars," upon which a black sword has been placed. Scars are worn either on the uniform's shoulders (below any unit and region patches) or on plain green epaulets; when not required to wear camouflaged or subdued uniforms, the scars are the color of the wearer's service branch. Enlisted personnel belonging to a planetary militia (instead of the Regular Army or a March Militia) have their scars edged in dark green, while the rest have theirs edged in black.

Recruit (Recruit)

Recruits are the lowest enlisted personnel and attain this rank upon enlisting in the AFFC. Recruits wear a black sword on a green field as their insignia of rank.

Private (Private)

Recruits are immediately promoted to the rank of private after graduating from a basic military training program. Privates continue at this rank throughout their advanced training courses and into regular service. Privates wear one Scar.

Private, First Class (Private, First Class)

The rank of private, first class was initially a holdover from both the AFFS and the LCAF, though it was only an unofficial step in the AFFS at the time. When the AFFC's new system was introduced, this rank was left out, apparently considered irrelevant. Training center commandants and unit commanders continued to award the rank as an indicator of excellent service, though the AFFC did not officially recognize it for more than a decade. This practice continues today, denoting basic training honor graduates and those privates who have shown themselves capable and worthy of additional responsibility. Privates, first class wear one Scar with a silver sword.

Corporal (Corporal)

After a private, first class has served for a short time, normally a year or so, and has shown his or her ability to follow orders and potential for leadership, he or she is promoted to corporal. Corporals lead infantry sections and small staffs or work details. The insignia for a corporal is two Scars.

Sergeant (Sergeant)

In many ways, the weight of the AFFC's chain of command rests squarely on the shoulders of its sergeants. These noncommissioned officers are truly the AFFC's first line of management, as they lead squads, command vehicles, lead technical and gunnery crews, and have responsibility over staffs throughout the AFFC's bureaucracy. Most graduates from a military academy also enter service as a sergeant. Except for academy graduates, who wear three curved Scars, sergeants wear three standard Scars.

Sergeant-Major (Sergeant-Major)

Sergeant-majors are among the AFFC's most capable and experienced enlisted personnel. These noncommissioned officers can lead entire lances, platoons, crews or staff sections, depending on their service and specialty. The standard insignia for a sergeant-major is a V-Scar, consisting of two downward-pointing chevrons with a disc upon which the black sword is centered. MechWarrior and other academy-graduate sergeant-majors wear the V-Scar with two curved chevrons. This is the highest rank an academy graduate can hope to achieve without attending an officer training school.

Command Sergeant-Major

The rank of command sergeant-major is another that was never officially adopted by the AFFS and so was not used in the AFFC, yet it lingered on unofficially. In today's AFFC, like private, first class, it is an unrecognized rank but used nonetheless. In general one sergeant-major per regiment is awarded the honorary rank of command sergeant-major. Command sergeants-major wear a V-Scar with three chevrons and a silver sword.

Officer Ranks

Officer rank insignia are mounted on epaulets in the shade of the wearer's service branch. The epaulets of general officers are further edged with gold piping.

Cadet/Subaltern (Cadet)

Cadets wear plain green epaulets during their stay at an academy, though they are authorized to wear the academy's cadet rank upon the epaulet. After graduation, those who have completed officer training are promoted to subaltern. The insignia for a subaltern is a plain green epaulet, trimmed in the color of their branch of service.

Leftenant

Subalterns are promoted to the rank of leftenant after successfully completing a six-month probationary period. Senior leftenants command 'Mech lances and armor and infantry platoons. Former sergeants-major who complete officer training are promoted directly to the rank of leftenant. The insignia for a leftenant is a branch-colored epaulet with a white border.

Captain (Hauptmann)

AFFC Captains command companies of soldiers and, like sergeants in the enlisted ranks, are collectively considered the backbone of the AFFC officer corps. Captains wear epaulets with one narrow white band across the base.

Major (Kommandant)

Field battalions and large staff organizations are commanded by majors. Because they are often required to operate cut off from the chain of command for extended periods of time, only those captains who show an unswerving dedication to duty and an ability to think on their feet are promoted to major. The insignia for a major is an epaulet with one broad band across its base.

Leftenant Colonel

While the rank of leftenant colonel existed in both the AFFS and the LCAF, the generals in charge of forming the AFFC chose to eliminate this rank to help streamline the new combined military. As with ranks such as command sergeant-major, the leftenant colonel lived on unofficially until it was recently returned to official use. A leftenant colonel is usually a senior battalion commander and/or a regimental executive officer, or, in many militia units, the commander of an infantry or armor regiment. In the Naval Service, these officers, known as light commodores, commonly command DropShips and Assault Ships. The insignia for this officer is two bands, one broad and one narrow, set upon the epaulet.

AFFC RANKINGS

Officers	Admiralty Equivalent
Prince	
Marshal of the Armies	
Field Marshal	Field Marshal
Marshal	Fleet Admiral
General	Admiral
Major General	Vice Admiral
Leftenant General	Rear Admiral
Colonel	Commodore
Leftenant Colonel	Light Commodore
Major	Major
Captain	Captain
Leftenant	Leftenant
Subaltern	Subaltern
Cadet	Cadet
Noncommissioned Officers	
Command	Sergeant-Major
Sergeant-Major	Sergeant-Major
Sergeant	Sergeant
Corporal	Corporal
Private, First Class	Private, First Class
Private	Private
Recruit	Recruit

Colonel (Leftenant General)

Colonels command entire regiments on the field of battle or departments within the AFFC bureaucracy. Within the Naval Service, these officers are known as commodores and commonly command DropShip squadrons or whole JumpShips. During the years of the combined AFFC, officers of this rank were known as leftenant generals but were never truly considered general officers and were still commonly referred to as colonels.

This rank represents the final step in the careers of many officers. Up to and including this rank, officers are expected to show loyalty to their men as well as imagination in their assignments but are still required to strictly follow their orders and work well within a staff. To have any hope of receiving a promotion to a leftenant general, however, the officer must be willing and able to take sole responsibility for a major project or command, a requirement to which many capable officers simply cannot make the adjustment. Colonels and commodores wear an epaulet with two broad bands.

Leftenant General

Though bearing the same name as the AFFC rank now known as colonel, today's leftenant generals command far more responsibility than did their AFFC predecessors. These officers commonly command infantry and armor brigades, entire planetary militias, and major staff sections and serve as command staff members in an RCT. In the Naval Service, rear admirals command multiple squadrons of DropShips or small JumpShip flotillas and serve as WarShip executive officers. The insignia for this officer is an epaulet with two broad bands and one narrow band.

Major General (Hauptmann General)

Leftenant generals who prove not only their command ability but also their tactical, strategic and political acumen can be promoted to the rank of major general. Officers elevated to this rank can expect to head important schools and institutions, command March Militias, serve as a key officer within a PDZ or RCT command staff, or, in the Naval Service as a vice admiral, command one of the AFFC's few WarShips. Major generals wear epaulets with three broad bands.

General (Marshal)

The rank of general is roughly equivalent to the AFFC marshal, or at least the "lower half" of that rank. Current AFFC generals often serve as the executive officers within PDZs and RCTs, though several have proved themselves enough that they are given full responsibility for these organizations. Generals can also head entire divisions within a military department or command the militia for one of the Federated Suns' largest planets. The naval equivalent is admiral, an officer who has responsibility for a small flotilla of ships formed around a single WarShip or for a naval repair yard. The insignia for a general is an epaulet with three broad bands and one narrow band.

Marshal

AFFC marshals command premier RCTs, major PDZs and the most prestigious divisions within a military department. The naval equivalent is a Fleet Admiral, who literally commands an entire naval fleet of WarShips, JumpShips and DropShips. These officers wear epaulets with four broad bands.

COMPARATIVE RANK TABLE

OFFICERS

Federated Commonwealth	Federated Suns	Lyran Alliance
Archon-Prince	Prince	Archon
Marshal of the Armies	Marshal of the Armies	General of the Armies
Field Marshal (Fld. Marsh.)	Field Marshal (Fld. Marsh.)	General (Gen.)
	Marshal (Marsh.)	Kommandant-General (Kmd.-Gen.)
Marshal (Marsh.)	General (Gen.)	Hauptmann-General (Hpt.-Gen.)
Hauptmann General (Hpt. Gen.)	Major General (Maj. Gen.)	Leutnant-General (Lut.-Gen.)
	Leftenant General (Lft. Gen.)	
Leftenant General (Lft. Gen.)	Colonel (Col.)	Colonel (Col.)
	Leftenant Colonel (Lft. Col.)	Leutnant-Colonel (Lut.-Col.)
		Hauptmann-Kommandant (Hpt.-Kmd.)
Kommandant (Kmd.)	Major (Maj.)	Kommandant (Kmd.)
Hauptmann (Hpt.)	Captain (Capt.)	Hauptmann (Hpt.)
Leftenant (Lft.)	Leftenant (Lft.)	First Leutnant (1st Lut.)
	Subaltern (Sub.)	Leutnant (Lut.)
	Cadet	Cadet

ENLISTED

Federated Commonwealth	Federated Suns	Lyran Alliance
	Command Sergeant-Major (Command Sgt.-Maj.)	Senior Sergeant-Major (Senior Sgt.-Maj.)
		Staff Sergeant-Major (Staff Sgt.-Maj.)
Sergeant-Major (Sgt.-Maj.)	Sergeant-Major (Sgt.-Maj.)	Sergeant-Major (Sgt.-Maj.)
		Staff Sergeant (Staff Sgt.)
Sergeant (Sgt.)	Sergeant (Sgt.)	Sergeant (Sgt.)
		Senior Corporal (Senior Cpl.)
Corporal (Cpl.)	Corporal (Cpl.)	Corporal (Cpl.)
	Private, First Class (Pfc.)	Private, First Class (Pfc.)
Private (Pvt.)	Private (Pvt.)	Private (Pvt.)
Recruit	Recruit	Recruit

Field Marshal (Field Marshal)

Field marshals command great responsibility within the AFFC. They lead entire Combat Theaters (or Marches, in the case of the most senior field marshals), military departments, the AFFC's entire Navy and major military organizations composed of multiple RCTs and brigades, such as the Davion Brigade of Guards. During wartime, these officers are named by the Prince or the Marshal of the Armies to lead major task forces. The insignia for this rank is a silver Federated Sunburst on the epaulet.

Marshal of the Armies (Marshal of the Armies)

The AFFC's Marshal of the Armies has full responsibility for the Federated Suns' entire military. This officer wears the same insignia of rank as the field marshals, only with a silver sword set upon the sunburst. When wearing a military uniform, the First Prince has the right to wear a golden version of this insignia.

AWARDS AND DECORATIONS

For the most part, the AFFC retained the AFFS tradition of awarding its people only a handful of different decorations for exemplary service or heroism under fire, even with the addition of five distinct decorations since the Fourth Succession War. The two individual State Commands, as well as the High Command, shared the responsibility for awarding these decorations. Each State retained its own unique awards, however, along with the authority to award them. Only the High Command or the Archon-Prince could award a decoration unique to one state or the other. The awarding authority for each is denoted in parentheses next to the decoration's name; AFFC indicates an award given by either state, while AFFS indicates an award given only by the Federated Suns State Command.

The Medal Excalibur (AFFS)

The Medal Excalibur is the highest honor the Federated Suns can bestow upon an individual and is awarded only to someone whose actions have resulted in some major enhancement to the Federated Suns. In the centuries since the fall of the Star League, it has been awarded only thirty-two times. The medal itself is a gold disc engraved with the image of King Arthur's mythical sword embedded in an anvil and set upon a Federated Sunburst. This award also confers with it the title of "Knight of the Realm" and an accompanying land grant.

The Sunburst Medal (AFFS)

The Sunburst Medal is awarded for exceptional duty, devotion and sacrifice in the name of the Federated Suns and exists in three ascending levels from lowest to highest: Silver, Gold and Diamond. The Starburst Medal can be awarded to both military personnel and civilians, as with the Medal Excalibur; the only requirement is that the recipient must perform a heroic act with statewide import. The medal is a black disc with a silver, gold or diamond sunburst on it.

Federated Commonwealth Medal of Honor (AFFC)

Awarded to those soldiers who show unflinching bravery at risk of life and limb in service not only to the Federated Commonwealth but also to their fellow men. A majority of these decorations are presented posthumously, as most of the recipients are killed while performing the actions that earned the award. Surviving awardees, and the immediate families of the deceased, receive a pension from the AFFC and preferential admissions to Federated Commonwealth academies and universities. The medal is a platinum disc upon which a Lyran Fist is wielding a Federated Sword, set upon an engraved sunburst.

Federated Commonwealth Star (AFFC)

One step lower than the F-C Medal of Honor, the Federated Commonwealth Star is awarded to those soldiers who show unquestioned bravery and honor under combat situations, and whose actions have a positive and measurable impact upon a major military endeavor. The F-C Star is a platinum medallion in the shape of a classic five-pointed star.

Order of Steiner-Davion (AFFC)

Originally named the Order of Davion, this prestigious award is given to those soldiers who show tremendous loyalty to the Steiner-Davion family and is therefore only awarded by the First Prince. Unfortunately, those few who have been awarded the order are themselves potential sources of conflict, as Archon Katherine has just added several AFFC officers to the order, following those Prince Victor added in the past decade. The badge of the order shows the Steiner-Davion family crest—a crouched fox with Steiner-blue eyes set upon a blood-red shield.

Knight of the Federated Suns (AFFS)

Membership in this select group is limited to the Federated Suns' hundred brightest and most heroic citizens. As with the Order of Steiner-Davion, membership in the KotFS is for life and includes a land grant and a generous yearly stipend. Members of this elite group are unquestioned leaders within the Federated Suns and are among the most vociferous opponents of Archon Katherine. The insignia of this order is a Federated Sunburst upon which the silhouette of a man (or woman, depending upon the gender of the recipient) holding a sword over his head is embossed.

Syrtis Medal of Honor (AFFS)

Only the commander of the Capellan March Regional Command has the authority to award this medal, which likewise can only be awarded to troops under his command. Field Marshal Duke George Hasek has recently awarded quite a few of these medals for actions taken in defense of the March and the St. Ives Compact against Capellan aggression. This diamond-shaped medal, made to look like the Hasek family crest, is made from gold, silver and highly polished 'Mech armor. In addition to the medal itself, recipients of this award are given the title "Knight of the Capellan People."

Robinson Medal of Valor (AFFS)

This award is given by the commander of the Draconis March to any of his troops who go beyond the call of duty in service to the March. Much as with the Syrtis Medal of Honor, this medal is as often given for purely political reasons as for legitimate military reasons. The medal consists of a burnished red disc emblazoned with a Federated Sword and Sunburst atop a Star of David.

Dragon Slayer's Ribbon (AFFC)

Though this ribbon has not been awarded for quite some time, many units still proudly bear the Dragon Slayer's Ribbon upon their standard. Originally an LCAF decoration, the AFFS began to award this ribbon just before the Fourth Succession War, when quite a few regiments won it following their successful campaigns along the Draconis Front. The ribbon itself is green and white, and it shows the symbol of the Draconis Combine, slashed with a black line.

Limp Sword Flag (AFFS)

Awarded for wildly successful action against the Capellan Confederation in which a severe blow has been struck against the Confederation, scores of individual regiments received this award following the Fourth Succession War. Fashioned to parody the Confederation's national symbol, the flag shows a drooping sword held by a smashed 'Mech hand, set against an inverted green triangle on a black background.

Regional Service Awards

Following the Fourth Succession War, both the AFFS and the LCAF instituted three awards denoting meritorious service under fire to a soldier's nation in one of three former Capellan regions. When the two militaries merged in 3042, they chose to keep these three awards active.

The Pride of Sarna Medal is awarded to soldiers who prove their mettle on the battlefields of the former Sarna March. The medal itself is a titanium medallion with a representation of the world of Sarna, framed in laurel, with a Federated Sword superimposed upon it.

Though no longer as frequently presented as the Pride of Sarna Medal, the Tikonov Sword is still awarded for gallantry on the worlds of the once-Capellan Tikonov Commonality. The sword itself is a rapier with an intricate sterling silver handguard and is authorized to be worn with the dress uniform.

The final regional award is the St. Ives Diamond. The diamond has become one of the most commonly presented awards in the AFFC, as the Capellan Confederation has attacked the St. Ives Compact numerous times over the past three decades. The St. Ives Diamond is a diamond-shaped silver medallion inset with eighteen small emeralds, representing the worlds of the St. Ives Compact.

AFFC PROFILES

DMI PROFILE CC62-RH5495U

Subject: George Michael Hasek
Position: Duke of New Syrtis, Field Marshal and Minister of the Capellan March
Born: 26 August 3030, Avalon City, New Avalon
AFFC Career:

- Warrior's Hall, New Syrtis (student): 3045–47
- War College of Goshen (OCS): 3047–48
- 1st Federated Suns Armored Cavalry (Lance & Company CO): 3048–51
- 6th Syrtis Fusiliers (Battalion & Regiment CO, RCT XO): 3051–54
- Sakhara Academy (Federation War College): 3054–55
- Capellan March Regional Command (Director of Logistics & Operations): 3055–59
- Capellan March Regional Command (Director): 3059–present
- Duke of New Syrtis and Minister of the Capellan March: 3060–present

The firstborn of Kym Hasek-Davion (nee Sorenson) and Morgan Hasek-Davion's four children, Duke George chose for himself a military life. Like his father before him, he demanded no special treatment. Instead, he chose to progress through his early career solely on the basis of his performance. The fact that he was heir to the Capellan March demanded he be put on the "fast track" to higher command within the AFFC however, and he transferred into the Capellan March Regional Command's HQ, assuming command some four years later. He inherited the dual titles of Duke of New Syrtis and Minister of the Capellan March following the assassination of his father.

Though a contemporary of Prince Victor Steiner-Davion (the two spent much of their early childhoods together as the Hasek-Davions lived for some time on New Avalon), Duke George is not an outspoken ally of his former Prince. On the surface, Duke George is a "child of the Capellan March," though unlike his grandfather, Duke George is wholly loyal to the men and women under his command. He is also a firm opponent of Archon Katherine, though he prefers to spar with her on a subtle battlefield, a fact that illustrates his intelligence and strategic acumen.

Based on his actions since assuming both military and political command of the Capellan March, it is obvious he is an officer with much ambition, one who possesses the loyalty of the majority of the men and women under his command.

DMI PROFILE DC62-RS22331U

Subject: James Dassert Sandoval
Position: Duke of Robinson, Field Marshal and Minister of the Draconis March
Born: 13 March 2994, Castle Sandoval, Robinson
AFFC Career:

- Robinson Battle Academy (student): 3010–14
- 1st Robinson Rangers (Lance, Company & Battalion CO): 3014–21
- 2nd Davion Guard (XO & CO): 3021–22
- NAIS CMS (Federation War College & Instructor): 3022–23
- Draconis March Coreward Combat Theater (CO): 3023–28
- 1st Federated Commonwealth RCT (CO): 3028–3033
- Federated Commonwealth Corps (CO): 3030–37
- Minister of the Draconis March: 3037–present
- Duke of Robinson: 3048–present

The product of a close-knit family, Duke James and his siblings formed a strong sense of duty at an early age, as well as a deep resentment toward the Draconis Combine. As befitting his position as firstborn son of the Duke of Robinson, James entered the Robinson Battle Academy at the earliest age possible, and later served within his father's First Robinson Rangers. His reputation as an outstanding motivator of men won him the command of the 2nd Davion Guards, while his tactical genius later earned him an instructor position as the CMS. His stock was so highly regarded that Prince Hanse gave him full latitude to build the F-C Corps as he saw fit. Though Prince Hanse expressed the desire to see him command the entire allied AFFC, Duke James could not ignore his family's position, assuming command of the Draconis March upon his father's retirement in

3037. He later inherited the ducal title upon his father's passing eleven years later.

Duke James is resolutely loyal to the Federated Suns and once felt the same towards the F-C alliance. Though outraged at Archon Katherine's secession, his feelings of loyalty to the Lyran people cooled as it became evident that few of them cared to continue the alliance that had revitalized their military and all but saved their nation from the Clan hordes.

Perhaps the greatest problem facing Duke James is the factious division of the AFFC. In an effort to keep his command together, he has of late unified his March under the traditional call to arms against a resurgent Draconis Combine, a call the Archon is unlikely to silence. Of course, as the most senior and respected Field Marshal within the AFFC, even a Prince would have difficulty challenging him.

DMI PROFILE FC62-RD7432U

Subect: Jackson Kennard Davion
Position: Marshal of the Armies
Born: 1 November 3000, Stirling, Argyle
AFFC Career:

Albion Military Academy (student): 3017–20
Davion Assault Guards (Lance, Company & Battalion CO, XO): 3020–26
Second Crucis Lancers (CO): 3026–32
1st Davion Guards (CO): 3032–36
Crucis March Coreward Combat Theater (CO): 3036–42
Capellan March Regional Command (Director): 3042–45
Federated Suns State Command (CO): 3045–60
Marshal of the Armies: 3060–present

Jackson Davion is the third of Yvonne Davion's seven children. All seven of the former Field Marshal and Prince's Champion's children entered AFFS service as soon as they could, each gaining some notoriety. Three of his siblings were killed in action throughout the years however, and though not accorded the same open opportunities as his older sister and brother, Jackson proved himself the most capable of Yvonne's children.

He quickly rose through the ranks, displacing his distant cousin, Field Marshal Stephen Davion, as commander of the First Davion Guards after the elder Davion finally fell out of favor. His service with the Guards eventually earned him command of the Capellan March, and later command of the entire Federated Suns State Command, following closely behind Duke Morgan Hasek-Davion's similarly meteoric rise.

Like most high-ranking AFFC officers who served under Prince Hanse, Jackson Davion is a true patriot, setting the needs of his nation ahead of his own. It is because of this view he is counted as one of the Archon's closest allies. While not a vocal supporter of Archon Katherine, he does view her as the Federated Suns' legal ruler and has sworn fealty to her. While likely an impossible task, he has devoted himself to keeping the AFFC united under one standard. Unfortunately, while his efforts have been largely unsuccessful, they have kept him busy enough to allow Field Marshal Gallagher to usurp some of his authority. Those efforts have apparently also blinded him to some of the less ethical instructions the Archon has given to the units she considers loyal to her.

DMI PROFILE XC62-CG82472U

Subject: Simon Jacob Fitch Gallagher
Position: Field Marshal, The Prince's Champion and Director of the Crucis March Regional Command
Born: 13 January 3001, Spring Fields, Timbuktu
AFFC Career:

University of Timbuktu (student): 3019–23
Rep-Dep Corps (LCAF): 3023–30
Department of the Quartermaster: 3030–33
412th Tamar Rifles (Logistics Chief): 3033–35
32nd Lyran Guard (Logistics Officer): 3035–41
NAIS CMS (Federation Command & Staff College) 3041 - 42
1st Argyle Lancers (Chief of Staff): 3042–43
Avalon Hussars (Director of Support Services): 3043–47
Crucis March Regional Command (Director of Logistics, XO): 3047–61
Crucis March Regional Command (CO): 3061–present

Field Marshal Gallagher is an officer who never sought out command, yet whose unmistakable talents demanded promotion through the AFFC. Gallagher has never been a combat officer, instead joining the Lyran Rep-Dep Corps directly after graduating from university. Throughout his career, he has shown himself to be an extremely capable logistician, a skill set that has kept many units throughout the old Federated Commonwealth operating in the face of adversity.

While he is an outstanding administrator, Gallagher is anything but a leader of men. During his brief assignment as Chief of Staff for the Argyle Lancers, the unit's morale dropped to an all-time low after Gallagher lost control of the staff sections operating under him. In his next two assignments his performance skyrocketed, though only because of the strong officers serving around him, keeping his staff in line while he concentrated on keeping lines of supply open.

Profiles indicate he is a weak-willed individual, but one with a strong sense of duty. Owing to his Lyran heritage and his personal loyalty to Melissa Steiner-Davion, he was one of the first senior officers within the Crucis March to accept Archon Katherine's leadership unconditionally. Since her assumption of control on New Avalon, the Archon has promoted Gallagher to commander of the Crucis March Regional Command and has given him the added title of the Prince's Champion. It is in those capacities that the Archon uses Gallagher as a virtual mouthpiece to direct the AFFC when she feels the need to bypass the Marshal of the Armies entirely. Hence, few senior AFFC officers feel anything but contempt for Gallagher.

SHARPENING THE CLAWS

The Federated Suns military has traditionally been considered the strongest in the Inner Sphere. The reason is simple: it chooses the right candidates and then invests heavily in their training. This training continues far longer than in most other militaries, and as a result, the men and women of the AFFC are far better prepared.

The training cycle does not end with graduation from the initial branch schools, however. All levels of the AFFC's leadership stress the importance of continued training throughout a soldier's career. Every AFFC unit, whether Regular Army or militia, participates in at least one major combat exercise on its homeworld every year, while Regular Army and March Militia units travel to a different world to do the same at least once every two years.

Likewise, individual soldiers and officers have multiple opportunities to advance their training throughout their tours of duty. Not only are they permitted, and even encouraged, to cross-train in other specialties while serving in their assignments, but the AFFC also offers a number of schools to provide them additional training opportunities. For example, all colonels and leftenant colonels are given the chance to attend the Federation Command and Staff College, a program they must graduate from before being considered for promotion to leftenant general.

The Department of Military Education (DME) is composed of four major directorates, each responsible for a distinct segment of the AFFC's training program. The members of these directorates, with the assistance of commanders in the field and the High Command, also devise the curricula and graduation requirements for each of the AFFC's schools and academies.

The AFFC's four major training routes are described below.

ENLISTMENT

The Directorate of Noncommissioned Training is tasked with the training of all of the AFFC's enlisted personnel. The directorate runs the facilities that train Regular Army and March Militia personnel. A liaison office coordinates with the schools outside the AFFC's direct control, such as those run by a planetary militia or some other outside concern, that wish to have AFFC accreditation. Without accreditation, given only to those schools that follow a strict curriculum requirement, graduates will not qualify for regular AFFC service without supplemental training.

BASIC MILITARY TRAINING

Enlistment accounts for some 85 percent of the recruits entering the AFFC every year, and more than 80 percent of those recruits receive their initial training at a state-run facility. These recruits receive an intensive course of instruction that prepares them for service in any branch or specialty within the AFFC. The specifics of each course vary between schools—indeed, some of the toughest courses are actually found at privately run schools and training camps.

Though the AFFC is in constant need of new recruits, it requires that its recruits meet certain standards. All recruits must pass a physical examination that tests their health and manual dexterity. They must also pass a rudimentary skills test before being accepted. Even with these precautions in place, some 10 percent of new recruits are mustered out of initial training before graduation, unable to handle the stress of military life.

SPECIALTY SCHOOLS

After graduating from the initial training camps, all AFFC recruits are screened for aptitudes and sent on to a Specialty School to learn their occupational specialty. Just as with the basic course, every PDZ hosts a Specialty School. Likewise, many private concerns also provide specialty training. In most cases, however, these facilities can train recruits in only the most basic specialties, such as infantry or cavalry. Only a few facilities in each March can train recruits in highly skill-intensive specialties.

After graduation from a Specialty School, recruits are assigned to a position within the AFFC based on their performance during their training and the recommendations of their instructors. A few quiet words from a patron (or enemy) within the AFFC bureaucracy have been known to make or break the potential career of a raw recruit.

These schools do not simply graduate classes of new recruits. Thousands of tenured officers and noncoms also cross-train in different specialties every year, providing the AFFC a far more flexible and skilled soldiery than almost any other military in the Inner Sphere. Most specialties are open for cross-training (usually with only the recommendation of an individual's commanding officer), and enough billets are set aside to advance whole classes of cross-training personnel. AFFC non-commissioned officers are eligible for this cross-training only after their initial four-year tour of duty is completed.

MILITARY ACADEMIES

The military academy has always been the preferred school of choice in the Inner Sphere for leaders to draw recruits from. Because these academies rarely spare any expense in training their students, the cost of a private academy education is prohibitive to all but the most affluent families. Those lucky few who gain admittance to a state-run academy are guaranteed an outstanding education at no cost, though they are required to serve a minimum time in the AFFC, usually six to eight years.

The AFFC draws recruits from almost a dozen primary academies scattered throughout the Federated Suns, as well as from a number of secondary institutions. Students graduating from one of the primary schools receive a better education and are often held in higher regard by many officers within the AFFC. Most of these academies also run specialized courses for existing AFFC officers and noncoms who wish to advance their education or cross-train in a different specialty.

STANDARDS OF TRAINING

The AFFC maintains a number of stringent requirements for each of its academies, monitored closely by the Department of Military Education. The AFFC also maintains reciprocal agreements with several military academies located throughout the Inner Sphere, such as the St. Ives Academy of Martial Sciences—though this has become strained with the ongoing war between St. Ives and the Capellan Confederation—allowing students from these institutions to enter the AFFC just as if they had graduated from one of the Federated Suns' own academies. Of course, attendees must still fulfill the curriculum requirements. Quite often, students who excel in their studies are offered the opportunity to participate in an exchange program with another academy to give them the opportunity to experience as much of the Inner Sphere as possible.

Enrollment

Because the AFFC does not directly run all the academies it draws recruits from, there is some variation in enrollment procedures. Since the end of the Fourth Succession War, however, AFFC has set up minimum admissions requirements that each institution must adhere to, though each is free to make the criteria more restrictive.

At a minimum, potential cadets must meet a far more grueling physical standard than those who simply enlist. Applicants must also show an outstanding academic ability and pass an extremely challenging entrance exam. Some 30 to 40 percent of all applicants are eliminated by these basic requirements every year; however, this does not preclude a candidate from enlisting or even entering a prep school, both of which could potentially earn them an academy berth later.

In addition to these standards, all academies require some sort of recommendation, often from a senior military officer. Though the Department of Military Education has done much to prevent graft, there are still a few academies where a well-placed bribe, or simply the right parentage, will ensure a position within the cadet corps.

Curriculum

All AFFC-accredited academies follow the same basic curricula, as directed by the Department of Military Education. Completion of such basic military courses, such as first aid, navigation and wilderness survival, are mandatory for all cadets, as are more academic subjects like the military campaigns of Caesar, Patton and Kerensky.

A cadet's first year at an academy is spent in this basic military training. Following this initial year, a cadet enters specialty field training. A few academies specialize in a particular field, or a narrow set of fields, offering training in only a few specialties. Most, however, offer training in quite a few fields.

The AFFC continually strives to accept only the most qualified candidates into its training programs. As a result, admission into many specialties is through academic testing and instructor recommendation only. Those who do not meet the admission or physical requirements for their chosen specialty can select a different specialty or even transfer to a different institution. Once excluded from a field, however, it is often impossible for a cadet to receive that training later.

When the academy is co-located with a civilian university, outstanding cadets may also take on a regular degree. Though this option is only offered to a few select cadets, the AFFC supports this program whenever possible.

Cadets who show outstanding leadership potential are offered the chance to attend an Officer Candidate School. In this one-year course, candidates are immersed in leadership theory and classical military theory.

Graduation

Cadets normally complete their academy training after three to six years, depending on their chosen specialty. Upon graduation, cadets entering the AFFC receive a promotion to sergeant or, if the cadet successfully completed an OCS program, to subaltern.

Graduates receive assignments in the AFFC based on their performance at the academy and their specialty. In some cases, a graduate's initial position is secured by a patron's political influence. Additionally, cadets who graduate from one of the AFFC's primary academies in a combat specialty commonly spend a single tour of duty in a training cadre made up of their classmates and commanded by veteran AFFC officers. Outstanding graduates are assigned directly to Regular Army units.

All graduates of a state-run academy are obligated to serve at least six years in the AFFC, while those who graduate from a privately run academy without the benefit of a scholarship are only obligated for two years.

ALBION MILITARY ACADEMY

The Albion Academy is the oldest and one of the most respected academies in the Federated Suns, dating back to the twenty-fifth century. Prior to the foundation of the NAIS, graduation from Albion almost ensured a successful career. Generally considered the most difficult academy in the Federated Suns, Albion fields the second largest cadet corps of any academy in the nation.

Located some thirty kilometers south of Avalon City on New Avalon, Albion has access to anything its cadets need, including a full combat training center. As many of the AFFC's most senior officers, including members of the Davion family, are alumni of Albion, the academy receives almost the same amount of support as does the NAIS and its attached College of Military Sciences. Because the NAIS receives the most attention and support, however, many of Albion's cadets feel some animosity toward their rival school.

Major General Frederick Hastings is Albion's commanding officer and superintendent.

Atmosphere

Considering its long history of graduating only the best cadets, Albion boasts the most physically and emotionally challenging program of any AFFC military academy. Excellence in all things military is demanded, and the cadets are often verbally assaulted or hazed during the course of their training. Though this practice is often decried by civilians, and the dropout rate is far higher than that of any other Federated Suns academy, Albion's graduates are also among the most decorated in the AFFC.

Because the academy is located so close to Avalon City, and by extension to the NAIS, Albion's cadets often interact with cadets from the CMS. The natural rivalry between the two schools continues with each new class of cadets, with the practical jokes perpetrated by each academy's cadets only feeding the fire of competition. This rivalry culminates in a yearly competition between each academy's cadet corps with a series of wargames on the Jasper Plain training field. The winners of these wargames are crowned "The Young Lions of New Avalon."

Special Notes

Because of its location at the heart of the Federated Suns, each of Albion's staff and candidates is thoroughly screened by the DMI to ensure they possess no potentially troubling political leanings or personal conflicts. Those who pass muster receive one of the best educations available in the Inner Sphere. The Albion Military Academy's training cadre can field two full BattleMech regiments, plus supporting elements.

NAIS COLLEGE OF MILITARY SCIENCES

While the Albion Military Academy is the Federated Suns' oldest training institution, the NAIS College of Military Sciences comes in a close second. The nearly forgotten origins of this institution lie with the New Avalon Military Academy, founded after the formation of the Star League. Overshadowed by the Albion Military Academy for many centuries, even to the point of many suggesting it be torn down or annexed by Albion, the NAMA barely survived the Succession Wars. It wasn't until Prince Ian Davion, himself a graduate of the beleaguered academy, came to power that the NAMA regained its prominence and again began to flourish.

Following the raid on Halstead Station, in which a huge Star League library was unearthed, Prince Hanse Davion built the New Avalon Institute of Science around the NAMA, giving the institution military and scientific prominence throughout the Inner Sphere. Renamed the NAIS College of Military Sciences, it quickly became known as the most lavishly equipped military academy in the Inner Sphere, with its huge network of advanced simulators, numerous Gauntlets and expansive live-fire training ranges.

General Ranier Wolfgram is the Commandant of the CMS.

Atmosphere

CMS cadets face the most difficult academic program of any military academy in the Inner Sphere. Space in each of the NAIS' programs, including the CMS, is limited, and as a result the atmosphere of the academy is highly competitive. As cadets are already under extreme pressure to excel in academic and military fields alike, a ban on hazing is actively enforced.

While below-average performance normally results in a transfer to a less demanding academy or even to enlistment, excellence is rewarded by increased responsibilities. Academically outstanding cadets can choose to study alongside civilian NAIS students to gain a degree in a non-military field, while cadets who master military skills often receive cross-training in one or more additional specialties or even early promotions into the training cadre.

Because the CMS is located in the heart of Avalon City, it often receives visits from the AFFC's most senior officers, as well as members of the royal family, further adding to its cadets' stress levels.

Special Notes

The cadets of the CMS have shown an unswerving loyalty to the Davion family for centuries, a feeling only amplified by the expansions during the reigns of Prince Ian and Prince Hanse. Of course, all NAIS and CMS attendees undergo an exhaustive background check by the DMI and/or MIIO. Like the Albion Military Academy, the CMS fields a large training cadre, including three full BattleMech regiments and a number of lesser supporting regiments.

WARRIOR'S HALL

Though it began life as little more than an officer's club with two training Gauntlets where SLDF and AFFS officers settled their disputes in mock combat, the Warrior's Hall on New Syrtis soon grew into a full-fledged military academy. By the end of the Star League era, it had added programs for nearly every military specialty and was graduating hundreds of cadets into the AFFS every year.

As the Warrior's Hall sits at the political center of the Capellan March, it has often served as the personal recruiting grounds of the Hasek family. Though that practice has all but disappeared over the past several decades, many of the cadets attending the Warrior's Hall owe much to the Haseks and continue to be life-long supporters of the family.

The Commandant of the Warrior's Hall is Leftenant General William Hasek.

Atmosphere

This academy's curriculum is very much focused on perfecting a cadet's military skills, often to the exclusion of any but the most basic academics. Since the formation of the AFFC some two decades ago, the Warrior's Hall has had an agreement with the University of Syrtis to allow cadets to pursue an education of more than just military skills. Less than 5 percent of Warrior's Hall cadets take advantage of this opportunity, however; many senior cadets and even some instructors argue that cadets should focus on their primary military studies.

The staff of the Warrior's Hall relies heavily on senior cadets to maintain strict discipline within the cadet ranks. Though this practice often results in excessive punishment for minor infractions, graduates of the Warrior's Hall are a skilled, if somewhat cold and calculating, group. The cadets of this academy also tend to harbor a deep hatred of the Capellan Confederation, especially for its Chancellor and armed forces.

Special Notes

The Duke of the Capellan March personally reviews the file of each potential cadet and has final approval of all applications. The regiments of the Syrtis Fusiliers often recruit the top graduates to fill vacant billets within their ranks.

ROBINSON BATTLE ACADEMY

Located on the capital world of the Draconis March, the Robinson Battle Academy has seen more than its share of good times and bad. Razed and rebuilt three times during the Succession Wars, the Battle Academy became a center for anti-Combine sentiments and a source of support for the Sandoval family. Following Prince Hanse's founding of the NAIS, the Sandoval family poured millions of C-bills into the Battle Academy in hopes of making it the equal of the CMS. Though the Albion Military Academy and the CMS are still considered superior schools, the cadets of the Battle Academy have access to an education and facilities equaled by few institutions in the Inner Sphere.

The Battle Academy experienced a sharp drop in attendance and prestige as detente grew between the Federated Commonwealth and the Draconis Combine following the Truce of Tukayyid. That trend has since reversed in light of the troubles brewing on the Draconis Front and the Federated Suns bankrolling the lion's share of Operation Bulldog and Task Force Serpent.

Hauptmann General Mai Fortuna, also commander of the First Robinson Rangers, is the Battle Academy's Commandant.

Atmosphere

The Battle Academy has always stressed a full curriculum of academic and military studies, reinforced by peer assistance and active guidance by staff counselors. Rather than base their curriculum solely on theory, the Academy's staff relies heavily on practical hands-on and field experience to train their charges, giving graduates a major advantage over many of their peers.

Because it is located at the heart of the Draconis March, the bulk of the Academy's cadet corps owes its allegiance to the Sandoval family. Likewise, no matter the prevailing sentiment throughout the rest of the AFFC, the cadets of the

Academy have never given up their deep hatred of the Draconis Combine.

Special Notes

The Duke of the Draconis March's power of approval over applicants to the Battle Academy often ensures a place for those who may not possess the aptitudes required for entry but profess loyalty to his family. The Robinson Rangers commonly recruits Battle Academy graduates, especially those who show unswerving loyalty to the Sandoval family and/or those with a borderline-psychotic hatred of the Draconis Combine.

SAKHARA ACADEMY

Though a small, privately run military academy, Sakhara nonetheless graduates some of the AFFC's most skilled personnel. The Academy is devoted to turning out the best combat soldiers possible, and as a result it only teaches a few specialty fields. All of the Academy's instructors are veteran AFFC officers, with some 80 percent being alumni of Sakhara who have returned to pass their knowledge on to a new generation of cadets.

Because it is privately run, the Academy must rely on the tuition it collects from its students to continue operating. The Academy does have a number of patrons inside the AFFC chain of command, as well as numerous corporate sponsors, ensuring its access to cutting-edge equipment for use in training its cadets.

Marshal Kilian Reason, AFFC retired and member of Sakhara's class of 3027, is the Academy's Commandant.

Atmosphere

As only the wealthiest families can afford to send a child to Sakhara, many outsiders consider the cadet corps a rather elitist group that looks down on anyone below their station. They are only half right.

The Sakhara "family" is a close-knit group, holding to many long-standing traditions. Many cadets hail from a long line of Sakhara graduates and often grow up with the children of other Sakhara alumni. Traditional hazing is not needed to maintain discipline; the fear of disappointing an instructor or a close relation who is an alumnus is usually more than enough to curtail any attitude problems. Cadets who experience difficulties with their studies can expect assistance from one of their classmates. Members of this "family" look out for each other and take care of their own problems. They do not look down on less fortunate people; they simply feel sorry for those who are not part of the family.

Special Notes

All Sakhara cadets complete officer training and enter the AFFC as leftenants after serving their two-year tours in the training cadre. Sakhara alumni in command positions are always looking to add more "family" members to their commands. Rarely has there been a year in which the requests for personnel have not exceeded the size of the graduating class by double or more.

ARMSTRONG FLIGHT ACADEMY

Unquestionably the premier flight school in the Federated Suns, and arguably in the Inner Sphere, the Armstrong Flight Academy is well into its third century of graduating the AFFC's most talented pilots. Co-located on Galax with Federated-Boeing Interstellar, the Inner Sphere's largest manufacturer of spacecraft, the Academy benefits from the experience of the company's brightest aerospace engineers, physicists and test pilots.

The Academy specializes in training pilots and aerospace and naval technicians. No other specialty fields are taught. In fact, the Flight Academy makes use of Galax's militia training camp to teach its cadets the required basic military skills. This allows the Academy's staff to concentrate on teaching their cadets to be the best pilots and technicians in existence.

Major General Christopher Saunders serves as the Academy's Commandant.

Atmosphere

Because many faculty and staff members are Federated-Boeing employees, Armstrong's atmosphere is, at least on the outside, the laxest of any AFFC academy. All the school's flight cadets realize that the next time they climb into the cockpit could be their last, however, and they show a great deal of personal discipline while conducting flight operations. "Hot-dogs" are tolerated neither by the faculty nor cadets and are quickly mustered out of the cadet corps. On the ground, however, Armstrong's cadets more than compensate for the stress in the cockpit or in the middle of a fusion engine.

Special Notes

Potential candidates are given far more rigorous physical testing than at any other academy. Because of this, slightly more than half of the Academy's applicants are eliminated before they even take an entrance examination.

POINT BARROW MILITARY ACADEMY

Once a flashpoint for terrorist activity, the Point Barrow Military Academy has purged the radical elements from its ranks and undergone a complete facelift. Though it was once considered merely adequate, apart from its technician and engineering programs, the Point Barrow Military Academy is today a shining example of the changes the AFFC wrought in both the Federated Suns and the Lyran Alliance.

Though it still specializes in technical fields, the Academy offers training in almost every military specialty. The cadets of the Academy owe a particular debt of gratitude to the late Archon Melissa Steiner-Davion, who helped shape the PBMA into one of the AFFC's primary military academies after the Fourth Succession War.

Hauptmann General Severine Brint is the Commandant of the Point Barrow Military Academy.

Atmosphere

Following its reformations in the 3030s, the PBMA turned into one of the Federated Suns' most demanding academies. Its students do not have access to the advanced simulators that Albion or CMS cadets do. Instead, they conduct nearly all of their training in or on the real thing. Because lives are on the line every time they conduct field training, failure is simply not an option. Dropouts because of health and mental problems brought on by extreme stress are not uncommon.

Special Notes

The Academy's staff and cadet corps feel they still owe a debt of gratitude to Archon Melissa. Many of PBMA's personnel have transferred that feeling to her daughter, giving Archon Katherine a bastion of support in the heart of the Federated Suns.

WAR COLLEGE OF GOSHEN

While not one of the AFFC's most notable academies, the War College of Goshen has been graduating skilled and talented soldiers for well over a century. Originally founded by a pair of retired AFFS generals who were tired of seeing the quality of the soldiers on the Capellan Front plummet, the school began life as the Goshen BattleMech Academy. As more and more students were attracted to the school, the founders added staff and courses. Within a decade, the newly renamed War College of Goshen was teaching almost every combat specialty to students from across the realm.

The War College is led by retired Marshal Lawrence Marlowe.

Atmosphere

Generals Alexander and Truscott set out to build an academy that taught cadets how to be not only great soldiers but also great thinkers. The College's curriculum includes a heavy course load of military history and classic and contemporary military strategy. Students spend far more time dissecting the methodologies of great military leaders than they do on their specialty studies. By the time cadets participate in their first live-fire exercises, they exhibit an almost Zen-like connection with their 'Mech or vehicle and a prescience for the battlefield.

Alumni scattered throughout the AFFC's bureaucracy ensure that the War College receives all the support it needs. Surprisingly, its commandant and staff require very little to keep the College operating. All of its primary training is accomplished using the actual 'Mechs, vehicles and weapons on live-fire ranges. The College does not even field an active training cadre; graduates are assigned directly into combat units.

Special Notes

The War College is often called the AFFC's best-kept secret because it does not publicly recruit potential candidates as do the AFFC's other academies. Instead, the College's halls are kept full by word of mouth. Much like the Sakhara Academy, attendance at the War College is often a "family affair," with generation after generation of soldiers sending their progeny there to receive what they consider the best education in the Inner Sphere.

KILBOURNE ACADEMY

Located in the Draconis March along the Federated Suns' border with the Outworlds Alliance, the Kilbourne Academy has never been considered one of the AFFC's premier military academies. On the other hand, it is one of the few institutes of higher learning in that region of space. Because of that, the Academy hosts a civilian university at its site, giving the inhabitants of the region the chance for an education and with it a better life. It may not be as flashy as the NAIS or have the long-held traditions of Sakhara, but the Kilbourne Academy provides a well-rounded education to all its students.

The Academy has recently become the target of a DMI investigation into sedition and treason within the region. Though no firm evidence has thus been uncovered and no suspects indicted, this ongoing investigation has caused enrollment, especially in the civilian disciplines, to plummet.

The Kilbourne Academy's commandant is Leftenant General Kara Rostov.

Atmosphere

As this is one of the few educational institutes in the region, competition for admission is usually stiff. Because of the DMI investigation, the Academy's atmosphere is tense, with many students fearing they will be the victim of an unfounded accusation or take the fall for whatever crime is being investigated.

Special Notes

The government of the Federated Suns provides scholarships for many of the civilians who attend the Kilbourne Academy; following their graduation, they are obligated to serve their nation for at least six years, as are its military cadets. Many Academy graduates find positions within the Draconis March bureaucracy or its Militia.

FILTVELT MILITARY ACADEMY

As many Filtvelt cadets like to say, their Academy is located "just this side of nowhere," a phrase that aptly describes the Academy's position along the Federated Suns' spinward reaches. Like the Kilbourne Academy, the Filtvelt Military Academy is devoted to educating both military cadets and civilian students from the region.

The Academy does not boast any high-tech simulators or multiple copies of the Helm Star League memory core. Its staff includes no Nobel Laureates or Safka Fellows. Its only mission is to train the extraordinarily undereducated population of the Federated Suns' peripheral worlds. Because of this, a degree from Filtvelt is often considered second-rate by those who attended more prestigious institutes; to most of its graduates, however, it is a ticket to a new and better life.

Major General Cluey Montserrat serves as the Academy's commandant.

Atmosphere

The atmosphere within the Academy is one of ravenous competition. Many civilians and would-be cadets see the Academy as their only hope of escaping their otherwise dead-end lives and will do anything to gain admittance to the school. Many students cannot handle its academic demands and are forced to drop out, perpetuating the Academy's high turnover rate. Those cadets that do excel academically are often given the opportunity to transfer to a different academy that will continue to challenge and motivate them.

Special Notes

Graduation from Filtvelt most commonly leads to a position within the Crucis March Militia, though the regiments of the Federated Commonwealth Corps have been known to snatch some of the Academy's outstanding graduates.

TIKONOV SCHOOL OF MILITARY DISCIPLINE

Once known as the Tikonov Martial Academy, one of the several Capellan Confederation military academies captured during the Fourth Succession War, the AFFS, and later the AFFC, rebuilt this academy from the ground up, both physically and academically. The new Tikonov School of Military Discipline became a beacon of light around which the people of the Sarna March could rally, giving any qualified citizen the opportunity to build a new life.

Though not as lavishly equipped as the CMS, the School boasts a modern simulator network, a number of Gauntlets and several large live-fire training areas. The School graduates hundreds of cadets from almost every major military specialty every year. Because of its long heritage as a part of the Capellan Confederation, the School is also at the heart of a political hurricane, one that shows no signs of weakening any time soon.

Leftenant General Craig Peltrov, a Tikonov native, commands the School.

Atmosphere

Normally, the School's cadet corps is evenly divided between citizens of the old Sarna and Tikonov Commonalities and the rest of the Federated Suns. Since the dissolution of the Chaos March and the increase in Capellan terrorist activities, however, the number of applicants from other regions has declined to almost nothing. Those Federated Suns cadets who choose to attend the School are in constant fear for their lives outside the School's grounds, and never travel in groups of less than four. Even so, a number of violent acts have been committed against the School's cadets, prompting the superintendent to mobilize the School's training cadre on numerous occasions to rescue their comrades.

Special Notes

The Republican Guard recruits heavily from the graduates of this academy, as does, surprisingly, the Free Capella movement.

REGIONAL TRAINING BATTALIONS

In addition to its enlistment centers and military academies, the AFFC has a third option open to qualified would-be MechWarriors. These so-called Training Battalions give recruits a chance to prove their mettle in a real-world, life-or-death environment.

Potential trainees must still complete their basic military training before being accepted into the Battalion. The AFFC's four Training Battalions have special agreements with local training camps, however, ensuring that potential cadets are put through a special course where Battalion instructors can observe their candidates and make final determinations as to who will be offered a position within the Battalion.

Once accepted into the Battalion, cadets enter the most hands-on training environment in the AFFC. Though all of the DME's required courses for the MechWarrior specialty are also taught, the instruction relies heavily on practical knowledge of the capabilities and inner workings of a BattleMech. There are no arcade-like banks of simulators or history professors in these Training Battalions. Classrooms often consist of a command tent in the field or a repair bay. Each cadet is assigned a senior cadet mentor and a staff advisor who help the cadet in his or her studies and evaluate the cadet's progress.

Battalion cadets do not graduate until they can demonstrate a mastery of all the requisite skills. Because of this, cadets are reassigned from the Battalion at an uneven rate. Some outstanding cadets have been known to graduate from a Training Battalion in less than two years; many take six or more years. When a cadet does graduate from a Training Battalion, he or she is not only a qualified MechWarrior but also an able technician. These MechWarriors are inducted into the AFFC at the rank of sergeant, but they may not become a commissioned officer without attending an academy OCS program.

Training Battalions attached to a specific academy display that academy's logo and color scheme where appropriate. All other Training Battalions simply display the Federated Commonwealth logo, though some have begun to sport the Federated Suns logo instead.

COMBAT TRAINING CENTERS

The Federated Suns' Combat Training Centers (CTCs) are not designed to train raw recruits in the art of warfare or teach them specialty skills. Instead, the CTCs give military units a place where they can train in the most realistic combat setting possible.

Though most worlds with a permanent garrison have some territory set aside for live-fire practice and simulated combat training, what sets the CTCs apart are their sophisticated computer networks, called GACTI (Ground-Aerospace Combat Tracking Instrumentation). Each BattleMech, vehicle, fighter and infantryman is fitted with a special transponder that is in constant communication with the GACTI system via sensor arrays spread out across the training area. The GACTI system records every simulated weapons attack and assigns hits against appropriate battlefield units. The sensors in the vehicles record these hits and, as they are connected through the vehicle's control systems, simulate battle damage.

Integral to the success of these CTCs are the combat units permanently assigned to each. Specially equipped and trained to simulate the combat formations employed by the Federated Suns' enemies, these units ensure the combat training is as realistic as possible.

The AFFC operates seven Combat Training Centers: on New Syrtis and Novaya Zemlya in the Capellan March, New Avalon and Defiance in the Crucis March, and Robinson and Woodbine in the Draconis March. The AFFC also jointly runs the CTC on Spica with the St. Ives Military Command; in light of the ongoing war in the St. Ives Compact, however, the SIMC has pulled all of its combat units off Spica, leaving the world's defense solely to the AFFC.

AFFC ROSTERS

This section describes the history, command structure, and available forces and composition, highlighting outstanding officers and trademark tactics where appropriate, of every major combat command traditionally associated with the Armed Forces of the Federated Suns. Units traditionally affiliated with the Lyran State and its Armed Forces are described in a companion volume. Though this means that a complete briefing on the units assigned to the AFFC is not contained within one volume, it is far more useful to see where each unit fits in its own historical context than to simply see where it has served for the past few years.

Each command follows the standard AFFC organizational and command structure unless otherwise noted. Each type of unit within a command is identified by one of the following symbols, indicating its predominant composition:

'MECH AEROSPACE ARMOR SPECIAL INFANTRY

Each command's entry includes its designation (Militia, RCT, and so on), insignia, size, quality and loyalty rating. Official nicknames, a long-standing tradition within the AFFC Regular Army, are included with those units that have them. Unit size ranges from a single squad or lance to one or more companies, battalions, regiments or wings. For more information on force composition, see *The Fox's Den*, pp. 13–31.

Quality ratings represent an estimate of a unit's overall skill and combat experience. Units given high ratings perform better on the whole than those with lower ratings, regardless of other factors. Quality ratings are assigned and regularly updated by the AFFC High Command's operations directorate.

Standard AFFC ratings are green, regular, veteran and elite. A green rating is given to units predominantly composed of new recruits or inexperienced soldiers. Regular units contain troops of average abilities and competent soldiers with some combat experience or advanced training. Veteran units employ troops with greater battle experience and skill and are generally reserved for vital or dangerous missions. Elite units are composed of the most highly trained, experienced and knowledge-

able troops. Because elite units are relatively rare, commanders reserve them for vital offenses and for defending key positions and persons.

The Department of Military Intelligence assigns loyalty ratings to every unit based on many factors, but primarily on the loyalty of the unit's key officers. This rating applies to loyalty solely toward the nation the unit is currently serving in. AFFC regulations prohibit revealing loyalty ratings to anyone outside the DMI and the High Command, though observant officers can often make educated guesses.

The standard loyalty ratings are questionable, reliable and fanatical. Questionable units are composed of potentially disloyal or seditious troops and bear close watching. The members of reliable units demonstrate consistent, loyal and appropriate conduct. Fanatical units display loyalty to their nation, whether the Federated Suns or the Lyran Alliance, that knows no bounds.

Over the past several years, the DMI has begun to use a secondary loyalty rating indicating the political leanings of the unit and its senior officers. These ratings have been attached to the deployment table following this section to better identify any potential problems. Units with a "Victor" rating tend to favor the leadership and views of Prince (now Precentor Martial and Commanding General) Victor Steiner-Davion. Those with a "Katherine" rating instead see Archon Katherine Steiner-Davion as their rightful ruler. Units with a "Neutral" rating have chosen not to involve themselves in the growing debate.

The Standard AFFC Nomenclature table lists abbreviations used throughout this section.

STANDARD AFFC NOMENCLATURE

Vehicles	**Regiments**	**Infantry**	**Regiments**
AB	Artillery Battalion	AI	Armored Infantry
ABY	Artillery Battery	BAB	Battle-Armor Battalion
AG	Artillery Group	CMR	Commando Regiment
ATR	Artillery Regiment	FR	Force Recon
AC(R)	Armored Cavalry (Regiment)	HI(R)	Heavy Infantry (Regiment)
AG	Armored Grenadiers	IR	Infantry Regiment
AR	Armor Regiment	JI(R)	Jump Infantry Regiment
ARR	Armored Regiment	JR	Jump Regiment
CR	Cavalry Regiment	LI(R)	Light Infantry Regiment
CA	Combat Auxiliary	MPR	Military Police Regiment
HA(R)	Heavy Armor (Regiment)	MDR	Mounted Rifles
HAY	Heavy Artillery	MI(R)	Mechanized Infantry (Regiment)
LA(R)	Light Armor (Regiment)	MR	Mechanized Rifles
LT	Light Tank	MRT	Mechanized Regiment
MG	Mounted Grenadiers	MTR	Motorized Regiment
PR	Panzer Regiment		
RR	Recon Regiment		
SR	Scout Regiment		
TR	Tank Regiment		

1ST ARAGON BORDERERS: THE PRIDE OF NEW ARAGON

Perched at the head of one of the Federated Suns' many salients into the Capellan Confederation, New Aragon has had to fight off Confederation raiding forces and Zhanzeng de Guang–backed terrorist activities for decades. Rather than assign even more Regular Army units to defend the world, the already overtaxed AFFS simply strengthened one of the planet's few militia units with two BattleMech battalions.

While the creation of the Borderers did not discourage the Confederation's continued attacks against New Aragon, it did result in fewer AFFS losses. Armed with BattleMechs rather than conventional armor, the world's defenders could finally fight the invaders on an even footing. That, coupled with the fervor of the men and women fighting to defend their home, cost the Confederation's many raiding forces dearly, though the price was often high for the Borderers as well.

Tied to New Aragon throughout most of the Third and Fourth Succession Wars, the Borderers nevertheless have had a number of occasions to prove their mettle elsewhere. Though instrumental in the capture of Hunan during the Fourth Succession War, the Borderers were still garrisoning their homeworld during the Clan invasion and were tied up with terrorist actions while the League and Confederation swept through the Sarna March. With the coming of Operation Bulldog, however, the Borderers had the opportunity to prove themselves in battle against Clan Smoke Jaguar.

After returning to their homeworld, the Borderers were surprised to find the Third F-C RCT still garrisoned on New Aragon, and even more surprised to learn of the Third's border-line insubordination. It seems the Third will be a continuing problem for New Aragon and its defenders, which also include the Twelfth Vegan Rangers' Alpha Regiment. With the entire Capellan March gripped with apprehension at the Capellan–St. Ives War and the growing dissent with Archon Katherine, there seems little chance of the tension on New Aragon evaporating any time soon.

The Borderers use a blue-and-white ceremonial paint scheme for their equipment. Their insignia is a black rectangle bisected by a white diagonal band, set upon a blue shield.

OFFICERS

Since the Borderers' return to New Aragon, Leftenant General Jimmie Kirston has taken on additional responsibilities as the director of New Aragon's defensive operations—a position mandated by New Aragon's laws and AFFC regulations. The Third's Hauptmann General Finnigan refuses to acknowledge Kirston's authority, however, and has instead arrayed his forces to gain complete control of two of New Aragon's three major land masses.

TACTICS

Primarily assigned to defensive duties, the Borderers prefer to win the battle before firing a shot. By anchoring a position with their heaviest units, the Borderers can force an enemy to come to them while the maneuver battalions outflank their enemy and force them to their doom.

1st Aragon Borderers
Regiment/Veteran/Reliable

CO: Leftenant General Jimmie Kirston

The Borderers have fielded a short BattleMech regiment throughout their long history, and though strengthened by salvage taken during the Fourth Succession War, the regiment experienced heavy losses while fighting Clan Smoke Jaguar. The Borderers currently field little more than two 'Mech battalions. Two of its companies consist solely of heavy- and assault-class 'Mechs, while the remainder operate mostly medium-class designs. The unit fields few new designs among its 'Mech complement, though it did salvage enough equipment during Operation Bulldog to rebuild nine Clan BattleMechs, including four OmniMechs.

1st Aragon Auxiliary
2 Battalions/Regular/Questionable

CO: Colonel Tab Daniels

The First Auxiliary was originally known as the Seventh Aragon Armoured Cavalry before the Borderers were created. Since that time, its numbers have dwindled, and its focus has shifted from heavy armor to a more mobile cavalry model. The Borderers' few mechanized infantry elements are assigned to the Auxiliary, as are its two aerospace fighter squadrons.

1ST ARGYLE LANCERS: THE FAST AND FURIOUS

Descended from one of the few SLDF units that chose not to accompany General Kerensky on his Exodus, the Argyle Lancers distinguished themselves throughout the Succession Wars in service to the Federated Suns. Their actions earned them not only the Federated Suns' most prestigious awards but also priority within the Department of the Quartermaster (DQ); the Lancers could count on access to the best equipment and personnel the AFFS could offer.

When Prince Hanse personally asked for volunteers to participate in an exchange program of sorts with the Lyran State, the Lancers' officers jumped at the chance to promote the AFFS in a different realm. The Lancers served a number of tours within the Lyran Commonwealth in the 3030s and 3040s, and they figured prominently in the Commonwealth's task force assigned to the 3039 invasion of the Draconis Combine.

By the time the Clans appeared, the Lancers were again stationed back within the Federated Suns. Though cleared for repositioning back to the Lyran State, the AFFC dragooned the Lancers' JumpShips, preventing the unit from reaching Lyran space until well after the Truce of Tukayyid. The Lancers nevertheless took up station along the border of the Jade Falcon OZ, giving a number of battle-weary AFFC units the opportunity to rest and refit.

Since the dissolution of the Federated Commonwealth, the Lancers have been caught in a political war between the Lyran Alliance and a number of traditional Federated Suns units. Though they have come to consider the Lyran Alliance's people as their own, the fact that Archon Katherine has been slowly crippling units loyal to her brother with supply shortages has soured them. Still stripped of their JumpShip support, the Lancers are virtual prisoners within a nation that is rapidly beginning to consider them an enemy of the state.

The insignia for the Lancers is a white letter "A" set against a red-and-black square. Likewise, the unit uses the ceremonial colors of black highlighted with red and white.

OFFICERS

Colonel Walter Scully is a Lyran native who graduated from the Sakhara Academy on an exchange program and entered the then-newly formed AFFC. Having served most of his career within the Federated Suns, Colonel Scully finds himself naturally within the pro-Victor camp. Like most Sakhara graduates, Scully is a masterful strategist, but his ideas are mostly ignored by his LAAF superiors. Chosen for command because of his Lyran heritage, Scully's superiors have since realized his true leanings and abandoned him and his unit.

TACTICS

The Lancers field a large quantity of fast BattleMechs and hovercraft and train extensively in a mobile style of warfare that relies heavily on its healthy aerospace support.

1st Argyle Lancers
Regiment/Veteran/Reliable
CO: Colonel Walter Scully

The Lancers' 'Mech regiment is operating at less than 100 percent strength, having fallen prey not to combat losses but to mechanical failures caused by lack of maintenance. The Lancers have been forced to cannibalize more than a company of 'Mechs to keep the rest functioning. While they have received a few under-the-table supply shipments from Arc-Royal, the Lancers need the shipments the LAAF keeps "accidentally rerouting" to its pro-Katherine units.

1st Argyle Auxiliary
Battalion/Regular/Unreliable
CO: Major Solomon Rogas

With repair parts for conventional vehicles far easier to come by than those for BattleMechs, the Auxiliary's hovercraft battalion is still functioning at near 100 percent. The unit is overworked, however, as it has taken on most of the patrol responsibilities normally assigned to the 'Mech regiment.

1st Argyle Air Wing
Wing/Elite/Unreliable
CO: Light Commodore Theodore Pallotine

Like the Lancers' 'Mech unit, their aerospace wing is mostly grounded, though in a pinch it could likely get three of its four squadrons airborne.

1ST CAPELLAN DRAGOONS: ROAR OF THE LION

The origins of the Dragoons are entwined with the Dukes of Kathil. Originally their private army, the Dragoons formed into a front-line combat regiment in 2835, following the start of the Second Succession War. Incorporating several members of the Capellan Confederation Armed Forces that had defected into the Dragoons following the First Succession War allowed them to achieve numerous victories against the Confederation.

However, beginning in 3005, the Duke of Kathil found himself enmeshed in the political war Michael Hasek-Davion was waging against Prince Hanse, and by 3009 his fortune was vanishing. With no other recourse, he sold the Dragoons to the Armed Forces of the Federated Suns, but with a proviso that he or his heirs could buy back the Dragoons at any time in the future.

Unwilling to trust the Dragoons because of their origins, the AFFS, and later the AFFC, posted the Dragoons on garrison worlds all along the Capellan Confederation border. Though they raided House Liao, they did not participate in any large-scale actions.

With the huge demands on the Kathil shipyards in the past few years to replace those JumpShips "hijacked" by the Lyran Alliance when it seceded, Duke VanLees of Kathil had amassed a fortune once more. In 3062, the Duke, son of the man who had originally sold the Dragoons, attempted to buy the unit back from Duke George Hasek. Duke Hasek refused but temporarily released the unit to the Duke of Kathil. Whether Duke Hasek had an ulterior motive for this strange action is unknown, but the Dragoons are nevertheless en route back to Kathil.

The Dragoons' insignia is a rearing lion set against a backdrop of the red-and-white crest of Kathil.

OFFICERS

Colonel Bobby Dunn was handpicked by Duke VanLees and sent to replace Leftenant General Ulso Dripe en route to Kathil. The Duke felt that Ulso, who had commanded the regiment since 3034, was too firmly entrenched in the Commonwealth's (or Katherine's) ways. A recent graduate of the Warrior's Hall on New Syrtis, Dunn was chosen not for his ability to command but because of his loyalty to Duke VanLees and Duke George Hasek.

Major Geoff Lee is the current XO and is the real leader of the Dragoons, with what amounts to a political appointee in command. A sound strategist and solid commander, Lee's no-nonsense approach and pragmatic neutrality in the face of a nation poised on the edge of civil war, have earned him the respect of the entire regiment and currently hold the Dragoons together.

TACTICS

The Dragoons have not fought in any large-scale engagement since the end of the Third Succession War. They have trained extensively in hit-and-run tactics and lightning raids.

1st Capellan Dragoons
Regiment/Regular/Questionable
CO: Colonel Robert "Bobby" Dunn
Aide: Major Geoff Lee

Built along the lines of a Star League Independent Dragoon regiment, the Dragoons field exclusively heavy and assault BattleMechs. However, due to their questionable loyalties, they have been at the bottom rung for resupply and are chronically short on parts, ammunition and consumables. Only Colonel General Bobby Dunn pilots a new BattleMech, a *Rakshasa* right off the General Motors assembly line on Kathil. This has caused additional resentment in the regiment.

1st Dragoon Wing
Wing/Veteran/Questionable
Wing Cmdr: Light Commodore Ceola Jackson

Called "The Crone" by the entire unit, the 74-year-old Jackson has commanded the unit since it was "sold" to the AFFS and does not appear to have retirement in mind. Though she commands her wing of medium aerospace fighters like the proverbial cold queen, there is no doubt that the unit's veteran status comes from her iron-handed leadership.

1st Armored Dragoons
Battalion/Regular/Questionable
Armor Cmdr: Major Michael Bauza

The Armored Dragoons are composed almost exclusively of heavy hovercraft such as the venerable Drillson and Condor designs, as well as newer hovercraft such as the Fulcrum Heavy Hovertank. The Armored Dragoons received these new vehicles through the efforts of the Duke of Kathil, who has consistently attempted to keep the unit mindful of its origins. Unable to assign new BattleMechs to the Dragoons because of the attention it would draw, the Duke settled for new vehicles.

1ST FEDERATED SUNS ARMORED CAVALRY: POINT OF THE SWORD

Formed in the earliest years of this century under the personal direction of Prince Ian Davion, the Federated Suns Armored Cavalry quickly became known as one of the AFFS' most potent units. Built around a paradigm of speed and firepower, the unit's core regiment consists primarily of medium-class BattleMechs. Coupled with its complement of seasoned veterans, the Cavalry has consistently shown itself capable of handling any situation.

Though first blooded in the final years of the Third Succession War, at first glance it would appear the Armored Cavalry did not play an active part in any of the Federated Suns' wars until Operation Bulldog. Nothing could be further from the truth. Poised at the head of the Suns' Ziliang salient since the end of the Third Succession War, the Cavalry has seen almost constant action against the Capellan Confederation, even during "peacetime." Though historians often overlook their deeds, the Cavalry have nevertheless been the reason many of the Suns' operations against the Confederation have succeeded. With their trademark lightning raids against key worlds, the Cavalry have often disrupted Confederation logistics lines and captured enough supplies to relieve the AFFS' supply difficulties.

The Armored Cavalry used these tactics to great success during the Fourth Succession War and, as has been recently declassified, during the Capellan-Andurien War as the AFFS built up toward a renewed Capellan invasion that ultimately never materialized. They continued this tradition during Operation Bulldog against Clan Smoke Jaguar's Zeta and Omega Galaxies, eliminating all opposition. Strangely enough, official AFFC documents insist that none of the Armored Cavalry's units have left Ziliang since returning from that operation. However, unconfirmed reports filtering out from the Confederation, in addition to supply and fuel requests far in excess of those generated even after a heavy training cycle, seem to indicate otherwise.

For ceremonial duties the Armored Cavalry commonly paints its equipment in the parade colors of Davion green with white highlights. The Armored Cavalry's insignia is a variation on the Federated Sunburst.

OFFICERS

Now entering his fifteenth year as the master of the Armored Cavalry, General Rudolph Chapman is one of the most respected officers in the AFFC, having earned both the Gold Sunburst and the Order of Steiner-Davion for classified actions. A master of maneuver and feint, Chapman has simultaneously served as the Director of Operations for the Capellan March's Edgeward Combat Theater for the past five years, planning and preparing the entire region's defense.

TACTICS

The Armored Cavalry's primary modus operandi is the lightning raid. Its speed and maneuverability give it the capability to hit fast and avoid contact with the heavy concentrations of enemy forces that could rip it to shreds.

1st Federated Suns Armored Cavalry
Reinforced Regiment/Elite/Fanatical
CO: General Rudolph Chapman
Aide: Leftenant General Tabitha Rennard

The Armored Cavalry is one of the few BattleMech regiments in the AFFC consisting almost entirely of equipment *younger* than its MechWarriors, due in part to its years of heavy combat. The unit is considered the Prince's chosen and is supplied as such. Furthermore, one company from each of the regiment's three battalions is equipped exclusively with OmniMechs and is trained to deliver the unit's Cavalier battlearmor battalion into battle.

1st Armored Cavalry Auxiliary
2 Battalions/Elite/Fanatical
CO: Leftenant Colonel Ephraim Talloch
4th Cavalry Battalion: Lft. Col. Ephraim Talloch
87th F-C BAB: Maj. Carlos Luckett

Though on paper the Auxiliary is an independent unit, in practice it operates as the regiment's fourth and fifth battalions. Tasked mainly with scouting and harassment missions, the Cavalry is made up solely of hovercraft, including two full platoons of the Confederation's new Regulators.

1st Cavalry AeroWarriors
Wing/Elite/Fanatical
CO: Commodore Choran Walters

In contrast to the Cavalry's predominantly medium-weight Mech regiment, the AeroWarriors, highly skilled in both air-to-air and air-to-ground combat, are equipped with the heaviest aerospace fighters in the AFFC.

1ST KESTREL GRENADIERS: ETERNAL FLAME

Smashed in a reconnaissance raid on the Capellan Confederation prior to the start of the Fourth Succession War, the Grenadiers were forced to sit out that war and almost suffered deactivation until the Duke of Kestrel stepped in to save his family's old guard unit. Prince Hanse, looking for a BattleMech regiment he could refit into a combined-arms unit in preparation for the creation of the allied AFFC, agreed to rebuild the Grenadiers.

By pulling MechWarriors from several independent regiments, as well as from three Capellan units that had defected to the Suns, Prince Hanse quickly built an experienced 'Mech regiment, but one with little combined-arms experience. Around this regiment he arrayed four conventional regiments and an aerospace fighter wing, turning the Grenadiers into a virtual RCT. After little more than three months of training, the Grenadiers went on to distinguish themselves in action against the Draconis Combine in the War of 3039. Later forced to retreat from New Canton during the Marik-Liao Invasion, the Grenadiers renewed their tarnished reputation during their Operation Bulldog assaults against Clan Smoke Jaguar.

The Grenadiers' insignia is a white torch with a blue flame, the same as that of Kestrel's ducal family.

OFFICERS

Marshal Agatha Stromp has commanded the Grenadiers for the past nine years; prior to that she served as the deputy commander for the Second Davion Guards. She is also a distant cousin of Kestrel's Duke Xerxes Cunningham, continuing the tradition of the Grenadiers' commander being a member of the ducal family. A highly competent officer, Marshal Stromp also professes a fanatical loyalty to Prince Victor and his family, a feeling she nurtures within her troops as well.

TACTICS

The Grenadiers have always been classified as a former household guard 'Mech regiment in the AFFC's official deployment tables, often making for a particularly nasty surprise for an enemy who is not familiar with the unit's recent history.

1st Kestrel Grenadiers
Reinforced Regiment/Elite/Fanatical
CO: Marshal Agatha Stromp
Aide: Colonel Lester Kuick

Often called upon to operate independently for lengthy periods of time, the Grenadiers' 'Mech regiment includes its own scouting and fire support elements. While the first two battalions consist primarily of heavy-class units, including a full company of *Longbows* and *Salamanders* in the first battalion, the third battalion is devoted to scout and cavalry missions.

1st Kestrel Armored Grenadiers
2 Regiments/Elite/Reliable
CO: Major General Timothy Van Norman
Aide: Major Oba Foulston
1st F-C AG: Maj. Gen. Timothy Van Norman
2nd F-C AG: Col. Nikki Pernacciaro
1st Kestrel Grenadiers AB: Maj. Dale Fusek

The First Armored actually fields four battalions of fast hovercraft and armor while the Second is made up of two battalions of the heaviest armor in the AFFC. Typical combat doctrine has the First locate and loosely encircle an enemy force while the mechanized infantry regiments set up ambushes along any possible escape routes. Meanwhile, the Second Armored and the 'Mech regiment operate in the open, keeping the enemy focused on their actions, until the trap can be sprung.

1st Kestrel Mounted Grenadiers
2 Regiments/Veteran/Fanatical
CO: Leftenant General Vojislav Stojsavljevich
Aide: Major Kichi Duecker
1st F-C MG: Col. Liang Kirsanova
2nd F-C MG: Col. Hector Gorenc
690th F-C BAB: Maj. Riel Khang

The Grenadiers' two mechanized regiments recently received a number of new Goblin and Maxim IFVs to replace their aging equipment, while the 690th is slated to begin receiving new transports within the next few months.

1st F-C Fighter Wing
Wing/Veteran/Fanatical
CO: Light Commodore Lyle Kuklinski

Outfitted with *Stukas*, *Chippewas* and *Eagles*, the First F-C, also known as the AeroGrenadiers, train extensively in Close Air Support (CAS) missions on whatever world they are assigned, allowing them to support their ground-based comrades in any terrain and weather.

1ST KITTERY BORDERERS: THE PRIDE OF KITTERY

Organized as a Regular Army unit for much the same reasons as the Aragon Borderers, the Kittery Borderers came into being less than a decade after Chancellor Dainmar Liao's surrender on St. Ives in 2862. By that time, the AFFS had consolidated its hold on Kittery and built a self-sufficient militia. Before the final occupation forces left the world, they built the Borderers from a group of Capellan front veterans.

Like the Aragon Borderers, the Kittery Borderers suffered a number of Capellan raids throughout their history. Situated at the head of the Suns' salient into the St. Ives Commonality, the St. Ives Armored Cavalry soon proved to be the Borderers' primary adversary. To their credit, the Borderers quickly learned their own weaknesses, and, when they knew they could not hope to defeat the highly skilled Lancers in a stand-up fight, they simply learned to use Kittery's terrain against the invaders.

This game of cat and mouse continued for many decades, throughout the Third Succession War. With the coming of the Fourth Succession War, the Borderers were naturally disappointed when no attack came and later voiced strenuous objections to the alliance with Duchess Liao's St. Ives Compact. Quite a number of the Borderers' personnel stepped down in protest, but by doing so they simply weakened the anti-St. Ives faction within the unit, as Prince Davion replaced the rabble-rousers with men and women who shared his own vision.

While there was little chance the Compact would launch a surprise invasion on the Federated Suns, the Borderers remained stationed on Kittery, where they could easily move to reinforce either the Compact or another Suns world along the Capellan border. Unfortunately, the years of garrison duty took their toll. Once considered an elite unit, the Borderers were soon downgraded to regular status, a slide that even weeks of intensive training on Spica each year could not reverse.

On the other hand, the Borderers' attitude toward the St. Ives Compact made a complete turnaround. As the Capellan–St. Ives War raged on, the Borderers served as Kuan-Yin Allard-Liao's unofficial guard unit, assisting her in delivering relief supplies from Kittery to the war-ravaged worlds of the St. Ives Compact.

The insignia for the First Kittery Borderers is a gold disk set against a stylized sunrise.

OFFICERS

Considered in the waning years of her career, Leftenant General Deedee Faulkner has bounced between a number of combat and support positions. She joined the Borderers right out of the Warrior's Hall on New Syrtis, and by all accounts requested a posting back with the Borderers, where she will end her lengthy service.

TACTICS

The Borderers know the terrain of Kittery better than anyone and remain more than a match for any unit that dares face them on their own soil. When taken out of their home environment, however, the Borderers lose much of their advantage.

1st Kittery Borderers
Regiment/Regular/Reliable
CO: Leftenant General Deedee Faulkner

A majority of the Borderers' mostly medium-class 'Mechs are simply refits of designs fielded since before the turn of the century.

19th Kittery Armored Regiment
Regiment/Green/Reliable
CO: Leftenant Colonel Pamela Li-Ting

Though elements of the Nineteenth Armored have operated with the Borderers throughout their history, they did not become an official part of the unit until after the Fourth Succession War. Now considered a Regular Army unit by the AFFC, the Nineteenth is still only supplied as if it were little more than a garrison unit.

12th Kittery Air Guard
2 Wings/Regular/Reliable
CO: Light Commodore Daut Jemeil

The Borderers officially absorbed the Twelfth Guard at the same time as the Nineteenth Armored. Originally little more than a conventional air unit, the Twelfth added its second wing, consisting of light to medium aerospace fighters, during the height of the Third Succession War.

AVALON HUSSARS

HONORED SERVICE

The Avalon Hussars are among the oldest professional military organizations in the Federated Suns, proudly tracing their lineage back to the Terran Alliance Marines. Stranded by the Alliance's Democratic Proclamation, the Marine Hussars remained on New Avalon to help defend their adopted homeworld. When President Lucien Davion brought the worlds of the Crucis Reach together under the aegis of the Federated Suns, the Hussars chose to sign on with President Davion rather than heed Admiral McKenna's recall order.

As one of the core units within the new Federated Peacekeeping Forces, the Avalon Hussars experienced tremendous growth and soon took responsibility for defending the Suns' entire border with the Combine. The Hussars later became one of the first organizations to field entire BattleMech regiments. During the Davion Civil War, the Hussars rapidly became divided—with different elements supporting Prince Alexander, Laura Davion and Dmitri Rostov—and lost a number of its best regiments.

The Succession Wars further whittled the Hussars' numbers down, though they did more than account for themselves. During the First War, their sacrifices helped hold the line against the Combine's advances into the Suns; during the Second War, their massed attacks brought more than half a dozen Capellan Worlds into the Federation. Nevertheless, the damage was done. By the end of the Fourth Succession War, only seven of the Hussars' original twenty-six 'Mech units were still operational, with a mere fourteen of thirty-two more conventional regiments still considered battleworthy. That number dwindled further following the Clan invasion and the subsequent events.

COMMANDING OFFICER

Marshal Roger William Waters commands the Avalon Hussars from the Command Center on New Avalon; he is one of the few AFFC officers who still maintains contacts within the LAAF and keeps another office on Poulsbo with his former command, the Forty-second Hussars. Having lost his wife to the machinations of a Lyran officer early in his career, Waters prefers to distance himself and his men from the political infighting unfortunately natural to any military organization. Unfortunately, Marshal Waters has been unable to keep the majority of his units from taking sides in the growing factionalism within the AFFC and the LAAF.

FORCES

As one of the Federated Suns' primary combat organizations, the Hussars have in theory almost unrestricted access to the DQ's supply chain. Since the breakup of the Federated Commonwealth and the ascension of Archon Katherine on New Avalon, however, many Hussars units are finding their access severely curtailed. While Marshal Waters has been able to clear up some of the administrative problems, several units have had to make their own under-the-table supply arrangements.

COLORS AND INSIGNIA

The Avalon Hussars prefer a ceremonial light gray paint scheme with highlights unique to each unit. The Hussars' insignia is a white sword set against a gold shield.

When Prince Victor rebuilt the Forty-first Avalon Hussars, he encouraged each Hussar regiment to embellish its logo; Victor hoped this act would cement their devotion to House Davion. The result was a mixed success.

11TH AVALON HUSSARS RCT: THE GALLOPING ELEVENTH

Until 3057 the Eleventh Avalon Hussars were stationed on Dustball, forming part of the defenses against Clan Jade Falcon. In the reorganization of units that followed Katherine Steiner-Davion's rise to power, the unit returned to the Federated Suns State Command. However, in their brief time on the Lyran border the Hussars had built a strong relationship with the locals and felt their withdrawal—ordered by Prince Victor—was a betrayal of their adopted people. When Archon Katherine took control of the Federated Commonwealth in 3060, the Eleventh were among her most vocal supporters and did much to promote acceptance of her rule.

The Eleventh uses a variant of the standard Avalon Hussars paint scheme, adding white highlights to the light gray base. The unit insignia is the white sword and gold shield motif of the Hussars, underneath which is the number 11.

OFFICERS

Though born on New Syrtis, Hauptmann General Leabo is a staunch supporter of Katherine Steiner-Davion. While at Warrior's Hall on New Syrtis, Leabo gained a reputation for being able to predict and counter the actions of his opponents. This skill applies as well to the political arena as it does on the battlefield, and Leabo has been able to dominate the pro-Victor but weak-willed Marquess Leon Therese on Brockway. Only Colonel Frederic Zaffson of Frederic's Gazelles, part of the Lexington Combat Group, has opposed Leabo, though his small force and mercenary status have limited his actions. However, the recent arrival of the Lexington Combat Group's remaining regiments has caused concern in the Eleventh, and the Hussars have undertaken low-key preparations for hostilities against the mercenaries.

TACTICS

The Hussars use their light and mobile 'Mechs and vehicles and mechanized infantry to outmaneuver their opponents, staging a series of lightning strikes and feints to throw the enemy forces off balance. The addition of a fourth battalion of 'Mechs in 3031, principally heavy and assault designs, gave the unit additional options in the field. The favored tactic has been to use this fourth battalion as an anvil, toward which hostile units are herded by the lighter elements of the RCT and then destroyed.

11th Avalon Hussars
Reinforced Regiment/Regular/Reliable
CO: Hauptmann General Justin Leabo
Aide: Leftenant General Saffron Gale
AeroSpace Brigade: Lft. Gen. Robin Kim

Like all the Avalon Hussars, the Eleventh Hussars are well-equipped with the most modern designs in the Lyran Alliance and the Federated Commonwealth. Though command of the regiment lies with Hauptmann General Leabo, day-to-day operations are overseen by Leftenant General Saffron Gale, a native of Tamar who joined the unit in 3053 after the destruction of her unit, the Twenty-fourth Arcturan Guards, on Dell in 3050. One of only nineteen survivors from the Twenty-fourth, Gale nurses a bitter hatred of the Clans, particularly the Wolves, and petitioned—unsuccessfully—for the Hussars to raid several Jade Falcon worlds.

The AeroSpace Brigade is composed of the 131st Recon and Twenty-second Assault Wings.

11th Avalon Hussars Armor Brigade
3 Regiments/Regular/Reliable
CO: Leftenant General Saffron Gale
Aide: Kommandant Larry Jarosik
21st Avalon Chasseurs: Lft. Gen. Trent Griffin
19th New Avalon RR: Lft. Gen. Louis Jones
326th Avalon AR: Lft. Gen. Raven Corrigan
98th Avalon AG: Kmd. Angelo Migliaccio

The Hussars' armor brigade is composed of fast, light vehicles such as the Striker and Centipede, supported by attack VTOLs such as the Warrior H8 and the Yellowjacket. Inidividually, the regiments are lack firepower, but they work well with the rest of the RCT, resulting in a combat effectiveness belied by their light weight.

11th Avalon Hussars Infantry Brigade
5 Regiments/Regular/Reliable
CO: Leftenant General Ayse Crosby
Aide: Kommandant Delphine Payne
421st Avalon LI: Lft. Gen. Ayse Crosby
422nd Avalon LI: Lft. Gen. Drew Watson
19th Avalon JI: Lft. Gen. Anne Ahmadi
834th Avalon Rifles: Lft. Gen. Geoff Klug
97th New Avalon MI: Lft. Gen. Lucya Bryant

Though many regard the infantry brigade as the RCT's weak point, the regiments have proved their worth time and again. Though only two regiments have integral transport assets—the 834th and Ninety-seventh—the brigade frequently surprises opponents with its speed and maneuverability. In recent exercises, they staged a forced march across the rugged Beautains Heights, traversing fifty kilometers overnight to catch a battalion of Frederic's Gazelles unawares in their bivouac area.

17TH AVALON HUSSARS RCT: DRAGON'S BANE

One of the more active elements of the Hussars' organization, the Seventeenth has been stationed along the Draconis front for much of the past two centuries. One of the units integral to Prince Hanse's invasion of the Combine in 3039, the Seventeenth made their way all the way to An Ting before being expelled from that important system by then-Gunji no Kanrei Theodore Kurita's counterattack.

When the Clans made their fateful appearance a decade later, the Seventeenth was still stationed within the Draconis March. Though Prince Hanse dispatched the Seventeenth as part of the AFFC's first wave of reinforcements, two-thirds of the RCT was stranded for several months following a misjump and did not reach their station within the Lyran State until a few weeks before the battle of Tukayyid. The Seventeenth spent the next five years jumping between systems along the Jade Falcon Occupation Zone in an attempt to keep the Falcons off guard while the AFFC rebuilt.

Returned to duty within the Draconis March three months before the Marik-Liao Invasion, the Seventeenth expressed outrage at Prince Victor's actions in respect to Joshua Marik and later at the signing of the Star League Constitution and the resulting military alliance with the Draconis Combine. A few weeks later, the High Command stripped the Seventeenth of its JumpShip group to assist in transporting the Bulldog and Serpent task forces; the JumpShip group has yet to be returned to the unit.

The Seventeenth makes use of the standard Hussars paint scheme with crimson highlights. The unit's insignia is that of the Hussars' organization with a "17" emblazoned below.

OFFICERS

Orphaned during the Third Succession War by a Combine raid on Benet III, Hauptmann General Kev Evans has vowed to repay that injustice in full. Though attitudes such as his once helped ensure a healthy career within the AFFC, Evans, once considered a prodigy, has been commanding the Seventeenth for more than a decade, with no apparent hope for promotion.

TACTICS

The Seventeenth practices regularly in both ground and airborne combat deployments and has often surprised enemies with its ability to maneuver even under the heaviest fire.

17th Avalon Hussars
Regiment/Regular/Reliable
CO: Hauptmann General Kev Evans
Aide: Leftenant General Cassandra Dzuiba
AeroSpace Brigade: Kmd. Jamie Doucette

Twice during their five years along the Occupation Zone, the Seventeenth surprised a Falcon raiding force by being where they weren't supposed to be. They eventually drove off the Clan forces both times, but only after suffering from the lack of experience they might have gotten from fighting in the Clan invasion.

The AeroSpace Brigade currently fields only the Forty-fourth Avalon Fighter Wing.

17th Avalon Hussars Armor Brigade
3 Regiments/Regular/Reliable
CO: Hauptmann General Trevor Cartmann
Aide: Kommandant Julia Simonsen
99th Avalon AR: Lft. Gen. J. E. Ortiguerra
198th Ipava Grenadiers: Lft. Gen. Josephus Schoenauer
445th Avalon RR: Lft. Gen. Lathan Kimayev
35th Avalon AG: Kmd. Pathinh Pi

17th Avalon Hussars Infantry Brigade
4 Regiments/Regular/Reliable
CO: Leftenant General Gandalf Darklight
Aide: Kommandant Dong Bong Kim
84th Avalon LI: Lft. Gen. Searlas Halliden
115th Crucis MI: Lft. Gen. Sonia Thao
405th Avalon MI: Lft. Gen. Helmer Fineas Wojahn
135th Federation JI: Lft. Gen. Darius Prohaska

The majority of the Infantry Brigade was using a different JumpShip as the Seventeenth RCT was underway to the Lyran Commonwealth in 3050 and thus arrived in the Commonwealth on time. The Brigade was held in reserve for several more weeks and then attached to the Ninth F-C for the assault on Twycross, where the 243rd Infantry was wiped out.

20TH AVALON HUSSARS RCT: THE SANDBAGGERS

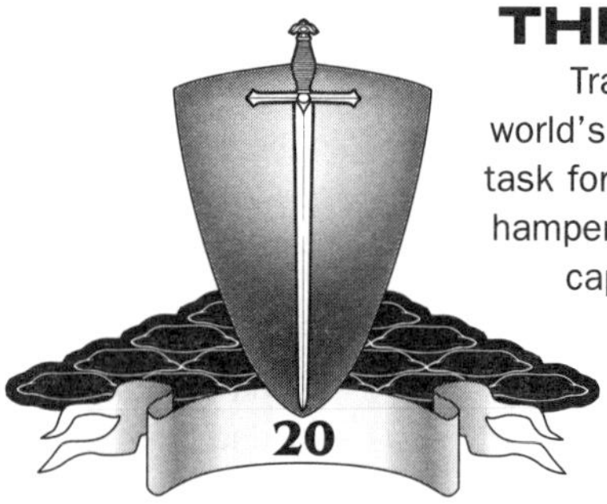

Traditionally considered a unit sent in after the primary invasion force to rebuild a newly captured world's infrastructure and military fortifications, the Twentieth Hussars took center stage as part of the task forces assigned to conquer the Capellan Confederation during the Fourth Succession War. While hampered by mechanical problems and adverse conditions, the Twentieth was actively involved in the capture of four worlds. After the war's conclusion, the Twentieth remained within the newly designated Sarna March, where its more traditional skills were needed.

Kept in place as a check against Capellan aggression during the Clan invasion, the Twentieth found itself in an untenable position when the Marik-Liao Offensive of 3057 propelled the Sarna March into total chaos. Cut off from its base of supply on Sarna and facing an increasingly hostile population, the Twentieth pulled back to regroup for the counteroffensive that never materialized. The majority of its JumpShips had been reassigned by the High Command years earlier, so the Twentieth commandeered two civilian JumpShips under the auspices of the Federated Suns' Emergency Warpowers Provisions to make their retreat. Though they were ultimately exonerated, the resulting legal battle effectively prevented the unit from supporting any of the AFFC's recent military endeavors.

The Twentieth Hussars tend to paint their equipment in the standard colors, with tan and green highlights. The Twentieth's insignia is the Hussars' shield and sword with the number 20 set below.

OFFICERS

General Jack Roberts' name is well-known within the AFFC and especially throughout the Capellan and former Sarna Marches. A graduate of Sakhara's class of '27, General Roberts willingly put his stellar career on hold when he turned down promotion after promotion to remain with the Twentieth Hussars, rather than take an assignment that would prevent him from commanding in the field. A master of battlefield strategy, Roberts has often put those skills to use in coordinating the massive reconstruction projects his Hussars have undertaken.

TACTICS

The Twentieth Hussars may be skilled at building and manning nearly impregnable defensive positions, but they have also used that well-known trait to their advantage by surprising an enemy with their mobility.

20th Avalon Hussars
Reinforced Regiment/Veteran/Fanatical
CO: General Jack Roberts
Aide: Leftenant General Yuriy McConeghy
AeroSpace Brigade: Cdre. Shemp Harrangue

Unlike many of his peers, General Roberts prefers to supervise his RCT from an area where he can gain a complete perspective, often giving Leftenant General McConeghy command of the Hussars' mostly medium and heavy BattleMechs.

The AeroSpace Brigade is composed of the Sixth Avalon Airlift, Eleventh Federation Fighter and 442nd Avalon Attack Wings.

20th Avalon Hussars Armor Brigade
3 Regiments/Regular/Reliable
CO: Major General Juan Villanova-Petain
Aide: Leftenant Colonel Catalina Jelinek
45th Avalon PR: Lft. Gen. Koko Nimkie
715th Federation HA: Lft. Gen. Darwin Gramstadt
19th Muskegon SR: Col. Soulou Iqbal
18th Argyle AB: Maj. Jodi Noonan

While the Forty-fifth and 715th are nominally designated heavy armor units, they also employ a number of fast hovercraft detachments.

20th Avalon Hussars Infantry Brigade
6 Regiments/Regular/Reliable
CO: Leftenant General Louizio Martine-Holm
Aide: Colonel K. Theodus Hasek-Osner
449th Argyle Infantry: Lft. Gen. Horatio Saveille
4th New Valencia Rangers: Col. Ren Hech
36th Federation MIR: Col. Sandra Zhen
200th Avalon JR: Col. Zenia Luciani
985th Combat Engineers: Col. Tariq Moen

In addition to the 985th Engineers, each of the Twentieth's infantry units fields one or more engineer companies.

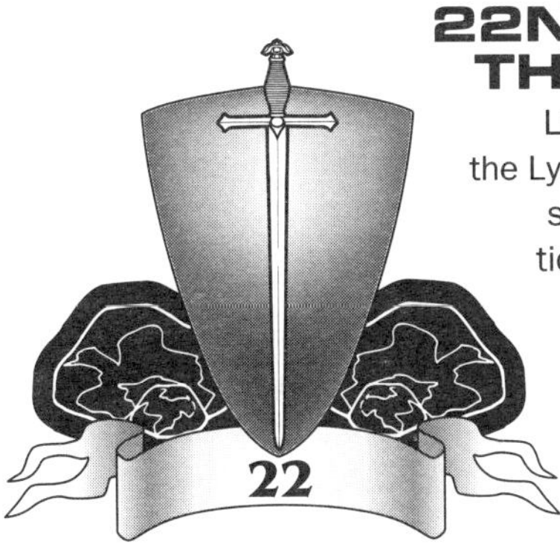

22ND AVALON HUSSARS RCT: THE STRENGTH OF TRADITION

Like the Eleventh, the Twenty-second Hussars were one of a number of units "repatriated" from the Lyran Alliance after the sundering of the Federated Commonwealth in 3057. The Hussars had been stationed along the Clan front since 3051, first as part of the mobile reserve and then as a tactical raiding unit following the Truce of Tukayyid. The unit took significant losses during this period, but their worst catastrophe occurred away from the front, when two companies were lost in a JumpShip accident in 3054.

The Twenty-second Hussars are staunch traditionalists, believing in loyalty to the established chain of command. As such, they stand by Katherine Steiner-Davion, who was legally made head of the Federated Commonwealth by Yvonne's decision to stand down in late 3060. This isn't to say they like her policies, but she is the head of state and thus deserves their loyalty.

The Twenty-second uses the standard Avalon Hussar paint scheme and insignia.

OFFICERS

Marshal Donna Iona has done much to rebuild the Twenty-second since she assumed command of the unit in 3053. Her promotion to marshal in early 3061 rewarded her efforts and long service with the AFFC—she joined the AFFS as a MechWarrior in 3018—but has confused the command structure within the Sirdar PDZ. Only her belief in following the established structure has prevented Iona from usurping Marshal Korsant's position, but as Korsant has shown signs of opposition to Katherine's rule, it is unclear how long this will continue.

Saule Rosenburg, commander of the Thirty-second Tactical Air Wing, gained a formidable reputation during the Clan invasion. Her ability to perform seeming miracles earned her the nickname "The White Witch," and she downed more than a dozen Clan aerospace fighters while serving with the Fifth Davion Guards against Clans Wolf and Jade Falcon. She narrowly escaped La Grave and spent more than a year recovering from her injuries before being reassigned to the Hussars in 3055.

TACTICS

The Twenty-second lacks any tactical specialties, preferring to be adept in a wide range of conditions and circumstances. Though generalization is often associated with mediocrity, the Twenty-second has built a formidable reputation for skill and tenacity. Their flexibility makes them ideal as a "firefighter" unit, able to be relocated wherever they are needed, and they have been called upon to deal with numerous raids by the Capellan Confederation and the Taurian Concordat.

22nd Avalon Hussars
Regiment/Veteran/Reliable
CO: Marshal Donna Iona
Aide: Leftenant General Tokiko Wallace
AeroSpace Brigade: Lft. Gen. Solomon D'Angelo

The Hussars' 'Mech regiment typifies their generalist approach, containing a wide variety of classes and designs that allow them to adapt to a wide range of circumstances. Almost all are less than twenty years old, including designs like the *Cataphract* and *Cestus,* though several date back to the early Succession Wars. These "mascot" 'Mechs, including a model 10 *Bombardier* and a model 733 *Highlander,* rarely see combat but serve as banner-carriers in the unit's parade formations.

The AeroSpace Brigade is composed of the Thirty-second Tactical and 192nd Air Wings.

22nd Avalon Hussars Armor Brigade
3 Regiments/Veteran/Reliable
CO: Leftenant General Bruce Vachon
Aide: Kommandant Adam Pearson
19th Crucis HA: Lft. Gen. Jessica Schlaeger
42nd RR: Lft. Gen. Max Holm
411th Avalon AR: Lft. Gen. Kim Delalande
8th Altair AB: Kmd. Timothy Cartwright

Leftenant General Vachon has struggled to keep the armor brigade in fighting trim. Spare parts for their predominantly Lyran tanks have been in short supply because of the friction between the Alliance and the Federated Commonwealth. The supply situation has improved in recent months, though the unit's technicians retain a "waste-not, want-not" mentality.

22nd Avalon Hussars Infantry Brigade
5 Regiments/Veteran/Reliable
CO: Leftenant General Keith Steel
Aide: Kommandant Rae Bond
192nd Tamar LI: Lft. Gen. Reynolds Duncan
3rd Avalon AI: Lft. Gen. Herbert J. Lang
42nd Tamar Rifles: Lft. Gen. Seth Racine
23rd Avalon JI: Lft. Gen. Pru Lauber
214th New Avalon MI: Lft. Gen. Weiman Harper

The Hussars' infantry brigade made good its losses against the Clans by recruiting from the local population on their former homeworld of Clermont and surrounding worlds. As a result, the brigade is the most pro-Lyran in the RCT.

39TH AVALON HUSSARS RCT: SWIFT WIND

The Thirty-ninth Hussars were one of the AFFS' first units assigned to the Lyran Commonwealth in the years after the Fourth Succession War. The unit proved invaluable when the LCAF began to integrate conventional regiments with its BattleMech units to form its own Regimental Combat Teams, assigning a number of its key officers to advise their Lyran counterparts. Meanwhile, the rest of the unit spent most of its first Lyran tour hunting Periphery pirate bands operating from the worlds of the former Rim Worlds Republic.

Ordered to hold their position during the Clan invasion, the Hussars could only watch as the Falcons gobbled up system after system. Though they pulled off several high-risk supply drops to resistance forces on Bone Norman, Black Earth and Somerset, those raids cost them dearly. In just two years, the Thirty-ninth lost a third of its DropShip contingent and two of its irreplaceable JumpShips. Rather than sacrifice the entire command to a lost cause, the High Command tasked the bulk of the Thirty-ninth with patrolling the length of the Falcon OZ border while the Infantry Brigade remained on station.

Today, the Thirty-ninth is back on Newtown Square along the Lyran Periphery. The unit has never been assigned transport elements to replace those lost in combat, however, even with the reported surplus following the Lyran secession. That fact, coupled with its refusal to turn over its only remaining JumpShip to assist in "the strategic repositioning of key LAAF assets," has fueled its growing feud with its Lyran masters.

The Thirty-ninth uses the standard Hussars paint scheme with blue and gold highlights, and its insignia shows the Hussars' shield upon which a "39" is superimposed.

OFFICERS

Major General Bella Bragg has found herself marooned in the middle of an increasingly hostile nation. With no chance for furthering her career and little hope that a return to the Federated Suns would do her or her unit much good, she has made it her goal to keep her unit alive while making life for her LAAF "superiors" as difficult as possible. Rumor has it she is receiving assistance in both areas from Arc-Royal.

TACTICS

Equipped with a preponderance of light and fast BattleMechs and vehicles, the Thirty-ninth can find its enemy long before its own main force is detected. The unit will then stage a number of false raids to keep the enemy looking in the wrong direction before finally making a full-scale assault.

39th Avalon Hussars
Regiment/Regular/Reliable
CO: Major General Bella Bragg
Aide: Colonel Andre Buetow
AeroSpace Brigade: Cdre. Prichard Seville Ardwright O'Donnell

Though "exiled" to the Periphery border, Major General Bragg has nevertheless managed to secure nearly a company of Lyran-built light and medium OmniMechs for her unit.

The Thirty-ninth AeroSpace Brigade fields only the Third New Damascus Recon Wing.

39th Avalon Hussars Armor Brigade
3 Regiments/Regular/Reliable
CO: Leftenant General Roland Hollenstein
Aide: Leftenant Colonel Ozzie Phifer
14th Federation LAR: Col. Keo Smith
27th Federation SR: Col. J. J. K. Illes
374th Marlette LA: Col. Suzanne Wiederhold
234th Federation AB: Maj. Frank Galfren

Each of the RCT's armor units fields a few Clan vehicle designs salvaged after several successful raids on the Falcon OZ.

39th Avalon Hussars Infantry Brigade
5 Regiments/Regular/Reliable
CO: Leftenant General Socrates Ukhanov
Aide: Major Fatima T'Lani
3rd Avalon FR: Lft. Gen. Gareth Ristau
74th Novaya Zemlya IR: Col. Olgert Parve-Jamison
19th Albion MI: Col. Somphone James
15th Avalon JI: Col. Kim Goke
34th Federation JI: Col. Leonid Sandoval

General Bragg has stripped the Nineteenth of almost all of its IFVs to exchange for other supplies and equipment to keep the entire RCT operational.

AVALON HUSSARS

41ST AVALON HUSSARS: RESTLESS SPIRIT

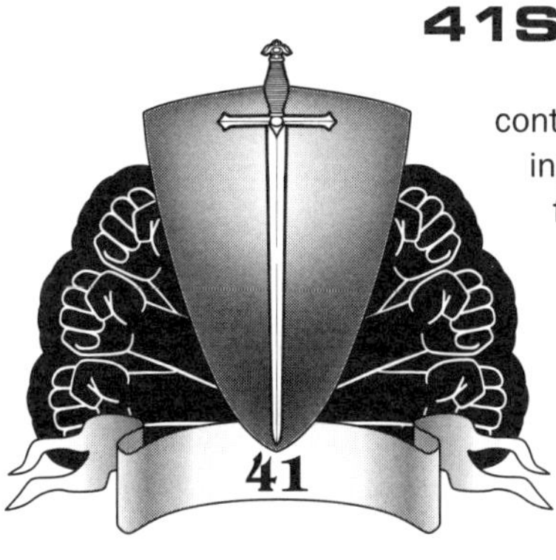

The Forty-first Avalon Hussars were once considered one of the luckiest units within the AFFC, continuing to field much of its complement of Star League–era equipment throughout the devastating centuries of the Succession Wars. Unfortunately, by the time of the Clan invasion, the Forty-first had apparently used up that luck. Assigned to the Lyran world of Planting, by all accounts they gave the Wolf invaders a difficult time but fell before Khan Garth Radick's concentrated efforts.

Following the Marik-Liao Invasion and Lyran secession in 3057, Prince Victor ordered three of his AFFC's BattleMech units rebuilt, including the Forty-first Avalon Hussars. Drawing heavily from the Federated Suns many academies, as well as the steady stream of expatriates returning from Lyran space, the reformed Forty-first entered service in late 3060. The unit then spent the next six months in intensive training at Robinson's Combat Training Center.

Since taking up station on Benet III, however, the Forty-first has run headlong into a number of roadblocks apparently set in their path by Archon Katherine's supporters in the AFFC bureaucracy. Though scheduled to re-form into an RCT, the infantry and armor regiments originally assigned to it have been rerouted to the Capellan March's Periphery border. Likewise, regular shipments of spare parts and other supplies have dried up, though the vociferously pro-Katherine Fifth Donegal Guards, also stationed onplanet, continues to receive its monthly shipments. In response, the men and women of the Forty-first Hussars have been involved in quite a few incidents with the Fifth Guards over the past several months. Though reprimands against both units have been issued from Robinson, the Forty-first continues to feel the weight of Archon Katherine's supporters bearing down on them.

Like the rest of the Hussars' units, the Forty-first uses the standard Hussars insignia with the number 41 set below the crest and paints its equipment in standard colors, with red and black highlights.

OFFICERS

General Laura Hamilton is a skilled administrator and trainer. In a relatively short period of time, she has brought a mixed group of academy graduates and bitter veterans together and turned them into a finely tuned fighting machine. Furthermore, she is well-connected within the AFFC, having served as the Deputy Commander for the Raman PDZ for many years, and has used those contacts to help build the Forty-first long after Archon Katherine's supporters closed down her official lines of communication.

Though an unsurpassed organizer of men and materiel, General Hamilton is considered by no one, including herself, to be a combat officer. Leftenant General Julius Gramstad, a veteran who fought with the Twenty-sixth Lyran Guards during the invasion, more than makes up for her lack of combat experience. While official reviews have often categorized Gramstad a mediocre officer, he has proved himself a quick thinker, a quality that has led him into trouble as often as to victory.

TACTICS

While the unit is still training together, under the leadership of Leftenant General Gramstad, the Forty-first has already proved to be an unpredictable and dangerous foe, at least on the training grounds of the Robinson CTC.

41st Avalon Hussars
Regiment/Regular/Reliable
CO: General Laura Hamilton
Aide: Leftenant General Julius Gramstad

As a part of Prince Victor's movement to enhance national pride in the Federated Suns, the Forty-first contains a high proportion of FedSuns BattleMechs, including the *Enforcer III* and *JagerMech III*. General Hamilton also managed to secure two full companies of OmniMechs for her unit, as well as a few Clan-based 'Mechs.

41st Avalon Hussars Armor Auxiliary
2 Battalions/Veteran/Unreliable
CO: Leftenant General Julius Gramstad
2717th Robinson AG: Lft. Gen. Jacobus Pillay
512th Federation ABY: Maj. Ivars Sodager

By using her contacts within the Draconis March, General Hamilton secured the temporary assignment of a battalion each of armor and artillery to her unit, an assignment that grows more permanent each day.

1001st Federation Aerospace Wing
Wing/Regular/Reliable
CO: Major Fred "Highball" Parneau

Considered by many to be little more than a collection of misfits, the 1001st has surprised everyone it has trained against with its skill.

42ND AVALON HUSSARS RCT: SONS OF THE FEDERATION

Deactivated after suffering horrendous casualties early in the First Succession War, the Forty-second Hussars were reborn in the final years of the Third Succession War. Organized from the survivors of the Thirty-fourth and Thirty-sixth Hussars RCTs, the new Forty-second was assigned to the Capellan border, where it soon cut its teeth on the series of raids that later became known as McCarron's War.

Archon Katrina Steiner was impressed with their trademark caution when the Forty-second was called into Lyran space during the Skye Crisis of 3034. She had them transferred to Poulsbo, where they became a far greater stabilizing force on that traditionally anti-Katrina world than any other Lyran unit could. The Forty-second has called Poulsbo home ever since.

For many years, the AFFC High Command kept the Forty-second staffed solely with men and women from the Federated Suns to help ensure the unit would not be susceptible to any influence from the late Duke Aldo Lestrade's retainers. Eventually, however, carefully screened Lyran-born personnel were allowed into the unit, further cementing its tie to the Lyran State. Today, although (or perhaps because) it has not taken a position in the growing rift between the supporters of Archon Katherine and Prince Victor, the Forty-second remains on excellent relations with both the LAAF and the AFFC.

The Forty-second uses green and blue highlights with the standard Hussar paint scheme. Its insignia is that of the Hussars' organization with a "42" superimposed.

OFFICERS

Leutnant-General Edwin May took command of the Forty-second after then-Hauptmann-General Waters accepted a promotion to head the Hussars' entire organization in late 3056. May often traveled with his parents (who own a trading concern based out of St. Ives) in his childhood and learned the fine art of diplomacy before reaching puberty. While he is adept at keeping the peace, he is also a skilled strategist who believes a victory is won long before reaching the battlefield.

TACTICS

While serving as executive officer, Leutnant-General May impressed on his staff the need for planning, something he continues to do to this day. Whether on the move or on the defensive, the Forty-second drives an enemy ahead of them by laying down extensive mine fields and building obstructions and/or fortifications.

42nd Avalon Hussars
Reinforced Regiment/Regular/Reliable
CO: Leutnant-General Edwin May
Aide: Colonel Brooke Hasey
AeroSpace Brigade: Lut.-Gen. Manoj Kujawa

The Forty-second maintains a full battalion of heavy- and assault-class fire support 'Mechs. The AeroSpace Brigade contains the Fortieth Federation Attack and 160th Crucis Interceptor Wings.

42nd Avalon Hussars Armor Brigade
5 Regiments/Regular/Reliable
CO: Leutnant-General Kirby Pethan
Aide: Colonel Jeremiah Lietzau
68th Avalon HAR: Col. Norbert Kozey
83rd Federation AR: Col. Reubena Prochnow
201st LAR: Col. Ernest McCluskey
222nd Lyran RR: Lut.-Col. L. Samuel Henes
7th F-C AR: Col. Reuben Caliteux

The Forty-second retains most of the armor it inherited at the beginning of the century, including one of the few artillery regiments in the AFFC.

42nd Avalon Hussars Infantry Brigade
7 Regiments/Regular/Reliable
CO: Leutnant-General Sara Steiner-Dauer
Aide: Colonel Kirk Suchenko
19th F-C BAB: Lut.-Gen. Hu Xin
61st Avalon Rangers: Col. Robert Proeber
13th Northfield Commandos: Col. Malachi Enge
109th Federation MI: Col. Nain Corriveau
111th Tamar MIR: Col. Brian Elvinn
49th Alliance JR: Lut.-Col. Lorenzo Grahovac
598th New Avalon MR: Lut.-Col. Amanda King

In addition to its full battle-armor battalion, two of the Forty-second's infantry units, the Thirteenth and the 161st, specialize in demolitions and other behind-the-lines operations.

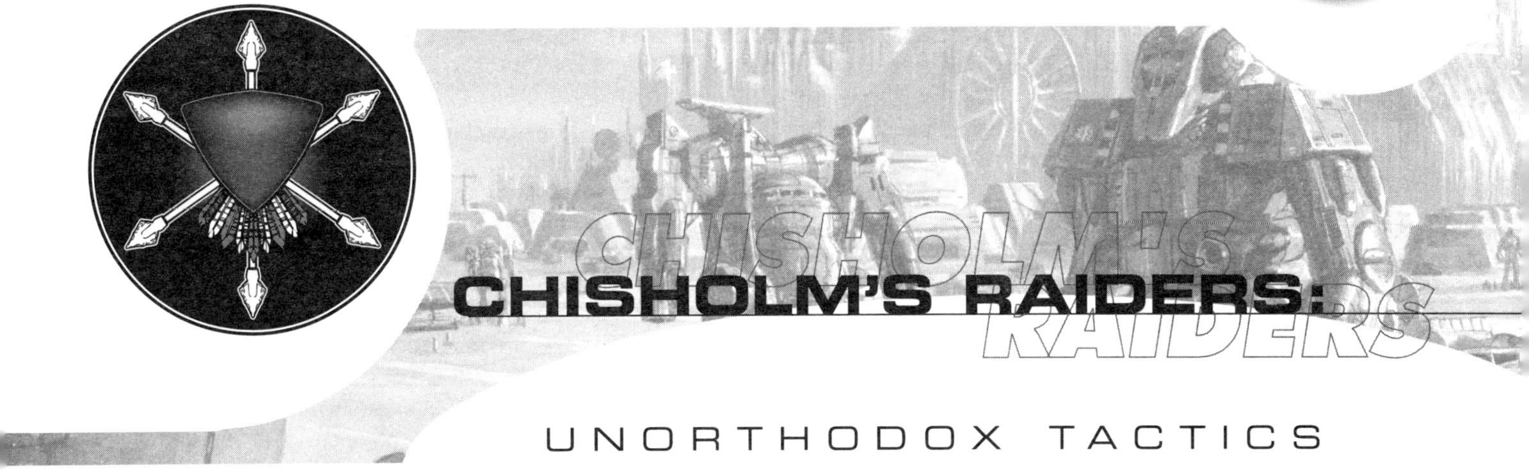

CHISHOLM'S RAIDERS:

UNORTHODOX TACTICS

In 2892, Marshal John Chisholm retired from the AFFS. Though he was regarded as an excellent commander, his constant criticism of the military establishment earned him many enemies and he was eventually forced out. Returning to his homeworld of Colorado, the marshal planned on living out his days in peace. However, after forty years in the military, he quickly became bored and set about finding outlets for his frustration. He took command of the planetary guards and began to instill in them the standards he staunchly believed in.

In 2897, reports of the unit's activities came to the attention of the AFFS, who dispatched a team of investigators. Though ill-equipped, the militia they found was a far cry from the ragtag mob they expected. Seizing his opportunity, Chisholm challenged the AFFS: his motley assortment of reservists, retirees and teenagers would fight any force the authorities cared to put forward, and win. The AFFS accepted.

Though ultimately defeated by the Second Crucis Lancers, the Colorado Militia performed well, outwitting the more experienced troops on several occasions. This convinced the AFFS that unorthodox tactics had a role to play in their efforts, and they designated the Colorado Militia the First Chisholm's Raiders and brought their equipment up to the standards of front-line AFFS units.

COMMANDING OFFICER

Tradition dictates that only those who have served in the Raiders can command both Chisholm RCTs. The current incumbent, Marshal Amelia Fitzwilliam, is no exception, having served with distinction in the First Raiders during the 3030s. What isn't clear is what she did after leaving the Raiders in 3038. It is widely believed that she spent five years with MI6, working with and later commanding House Davion's elite Special Forces units. However, any documents that would clarify the situation remain sealed under the AFFC's Military Secrets Protocol.

FORCES

The Chisholm's Raiders RCTs fall outside the regular AFFC chain of command, reporting directly to the High Command rather than to their PDZ, Theater and March commanders. With factionalism rife in the Federated Commonwealth, this non-standard command structure may allow the Raiders to remain operational when other units are bogged down in red tape. However, the regiments are at the mercy of the established AFFC structures for supplies and materiel, and Marshal Fitzwilliam has had to work hard to keep her forces in fighting trim.

COLORS AND INSIGNIA

The insignia of Chisholm's Raiders is a blue disc on which is a black triangle representing a Native American shield. Radiating out from the shield are six white arrows, each pointing in a different direction. Recently, a fan of feathers was added to the base of the shield.

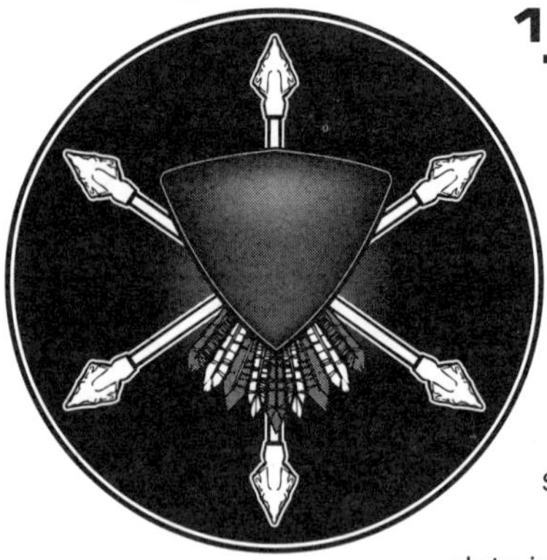

1ST CHISHOLM'S RAIDERS RCT: THE GREEN GIANT

Since the formation of Chisholm's Raiders in the late twenty-ninth century, they have come to epitomize flexibility and adaptability in a military unit. However, the First Raiders were effectively destroyed on Deshler by the Draconis Combine during the Fourth Succession War. Hanse Davion recognized their immense contribution—and their sacrifice for the Federated Suns—by ordering the AFFC to rebuild the unit. The material losses were made good by 3035, but the unit has struggled to regain its former ability. Indeed, its troubles have been so bad that even before the Clan invasion, the AFFC had designated the First Raiders a "sink" unit and used it as a repository for troublemakers and malcontents. The unit was in such poor shape by 3053 that they were removed from the AFFC rosters.

The need for troops after the Lyran secession led to the Raiders being returned to active duty in early 3058, though little was done to improve their personnel or equipment. They were indifferent to Archon Katherine's takeover of the Federated Commonwealth, but her promises of aid have brought the unit grudgingly into her camp. However, the lack of progress in the nearly two years since her accession has led to increasing disillusionment with the Archon's promises.

The Raiders' 'Mechs and vehicles lack a cohesive paint scheme, but all bear the Chisholm's Raiders insignia. The quality of the painting, done by the pilots, varies considerably, lending to the slapdash image of the unit.

OFFICERS

Hauptmann General Arisota Neece has gained a reputation for single-minded determination in her five decades of service, first with the AFFS and then the AFFC. She earned the nickname "Try Again" for her persistence, a trait that less charitable members of the AFFC credit for her rise to command despite her poor tactical skills. However, for all her failings, Neece's stubbornness and her infamous temper seem the best way of keeping her often rowdy troops in line.

Hauptmann Harold Kerr serves as Neece's enforcer within the unit. Nicknamed "Hook," Kerr's massive bulk and willingness to use violence to enforce Neece's orders serves to keep many of the less cooperative members in line.

TACTICS

Prior to the Fourth Succession War, the First Raiders were proficient at adapting their tactics to the terrain and the enemy. Today, according to the regiment's critics, they would struggle to cope with a change in meal times. The unit certainly lacks cohesion and tactics, though the abilities of individual warriors are superior to those normally found in a green regiment.

1st Chisholm's Raiders
Regiment/Green/Questionable
CO: Hauptmann General Arisota Neece
Aide: Leftenant General Jackson Tree

The 'Mechs used by the First Chisholm's Raiders are the cast-offs of the AFFC, too old or unreliable to be used by other units. None is newer than thirty years old, and most date back a century or more. A few have been upgraded with Star League technology, but Neece carefully doles out such equipment to her supporters as a reward to ensure their loyalty.

1st Chisholm Armor Brigade
2 Regiments/Regular/Reliable
CO: Leftenant General Albert Vega
Aide: Kommandant Mike Hisashi
19th Breed HA: Lft. Gen. Albert Hook
23rd Breed Recon: Lft. Gen. Fatima Vance
225th Breed ABY: Kmd. Annabelle Feff

Formerly part of the Breed Planetary Militia, the First Chisholm Armor Brigade has a difficult relationship with its "parent" regiment. In a reversal of the normal relations between militia and front-line troops, Leftenant General Albert Vega looks down on the Raiders' 'Mech regiment. He regards Neece's methods as crude and her troops as grossly unprofessional. In his opinion it is the Armor Brigade that is the true heart of the RCT, a view that has prompted frequent conflicts with the Hauptmann General.

1st Chisholm Infantry Brigade
2 Battalions/Green/Reliable
CO: Leftenant General Greg Prescott
Aide: Kommandant Ellen Forbes

The Chisholm Infantry Brigade—in reality the First Chisholm Light Infantry—is the newest addition to the RCT, having only been in existence since early 3062. The core of the unit, Stormhammer Company, was transferred from the Second Raiders, but the bulk of the unit consists of new recruits. At present the unit lacks transport assets and has no weapons heavier than a gyrojet rifle.

CHISHOLM'S RAIDERS

2ND CHISHOLM'S RAIDERS RCT: THE APACHES

Though originally little more than a mobile reserve and feeder regiment for the First Raiders, the Second Chisholm's Raiders were raised to full RCT status in 3055. Prince Davion wished to retain the unorthodox methods and adaptability of the Raiders, and with the First Raiders deemed unsalvageable, upgrading the Second Regiment was the most logical course. Though militarily sound, the move had a number of political ramifications. The former officers of the recently dissolved First Raiders protested the move, claiming the resources dedicated to upgrading the Second Regiment would have been better used returning their RCT to operational status. Similar protests were made by units struggling to rebuild after the Clan invasion, particularly Lyran units that cited the move as further evidence of Davion favoritism.

The Second Raiders use a green and gray color scheme, with color banding distinguishing battalions and other sub-units. In addition to the regular Raider insignia, the 'Mechs, vehicles and fighters of the Second bear a stylized Native American headdress.

OFFICERS

Helen Eisner has spent almost her entire career with the Second, joining the regiment as a captain shortly before the War of '39 and rising to command the regiment in 3046. An inspirational speaker and formidable tactician, the hauptmann general is an excellent leader and teacher. However, the expansion of the unit to RCT status has taxed her administrative and political skills, forcing her to rely increasingly on her deputy, Ryan Davion-Cole.

A very distant relative of the ruling family, Davion-Cole's skills complement those of his CO, and the pair make a formidable team. However, Eisner fears the partnership will soon be broken up as Davion-Cole is in line for his own command.

TACTICS

The Second Raiders are proficient in highly mobile warfare, using speed and flexibility to devastating effect, particularly in flanking and enveloping maneuvers. Almost all the 'Mechs and vehicles used in the RCT's constituent regiments are capable of sustained speeds over 50 kph and almost a third can exceed 100 kph for short periods.

2nd Chisholm's Raiders
Regiment/Veteran/Reliable
CO: Hauptmann General Helen Eisner
Aide: Leftenant General Ryan Davion-Cole
Chisholm's Flyers: Lft. Gen. Douglas Brennan

The Second Raiders retain the decentralized command structure they used while still an auxiliary unit to the First RCT, placing considerable emphasis on individual initiative and responsibility. This allows the unit to operate well on a wide range of levels, from individual lances to the entire RCT. As a consequence, elements of the RCT are often sent on detached duty, such as pirate hunting or tactical raiding. The unit has been particularly active since Sun-Tzu Liao's invasion of the St. Ives Compact, staging extensive patrols along the Capellan border and throughout the Sirdar PDZ's Ziliang Salient.

The regiment makes heavy use of new BattleMech designs. Many Free Worlds League designs such as the *Wraith* and *Falcon Hawk* are common.

Chisholm's Flyers field two Wings.

2nd Chisholm Armor Brigade
2 Regiments/Regular/Reliable
CO: Leftenant General Anders Crow
Aide: Kommandant Suk Mey Meng
1st Raider RR: Lft. Gen. Anders Crow
4th Chisholm LT: Lft. Gen. Toren Cage
1st Chisholm HAY: Kmd. Kurt Farchione

The Second Chisholm Armored Brigade contains virtually no tracked vehicles, preferring to focus on swift wheeled and hover vehicles like the Striker and Pegasus. Heavy firepower is provided by an "assault company" dominated by the new Pilum heavy-wheeled tank. General Cage's pro-Steiner leanings have also allowed the unit to acquire several Fulcrum hovertanks.

2nd Chisholm Infantry Brigade
3 Regiments/Regular/Reliable
CO: Leftenant General Curtis Sozzi
Aide: Kommandant Nikki Kreviazuk
1st Airmobile Regiment: Lft. Gen. Alain Novak
3rd Chisholm JI: Lft. Gen. Ivan Messenger
9th MI: Lft. Gen. Yei Munsterberg

All three regiments of the Infantry Brigade are highly mobile, equipped with a wide range of integral transport. The airmobile regiment makes heavy use of the Karnov UR VTOL transport as well as the swifter Planetlifter. Combat VTOLs escort these vulnerable transports and provide grounded troops with supporting fire. The mechanized infantry use a mix of APCs and light personal or squad transports such as motorcycles and jeeps.

CETI HUSSARS

GRAND EXPERIMENT

One of the first true "experimental" units in the AFFS, the first Ceti Hussars unit was organized in the wake of the Davion War of Succession in the early twenty-eighth century. Based loosely on the SLDF's Regimental Combat Teams, the concept of the Hussars proved to be an idea before its time. Rather than organize an RCT or brigade in support of one or more BattleMech regiments, the Hussars divided all of its forces into three sub-units, or Combat Commands, leaving only a few specialized units, such as its artillery, independent. Though on paper the aerospace assets of the Hussars were attached to each Combat Command Alpha, in practice they also operated independently. The forces within each of these Combat Commands were then assigned to battalions based on their mission and capability. Though this process often segregated infantry units, in many cases armored vehicles and BattleMechs operated side-by-side in battalion- and company-sized formations.

Organized primarily for raiding and other fast assaults, the Hussars cut their teeth in the years of inter-League territorial disputes and border wars in the mid-twenty-eighth century. The Hussars more than vindicated themselves in action against the Capellan Confederation, but many AFFS officers had trouble understanding their purpose, and support was intermittent at best. The Ceti Hussars would likely have been disbanded if not for the Usurper Amaris and General Kerensky. Following Kerensky's Exodus, the Inner Sphere plunged into war. The Federated Suns soon needed every military unit it could get, so Prince John organized two more Hussars RCTs from the elements of several SLDF units he persuaded to remain.

The Hussars proved the value of the combined-arms concept time after time throughout the Succession Wars. In fact, it was partly because of their successes that Princess Melissa Davion formed RCTs around many of her key BattleMech units in the late twenty-ninth century. Though she stopped short of integrating all of her RCTs as radically as in the Hussars, two centuries later the St. Ives Military Command attempted to duplicate the Hussars organization with the Janissary Brigades.

COMMANDING OFFICER

Marshal Nebuchadnezzar Talles has commanded the Hussars for more than twenty years. Well past the age most officers retire, Marshal Talles not only continues to direct his three RCTs but is also a professor of military history at the NAIS College of Military Sciences. With a capable staff under him that issues his directions to the Hussars and keeps them operating, as well as three extremely skilled officers commanding his RCTs, Marshal Talles has enough time for the two advanced courses he teaches every semester. He often brings his students into the Hussars' headquarters to teach them the practical lessons of command.

FORCES

The Hussars may have been the first successful combined-arms paradigm within the Federated Suns' military, but the organization has nevertheless been relegated to a status well below the Royal Brigade or the Avalon Hussars. The situation improved somewhat when Marshal Talles moved the Hussars' headquarters from the Emporia system to New Avalon two decades ago, but the Hussars' tactical organization continues to draw criticism from traditionalists in the AFFC bureaucracy, even in light of their many successes.

COLORS AND INSIGNIA

For ceremonial occasions, the unit paints its equipment in dark blue with light blue and red highlights. The unit's insignia is a stylized rendition of a JumpShip with a DropShip.

1ST CETI HUSSARS RCT: THREE HEADED SERPENT

The First Ceti Hussars were nicknamed "The Grand Experiment" by Prince John Davion's staff, but pundits in the AFFS simply referred to them as "Humpty Dumpty," alluding to a popular child's rhyme. Even after the First Hussars dealt the Capellan Confederation a stunning defeat during Prince John's incursion in 2760, few supporters rallied behind the Prince's new unit. The Hussars performed well enough to persuade the Prince to form two more units several years later, however.

The First Hussars have called the Draconis March their home for many decades, sojourning out only while engaged in action against the Combine, as has occurred a number of times this century. They were assigned for some time to help shore up the Lyran border with the Jade Falcon OZ, though once Archon Katherine made her Secession Proclamation, the unit returned to the Federated Suns rather than allow themselves to be subject to the Archon.

The First Hussars' insignia shows a Hydra, its wings spread and heads ready for attack, next to the number "1".

OFFICERS

Marshal Vance Lamont has commanded the First Hussars for more than a dozen years. Though he has been offered a number of promotions, he is unwilling to take any post that would end his career in the field, though he has professed a desire for Marshal Talles' position once Talles retires. An opinionated officer who speaks his mind, often very publicly, Marshal Lamont is also an outstanding, if rather unorthodox, tactician. In fact, Lamont has served as director of operations in every PDZ his unit has been permanently assigned to during his command. When Marshal Petra Nichols took command of the Bryceland PDZ three years ago, however, she relieved him of that responsibility. The two officers have since shown a marked animosity toward each other. As Lamont is a vocal supporter of Prince Victor, while Nichols is a strong proponent of Archon Katherine, the open hostility could simply stem from the difference of opinion, though one DMI report suggests an affair dating back to their concurrent attendance at Albion could be the genesis of this feud.

TACTICS

Under Marshal Lamont's leadership, the First Hussars have embraced a highly mobile method of warfare. While they can often catch an enemy off-guard with their punishing "leapfrog" assaults, they also tend to leave their flanks wide open, giving a watchful enemy an opportunity to deliver a swift and deadly blow.

Combat Command Alpha
3 Regiments/Regular/Fanatical
CO: Major General Clayton Intravaia
Aide: Colonel Va Yee La Barbara
AeroSpace Group: Cdre. Lay Jesse

Combat Command Alpha serves as the First's maneuver element. In addition to its two companies of 'Mechs, it fields equal numbers of fast armor and mechanized infantry platoons.

The First Ceti Hussars AeroSpace Group fields two Wings.

Combat Command Bravo
3 Regiments/Green/Fanatical
CO: Leftenant General Alexander Davion Malen
Aide: Leftenant General Kimberly Lague

CC Bravo includes the RCT's only battalion of battle armor.

Combat Command Charlie
3 Regiments/Regular/Reliable
CO: Major General Laszlo Nunez
Aide: Leftenant General Erasmus D. L. Lamprakos

CC Charlie is the First's anchor, fielding almost an entire battalion's complement of assault BattleMechs in addition to its massive vehicle and ranger force.

1st Ceti Hussars Artillery Group
2 Battalions/Elite/Reliable
CO: Colonel Rashidi Kilton

As in the other Hussars units, the First Hussars' artillery group is assigned by the RCT headquarters, allowing a concentration of fire where it is needed most.

2ND CETI HUSSARS RCT: THREE OF DIAMONDS

Formed from the SLDF's Eighth Striker and 1894th Light Horse, the Second Hussars quickly became a significant force in the AFFS. They were central to the Suns' offensive against the Capellan Confederation during the Second Succession War—an action notable as the Hussars' most public failure. Assigned to bring the planet Tikonov into the Federated Suns, the Second twice failed to pacify the world before Capellan reinforcements forced their retreat. These incidents gave the entire Hussars' organization a worse reputation within the AFFS than it already had, even though the Second Hussars experienced wild success in all aspects of the operations except the one mission they were neither trained nor equipped for: an extended siege.

The Second's perceived failures haunted them for years, and some hold those failures against the unit even now, two centuries later. Whether for that reason, or simply because it was needed where it was, the Second remained stationed along the Capellan border during the years of the Clan invasion and today. According to all reports, the Second has not been involved in any action against the Confederation during its war against the St. Ives Compact, though Field Marshal Hasek has recently repositioned the unit, apparently in hopes of keeping the Confederation off balance.

The Second Hussars' insignia is a fan of playing cards with a three of diamonds showing, next to the number "2".

OFFICERS

Hauptmann General Oscar Carlson has spent his entire career in the Ceti Hussars. After enlisting, Carlson became one of the organization's many supply clerks. He proved an incredible organizer and thinker, however, and was recommended to the Sakhara Academy, where he graduated with honors and a degree in engineering. He returned to the Hussars, where he proved flexible enough to take command of one of the three most disorganized units (at least to an outsider) in the AFFC. Carlson is a conservative commander, both operationally and politically, and has refused to allow his unit to take on the traditional AFFS colors and insignia, even though a majority of the other units in the Capellan March already have.

TACTICS

The Second Hussars are often compared to a predatory feline: slow, methodical and utterly deadly. In many ways the First Hussars' diametrical opposite, the Second is often accused of not taking the initiative. On the other hand, its officers commonly study every piece of information available on an upcoming situation and thus are rarely surprised.

Combat Command Alpha
3 Regiments/Veteran/Reliable
CO: Hauptmann General Sebastian Jones
Aide: Leftenant General Dennis Malloy
AeroSpace Group: Kmd. Mary Rothe

CC Alpha fields the majority of the RCT's infantry forces as well as a concentration of slow, heavy armor.

The Second Ceti Hussars AeroSpace Group currently fields only one Wing.

Combat Command Bravo
2 Regiments/Elite/Fanatical
CO: Hauptmann General Marquis de Lafayette
Aide: Leftenant General Edward James Niemi

Hauptmann General de Lafayette is one of the most energetic and loyal officers in the AFFC, qualities that have rubbed off on the vast majority of his soldiers. Unfortunately, that enthusiasm often sets his Combat Command at odds with General Carlson, who frequently has trouble controlling this "renegade" unit.

CC Bravo does not deploy any infantry forces, instead concentrating on its massed 'Mech and armor formations.

Combat Command Charlie
3 Regiments/Veteran/Reliable
CO: Leftenant General Christian Peterson
Aide: Leftenant General Dominic Raabe

When he takes to the field, Hauptmann General Carlson assumes direct command of Combat Command Charlie. Fielding some 60 percent of the RCT's BattleMech complement, the Charlie Command represents the majority of the Second's ground-based firepower.

2nd Ceti Hussars Artillery Group
Battalion/Regular/Reliable
CO: Kommandant Quint Rabideaux

The Second's Artillery Group is one of the lightest in service with any RCT, fielding only twenty-four of the lightest-caliber tubes in the AFFC. Its six Arrow IV missile platforms help to make up for that deficiency, however.

3RD CETI HUSSARS RCT: THREE OF SPADES

The Third Ceti Hussars were formed, much like the Hussars' Second RCT, in the days following General Kerensky's Exodus. Composed of members from the SLDF's 199th and 396th Dragoon Regiments and the 250th BattleMech Division, the Third Hussars quickly proved one of the AFFS' most potent units, though like its two sister units the Third paid dearly for every victory in soldiers and equipment. The unit's continued successes throughout the Succession Wars were enough to persuade the High Command to keep the Third RCT operational, though the unit received barely enough support to maintain its combat readiness.

While they have made dozens of major raids on the Confederation's Sian and Victoria Commonalities in this century alone, the Third Hussars have not become involved in the ongoing war between the Confederation and the St. Ives Compact. With the unit showing no particular political leanings, Field Marshal Hasek is unwilling to commit it to an action that requires the utmost fidelity.

The insignia for the Third Hussars is a three of spades playing card with a bullet hole in it, sitting atop crossed pistols, next to the number "3".

OFFICERS

Major General Kellie Lee-Merrow is often considered a bit odd, even for a commander of a Ceti Hussars RCT. Rumors about her childhood and early career circulate through the unit. Many are contradictory and far too outrageous to be true; however, they all agree on one point: her youth was wildly misspent. Further adding to the mystery is the fact that Kellie Lee does not seem to have existed prior to 3032, when she was promoted from leftenant to captain while serving on New Avalon. Married, with four children and quite a few grandchildren, General Lee-Merrow speaks plainly and often acts with the impetuousness of youth; nevertheless, she has forged a successful career.

TACTICS

Many an enemy has mistakenly attributed Major General Lee-Merrow's reputation to her entire unit. While she gives the unit overall direction and formulates its grand strategies, her staff turns those general goals into detailed operational plans. Each of the RCT's Combat Commands often divides itself into smaller battalion- and regimental-sized units, allowing the Third to temporarily form one or more additional CCs and mount surprise attacks.

Combat Command Alpha
3 Regiments/Regular/Reliable
CO: Leftenant General Roman Do Linh
Aide: Leftenant General Cyrus Guthman
AeroSpace Group: Lft. Col. Rene Donahue

The Third RCT's Alpha Command is often split into four sub-commands, each with a different mission. Its largest group consists of almost a regiment of heavy-fire support BattleMechs and vehicles. Other smaller commands include independent scout, battle-armor and mechanized infantry units.

The Third Ceti Hussars AeroSpace Group fields two Wings.

Combat Command Bravo
3 Regiments/Veteran/Unreliable
CO: Major General Femi Naeem
Aide: Leftenant General Brent Jameson

The Bravo Command fields a well-integrated mix of medium BattleMechs, armor and mechanized infantry, trained extensively for mutual support even in the face of heavy opposition.

Combat Command Charlie
2 Regiments/Regular/Reliable
CO: Leftenant General Billy Rickgauer
Aide: Colonel Miriam Olson

CC Charlie fields the same mix of forces as the Bravo Command, though action against the Confederation in the Third Succession War permanently reduced its numbers.

3rd Ceti Hussars Artillery Group
3 Battalions/Regular/Reliable
CO: Colonel Lucien Lewis-Peron
Aide: Captain Diane Shaul

Fielding more than a hundred artillery tubes, in addition to other missile and air-defense platforms, the Third RCT can lay down precise artillery barrages that have shattered the resolve of entire divisions.

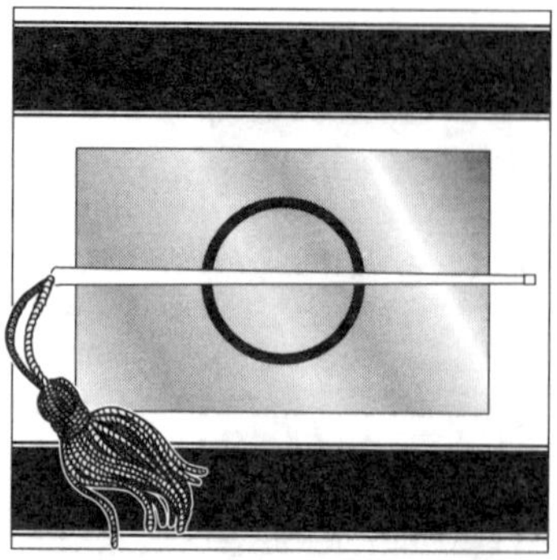

CRUCIS LANCERS

PRIDE OF THE PEOPLE

Like several other units in the AFFC, the origin of the Crucis Lancers lies with those SLDF troops who chose not to accompany General Kerensky on his Exodus. Most of the few units that hired on with the Federated Suns were incorporated into other organizations, such as the Ceti Hussars or the Deneb Light Cavalry. A majority of the SLDF troops that joined the AFFS no longer belonged to a coherent unit, however. While he used some of these hardened veterans to shore up numbers in his understrength units, Prince John Davion formed two complete Regimental Combat Teams around a core group of these former SLDF soldiers.

Equipped with weapons and vehicles from Star League storehouses, the first two Lancers units were not rushed immediately into combat. Instead they actively trained for several years while assisting in the formation of several more Crucis Lancers units. When the High Command finally released the Lancers for combat duty, they surprised the Capellan Confederation and the Draconis Combine, which thought them little more than glorified garrison forces.

The Lancers experienced a shake-up in the early thirtieth century, as a result of Princess Melissa Davion's reorganization of the AFFS, begun in the final years of the twenty-ninth century. Mostly administrative in nature, this reorganization nearly broke the Lancers apart. Ultimately, however, it strengthened the entire organization, allowing the Lancers to thrive through the coming years, which were to bring the bloodiest wars the Inner Sphere has seen.

COMMANDING OFFICER

A native of Tikonov, Marshal Ivor Wasjinji made passage into the Federated Suns working as a longshoreman and joined the AFFS before the Fourth Succession War. He left the AFFS to join Pavel Ridzik's Tikonov Free Republic Army, only to be incorporated into the AFFS when the TFR voted to enter the Federated Suns. Marshal Wasjinji has harbored a grudge against the Capellan Confederation his entire career and has done everything within his power to fight the Confederation or support those who were. Since his youth he has nurtured contacts across the Inner Sphere, both legitimate and not, which he uses to keep his units fully operational and to funnel support under the table to the St. Ives Compact and organizations like Free Capella that continue to fight the Capellan Confederation.

FORCES

While the Crucis Lancers owe their existence to SLDF expatriates, each of the organization's eight units took on the name of a different administrative sector within the Crucis March and eventually began to recruit its members almost exclusively from these regions. Because of this, the members of each unit tend to form a tighter-knit group than in most other units.

The eight Lancer units have been stationed throughout the Federated Suns and Lyran nations for the past three decades, resulting in an eclectic mix of equipment. While the Lancers have not been at the top of the supply chain for many years, they nevertheless field an impressive amount of upgraded technologies.

COLORS AND INSIGNIA

The Lancers use the parade colors of Davion green, highlighted with the colors of whatever Combat Region they consider home. The organization's insignia is a stylized general's baton set against a yellow flag.

1ST CRUCIS LANCERS RCT: NEW AVALON LANCERS

Though originally made up of former SLDF personnel, as combat losses increased during the First Succession War the First Crucis Lancers drew heavily on the population of the New Avalon Combat Region. As the years rolled on and the First Lancers continued to prove their loyalty to the Federated Suns and to the royal family in particular, the Davion Princes rewarded the unit handsomely.

The Lancers were heavily involved in the Fourth Succession War and the War of '39, but they did not reach the Lyran Theater until the late stages of the Clan invasion. The unit remained along the Jade Falcon border until Archon Katherine ordered them out of the Lyran Alliance following the Marik-Liao Offensive, after which the unit returned to the Draconis March.

Owing to years of heavy combat losses and the intense scrutiny of the DMI, the First Lancers have rarely fielded their full strength over the past two centuries. The unit is currently running at slightly less than 90 percent strength.

The First Lancers use the standard Crucis Lancers colors with red, blue and white highlights. Its insignia is that of the Crucis Lancers organization with the initials "N.A.L.," signifying New Avalon Lancers, set below the flag.

OFFICERS

While Marshal Wasjinji is the titular commander of the First Crucis Lancers, his duties as commander of the entire organization often keep him away. After taking charge of the organization, Wasjinji groomed a young leftenant general he had encountered in the Sarna March to act as his deputy. Eventually, Wasjinji turned over the day-to-day operations of the First Lancers to this upstart, who quickly proved himself more than worthy of his surname. Mordecai Rand-Davion has never played upon his connection to the royal family, often leaving the "Davion" off to further distance himself. He is, however, a distant grandson of Prince Richard Davion and the Duke of Abbeville. Marshal Rand-Davion is an outstanding officer, nurturing a deep esprit de corps within his unit and displaying a strategic aptitude worthy of the Davion name.

TACTICS

The First Lancers are one of the very few units that has ever been authorized to carry the Davion family standard into battle. The unit has fought with an almost unequaled vigor throughout its history, especially against enemies of the Davion family, never allowing its own standard or that of the royal family to fall in battle.

1st Crucis Lancers
Regiment/Regular/Fanatical
CO: Marshal Mordecai Rand-Davion
Aide: Leftenant Colonel Kathryn Louise Murphy
Aerospace Brigade: Vice Adm. Ho Van Luu

As one of the Davion family's favored units, a majority of the First Lancers' mostly heavy-class 'Mechs are less than eight years old.

The First Crucis Lancers AeroSpace Brigade fields the First and Second Crucis Lancers Aero Wings, along with the 315th Federation Independent Air Wing.

1st Crucis Lancers Armor Brigade
3 Regiments/Regular/Reliable
CO: Major General Gerald Kroeger
Aide: Leftenant Colonel Andrea Marcek
1st Crucis Lancers ARR: Lft. Gen. K. Richard Miller
2nd Crucis Lancers ARR: Col. Persephone Jefferson
3rd Crucis Lancers ARR: Col. Carolyn Rode
1st Crucis Lancers AR: Maj. Lorenzo Sammataro

Though it deploys mostly fast armor and hovercraft vehicles, the First Armor Brigade also commands an entire regiment of heavy artillery.

1st Crucis Lancers Infantry Brigade
8 Regiments/Veteran/Fanatical
CO: Major General Natascha Tamblyn
Aide: Leftenant General Willie Ragan
1st Crucis Lancers IR: Lft. Gen. Wolfgang Mattson
2nd Crucis Lancers IR: Col. Elihu Lang
1st Crucis Lancers MIR: Col. Melesse Javadi
2nd Crucis Lancers MIR: Col. E. J. Hannah
1st Crucis Lancers JIR: Col. Lucifer "The Black Devil" Roussa
11th Freisland MR: Col. Constantin Kolesnikov
80th F-C BAB: Lft. Gen. Michael James Lewandowski
2nd Argyle Rangers: Col. Elric Lynch

Each infantry unit fields its own organic transportation assets, while the Second Rangers also makes use of a full battalion of VTOLs.

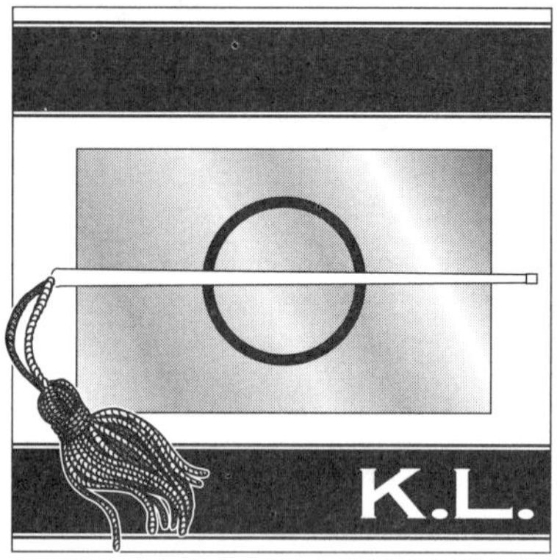

2ND CRUCIS LANCERS RCT: KESTREL LANCERS

The Second Crucis Lancers owe their fealty and prosperity to the Davion family, just as do the First Lancers, and though overshadowed by their more prominent sister unit, the Second Lancers are no less capable or loyal. After they suffered tremendous losses in the Succession Wars, the High Command temporarily deactivated the Second Lancers while the unit rebuilt.

Reactivated shortly after the formation of the AFFC, the Second was repositioned along the Falcon line of attack when the Clan juggernaut appeared. Tasked to aid survivors and other rebel groups operating behind the Falcons' lines, by the time the invasion ended, the Second Lancers had been reduced to little more than a reinforced brigade.

The AFFC rebuilt the Second as best it could, but that ended with the secession of the Lyran Alliance. After reassigning the unit to the "important" station of Timbiqui, the LAAF has apparently forgotten about the Second Lancers, giving it barely enough supplies to stay operational. In a stroke of genius, General Sung agreed to work with the Raasch Brewhaus, brewers of Timbiqui Dark, to produce a continuing series of advertisements hailing the drink as the "Brew of the ilKhan." While not condoned by the LAAF, these ads provide the Second enough money to maintain its combat edge.

The Second uses the standard Lancers colors, highlighted in light green. The initials "K.L." are displayed below the Lancers insignia.

OFFICERS

While devoted wholeheartedly to the Davion family and the Federated Suns, General Anne Sung is above all a realist. Realizing no support for the Second Lancers would be forthcoming, she agreed to the deal with the Raasch Brewhaus, one that is specifically prohibited by AFFC and LAAF regulations. Rumor has it she has also secretly hired out portions of her command on security assignments to add more funds to the unit's depleted coffers.

TACTICS

The Second Lancers learned the art of guerrilla warfare during their extended assignment along the Falcons' OZ. The Second often breaks its component regiments down into small combined-arms battlegroups that are as proficient at stealthy overland movement as they are at orbital drops.

2nd Crucis Lancers
Regiment/Veteran/Reliable
CO: General Anne Sung
Aide: Leftenant Colonel Bruno J. Groskopf
AeroSpace Brigade: Lgt. Cdre. Schoener Liljestraand

Still reeling from losses during the Clan invasion, the Second's 'Mech unit is more than twenty 'Mechs short of full strength. Slightly more than a dozen Clan-based 'Mechs are scattered among its companies, however, giving it a slight advantage.

The Third and Fourth Crucis Lancers Aero Wings make up the Second Crucis Lancers AeroSpace Brigade.

2nd Crucis Lancers Armor Brigade
3 Regiments/Veteran/Reliable
CO: Major General Edward Cook
Aide: Leftenant Colonel J. G. Julian
4th Crucis Lancers ARR: Col. Anna Ortiz-Sandoval
6th Crucis Lancers ARR: Col. La Grange Ingram
2nd Crucis Lancers AB: Lft. Col. Randy Palen

The Sixth Armored, plus a full lance of Clan-based 'Mechs painted to look like ilKhan Ulric Kerensky's Golden Keshik, are based at the Raasch Brewhaus and its accompanying airfield.

2nd Crucis Lancers Infantry Brigade
4 Regiments/Green/Reliable
CO: Major General Charity Jessman
Aide: Major Marc Geary
4th Crucis Lancers IR: Col. Don Schmidt
3rd Crucis Lancers MIR: Col. Joelle La Mere
4th Crucis Lancers MIR: Col. Lyons Freeman
2nd Crucis Lancers JIR: Lft. Col. Karl Lim

After rebuilding its Infantry Brigade, shattered in the wake of the Succession Wars, the Second Lancers again lost most of its infantry strength against the Jade Falcons. The Third Infantry is still inactive, while the remaining four regiments are just now reaching full strength. The Second Jump does field two companies of the F-C standard Cavalier battle armor.

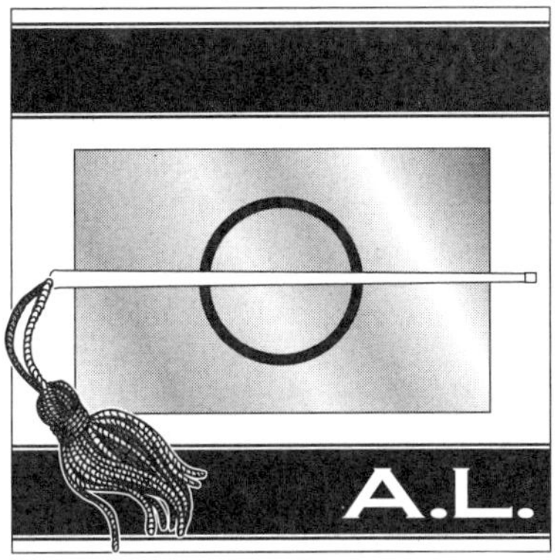

3RD CRUCIS LANCERS RCT: ACHERNAR LANCERS

Stationed within the Kentares PDZ at the onset of the Clan invasion, the Third Crucis Lancers were among the first units repositioned to buttress the Lyran State's defenses. The Third found itself in Field Marshal Nondi Steiner's strategic reserve until Marshal of the Armies Hasek-Davion redeployed the reserve into positions where it could slow or stop the Falcons' advance. While assigned to Babaeski, the Third Lancers took part in a number of raids against the Falcons, especially against their bases on Sudeten.

Had the Third remained there, they would have been in the perfect position to contest the Falcon incursion of 3058. Instead, following the Lyran secession a year earlier, the Third returned to the Federated Suns, though not before turning over the defense of Babaeski to Duke Morgan Kell and his Arc-Royal Defense Cordon.

The Third uses yellow highlights with the standard Lancers colors. The initials "A.L." are displayed below the Lancers insignia.

OFFICERS

General Acabee Zardetto is one of the few non-MechWarriors to command an RCT, especially one as prominent as the Third Crucis Lancers. A Tharkad native, Zardetto was one of the first Lyran cadets to enter the NAIS. His older brother Paul, now commanding the Third Lyran Guards, entered Tharkad's Sanglamore, leading to a rift that has kept the brothers apart for more than a decade. General Zardetto is a popular commander who has forged close alliances with officers and political figures throughout the former Federated Commonwealth. He is also a devout critic of Archon Katherine, a fact that has somewhat sullied his reputation within the Lyran State and further distanced him from his brother.

TACTICS

The officers of the Third Lancers prefer the offensive, where they use their overwhelming numbers of 'Mechs and armored vehicles to drive the enemy before them. Whether on the offensive or the defensive, the Third places its infantry units in prepared or fortified positions from which they can launch surprise attacks.

3rd Crucis Lancers
Reinforced Regiment/Elite/Reliable
CO: Leftenant General Katriel Macon
Aide: Leftenant Colonel Kevin Kirschner
AeroSpace Brigade: Rear Adm. Dmitri Mollerskov

The Third fields four full battalions of medium to heavy BattleMechs, including eight lances of OmniMechs and two more of Clan-tech 'Mechs. Inexplicably, General Zardetto has managed to acquire two brand-new *Hauptmanns* for his MechWarriors.

The Third Crucis Lancers AeroSpace Brigade fields the Fifth and Sixth Crucis Lancers Aero Wings.

3rd Crucis Lancers Armor Brigade
4 Regiments/Elite/Fanatical
CO: General Acabee Zardetto
Aide: Leftenant General Lloyd Ra
7th Crucis Lancers ARR: Col. Isis Jaske
8th Crucis Lancers ARR: Col. Terri Kuhn
9th Crucis Lancers ARR: Lft. Gen. Jennifer Boyd
8th Elbar LA: Col. Albin Jancar
3rd Crucis Lancers AB: Lft. Col. Theo P. Reaves III

As with his 'Mech unit, General Zardetto has made certain the armor brigade is equipped with the best. A full battalion each of Alacorns and Challenger Xs are scattered among his four armored regiments, as are a handful of Clan-tech-equipped vehicles.

3rd Crucis Infantry Brigade
5 Regiments/Veteran/Reliable
CO: Major General Tina Dorsey
Aide: Colonel James Mingjun
5th Crucis Lancers IR: Lft. Gen. Raghib Li
6th Crucis Lancers IR: Col. Emma Valentino
5th Crucis Lancers MIR: Col. Christopher Snow
6th Crucis Lancers MIR: Col. Lei Ntwana
3rd Crucis Lancers JIR: Col. Ronald Montrose

The Fifth Infantry pulls double duty as the RCT's engineer regiment, operating its heavy construction machinery. The Sixth Infantry has been known to ravage entire 'Mech battalions when given the chance to deploy their large quantity of heavy towed weapons.

4TH CRUCIS LANCERS RCT: POINT BARROW LANCERS

The Fourth Lancers began their long career within the Lyran State when they were called upon to help the LCAF form RCTs around several of its premier units in the late 3030s. As part of General Nondi Steiner's attack force, the Fourth Lancers acquitted themselves well against the Draconis Combine during the War of 3039, though they were ultimately driven from Konstance by the Third Dieron Regulars.

The Fourth has faithfully served the Lyran State since, stationed primarily along its border with the Free Worlds League. This long association with the Lyran people is causing quite a bit of strife within the unit, however. While a great majority of the unit's personnel hail from the Point Barrow Combat Region, most of its dependents have a Lyran heritage. A few families have split as a result of the growing schism between Archon Katherine and Prince Victor. Rather than tear the unit apart, most of its personnel are refusing to choose sides, though many are keeping a close eye on the situation in their homeland.

The Fourth Lancers add yellow and blue highlights to the standard Lancers colors and the initials "P.B.L." below the Lancers insignia.

OFFICERS

A tenth-generation Federated Suns officer, Leutnant-General Andrew Giggins began his career at the Point Barrow Military Academy, then experiencing a number of difficulties with anti-Davion radicals. Giggins nevertheless made a name for himself in the AFFS and later attended the CMS Federation Command and Staff College, graduating with top honors. All AFFC profiles show that Giggins is a dedicated officer who holds service to his nation above all.

TACTICS

The Fourth Lancers are often considered the "black sheep" of the Crucis Lancers, fighting in places and against foes that few other units would consider. The Fourth will, almost as a rule, take the path less trodden, surprising and dislodging even the most entrenched enemy. In contrast, the unit functions barely above average on an open battlefield.

4th Crucis Lancers
Regiment/Veteran/Reliable
CO: Leutnant-General Andrew Giggins
Aide: Leutnant-Colonel Sakhone Phoun
AeroSpace Brigade: Lut.-Kdre. "Wild" Willie Brueckner

The Fourth's 'Mech regiment fields almost exclusively medium-class BattleMechs, with nearly the entire regiment jump-capable. Its one Assault company, all retrofitted with jump jets, mounts the heavy autocannon and Gauss rifles needed to break through the strongest enemy fortifications.

The Fourth Crucis Lancers AeroSpace Brigade is oversized, fielding the Seventh and Eight Crucis Lancers Aero Wings, along with the Fifth Baxley Aerial Recon and 120th Minette Light Wings.

4th Crucis Lancers Armor Brigade
3 Regiments/Elite/Reliable
CO: Leutnant-General Adonias Felice
Aide: Kommandant Jamie Rand
10th Crucis Lancers ARR: Col. Arcadio Benz
11th Crucis Lancers ARR: Col. Ramond Rameshkumar
12th Crucis Lancers ARR: Lut.-Col. Marshal Kail
4th Crucis Lancers AB: Hpt.-Kmd. Rob Roberts

The Tenth Armored is designated heavy armor and often fights alongside the RCT's few heavy 'Mech elements. The Fourth RCT's artillery double-strength battalion is used to demolish fortifications its more mobile cousins cannot.

4th Crucis Lancers Infantry Brigade
3 Regiments/Veteran/Fanatical
CO: Leutnant-General Danielle Hartman
Aide: Leutnant-Colonel Apollo Makar
7th Crucis Lancers MIR: Lut.-Col. Agueda Lopez-Repp
8th Crucis Lancers MIR: Lut.-Col. P. N. Meyers Jr.
4th Crucis Lancers JIR: Col. Heather Soden
1430th Federation Ranger Batt: Kmd. Lex Bradford

The Fourth Lancers lost the majority of their conventional infantry units more than a century ago, but they have compensated by equipping their mechanized elements with a variety of towed weapons.

5TH CRUCIS LANCERS RCT: REMAGEN LANCERS

Assigned to the Sarna March when the Marik-Liao Offensive of '57 hit, the Fifth Lancers were caught off guard and forced from the region by the combined efforts of the CCAF and the Zhanzeng de Guang terrorist cells. Shamed by their defeat, the Fifth continued to fight an unauthorized campaign on several worlds nominally in Capellan hands, primarily against the Zhanzeng de Guang organization. Four years and nine worlds later, the Fifth has scaled back its operations, having wiped out seven of the terrorist organization's major groups and further tied up several CCAF units that otherwise would have been assigned to the war in St. Ives. Lately the Fifth has turned its frustrations against the pro-Katherine Fifteenth Deneb Light Cavalry.

The Fifth Lancers use white highlights over the standard Lancers colors and display the initials "R.L." below the Lancers insignia.

OFFICERS

General Olaf Richardson has commanded the Fifth Lancers for almost fifteen years, during which time he has built a reputation as a freelancer. Once considered for promotion to New Avalon, his long years of service in the former Sarna March have given him a penchant for taking matters into his own hands. Though he extricated his unit from the rapidly degenerating Chaos March relatively intact, he garnered a severe reprimand from New Avalon over his recent actions within Capellan territory. Psych reports indicate the general has recently begun to experience deepening paranoia and schizophrenia.

TACTICS

The Fifth Lancers do not like to stay in one place for any length of time. Whether on the attack or the defensive, the unit is always on the move, with elements often seemingly operating independently from one another. General Richardson keeps complete track of his RCT from the cockpit of his ancient *Cyclops,* however, coordinating all the frenetic actions of his subordinate units.

5th Crucis Lancers
2 Battalions/Veteran/Fanatical
CO: General Olaf Richardson
Aide: Major Rett Brody
AeroSpace Brigade: Cdre. Leroy Oliver

Destroyed in the final years of the Third Succession War, the Fifth's 'Mech regiment was reconstituted after the formation of the unified AFFC in 3042. Though the unit survived the best the CCAF could throw at them over the past several years, almost two-thirds of the command perished just a few months ago while training in Demeter's Kuriak Mountains. After losing contact with the Second and Third battalions for almost three days, General Richardson dispatched a recon force from the 204th Scout. A day later he led four regiments into the mountains, where he found all his missing equipment but the remains of only a few of the 204th's soldiers. DMI investigators later uncovered a mass grave and concluded a Thuggee attack had killed Richardson's troops. To date, though he has enough equipment to field a full 'Mech regiment, Richardson has only enough rated pilots for two battalions.

The Fifth Crucis Lancers AeroSpace Brigade fields the Ninth and Tenth Crucis Lancers Aero Wings.

5th Crucis Lancers Armor Brigade
6 Regiments/Elite/Fanatical
CO: Major General Meshach Felsner
Aide: Colonel Michele Ku
13th Crucis Lancers ARR: Lft. Gen. Donato Guiseppi
14th Crucis Lancers ARR: Col. Elson Mortazavi
15th Crucis Lancers ARR: Col. Lisa Ravenwater
74th Remagen AC: Col. Joichiro Tanaka
587th Federation HAR: Maj. Gen. Meshach Felsner
204th Defiance SR: Col. Paul Matthews
5th Crucis Lancers AB: Maj. Emil Mosthov

The 204th is back up to full strength, though the soldiers crewing the two companies of recovered vehicles are all exhibiting various health problems, as are the new members of the second 'Mech battalion.

5th Crucis Lancers Infantry Brigade
6 Regiments/Veteran/Fanatical
CO: Leftenant General Melanie Sode
Aide: Major Kwinn Illari
9th Crucis Lancers IR: Lft. Gen. Andrei Gorobtsov
10th Crucis Lancers IR: Col. Quan-Lu D'Aquisto
9th Crucis Lancers MIR: Col. Craig Peterson
10th Crucis Lancers MIR: Col. Babette Kiska
5th Crucis Lancers JIR: Col. Halsted Sharpe
98th Crucis JI: Col. Ping Tam

6TH CRUCIS LANCERS RCT: TSAMMA LANCERS

Often called the "Flower Lancers" because of the way the unit decorates its combat equipment, a tradition that hails from an annual festival celebrated on Tsamma, the Sixth Lancers are as capable as any of their sister RCTs. The Sixth Lancers have been targeted by Capellan terrorists since their bloody capture of Mandate during the Fourth Succession War. The High Command even reassigned the unit to the Lyran State in the early 3030s to ease the pressure placed on the Sixth by Capellan patriots. The unit gained a respite from the constant harassment until Chancellor Liao's Xin Sheng movement stirred up traditional Capellan feelings again. Since the beginning of the Capellan-St. Ives War, the Sixth Lancers have endured almost a dozen terrorist attacks aimed at crippling or killing its command staff.

The Sixth Lancers use yellow and light green highlights on the standard Lancers colors. In addition, the members of the Sixth Lancers traditionally decorate their equipment with actual flowers or, when those are impractical, painted ones. The Sixth adds the initials "T.L." below the Lancers insignia.

OFFICERS

The Sixth Lancers are rapidly unraveling. Once considered an outstanding officer, the stress of continued terrorist attacks has turned Patricia Vineman into a totalitarian leader. She has taken de facto control of Langhorne and has turned her RCT into a police force dedicated to ridding Langhorne, and if necessary surrounding worlds, of terrorists. She received a severe reprimand and a demotion to colonel after ordering her troops to put down a riot, even though, although hundreds were killed, it would likely have resulted in the destruction of much of Langhorne's capital city of Cherone.

TACTICS

After taking part in so many police actions over the past several years, the members of the Sixth Lancers have become quite adept at fighting within the confines of a city against a well-entrenched foot-infantry (in this case terrorist) force.

6th Crucis Lancers
Regiment/Veteran/Reliable
CO: Colonel Patricia Vineman
Aide: Hauptmann-Kommandant Albert Jehlik
AeroSpace Brigade: Lut.-Kapt. Deuce MacAdams

Since beginning to train her unit in police and city-defense tactics, Colonel Vineman has used her few contacts within the LAAF to trade the majority of her heavy and assault-class BattleMechs for more mobile lights and mediums. She retained only two companies of heavy 'Mechs for her first battalion.

The Eleventh and Twelfth Crucis Lancers Aero Wings make up the Sixth Crucis Lancers AeroSpace Brigade.

6th Crucis Lancers Armor Brigade
2 Regiments/Regular/Unreliable
CO: Colonel Emmanuel O'Shea
Aide: Kommandant Enrico Tedesco
16th Crucis Lancers ARR: Col. Dick Ziogas
18th Crucis Lancers ARR: Col. Pinnard Smith
6th Crucis Lancers AB: Lut.-Col. Craig Mackie

Both the Sixteenth and Eighteenth field a mix of heavy tanks, like the venerable Patton, and lighter armored vehicles, such as the Bulldog and Plainsman.

6th Crucis Lancers Infantry Brigade
7 Regiments/Veteran/Reliable
CO: Colonel Ilena Soden
Aide: Hauptmann-Kommandant Amy Arrizon
11th Crucis Lancers IR: Col. Caridad Hasek-Goros
12th Crucis Lancers IR: Lut.-Col. Peng Popovic
11th Crucis Lancers MIR: Col. Omar Herzog
12th Crucis Lancers MIR: Col. Dana Miguel De Jesus
6th Crucis Lancers JIR: Col. Sam Worthington
Gambier Lancers JI: Hpt.-Kmd. Calypso Kaegi
1st Langhorne MPR: Col. Jessup Wilmer

The First Langhorne is a recent addition to the Sixth Lancers, consisting of seven full battalions of military police, many recruited from Langhorne's police forces. The seventh battalion is tasked with undercover investigations and other covert anti-terrorist operations.

7TH CRUCIS LANCERS: BROKEN WHEEL CHARIOTEERS

Reduced to little more than a paper regiment while defending Kentares IV during the First Succession War, the Seventh Lancers were slowly rebuilt during the early years of the Third Succession War. The unit distinguished itself during the Third and Fourth Succession Wars but was blind-sided by then *Gunji-no-Kanrei* Theodore Kurita's counterattack during the War of 3039. The Seventh Lancers withdrew from Klathandu IV, broken but not dispirited.

The Seventh quickly rebounded from their rout, but stripped of their Naval Transport Squadron at the outset of the Clan invasion, the unit saw no action against the Clans, virtually trapped as they were along the Lyran Periphery. When the LAAF finally reassigned permanent transport to the unit in 3058, the Seventh RCT engaged in a long series of independent exercises in the Periphery. The elements of the RCT alternated between garrison duty on Winter and training on worlds in the near Periphery.

This continued for a little more than two years before the unthinkable happened. Leutnant-General Zibler remained on Winter with his 'Mech regiment and a few supporting elements while the rest of the RCT made their way to an uninhabited world in the Periphery system known as Savarin 223-74. When the detachment had not reported back after six weeks, Leutnant-General Zibler dispatched a scouting force to the system but found no trace of the detachment. The LAAF and even ComStar sent a number of expeditions into the Periphery, but after almost a year of searching, no evidence of the detachment's fate has turned up. Intelligence analysts have heard no evidence suggesting the detachment fell prey to a Clan task force. The unit's four JumpShips, sixteen DropShips and some eleven thousand personnel have apparently become the latest victims of the often mysterious and dangerous Periphery.

The Seventh highlights the standard Lancers colors with red and blue. The unit's insignia is the Lancers flag with the initials "B.W.C." inscribed beneath.

OFFICERS

Leutnant-General Jasper Zibler worked his way up through the ranks of the Seventh Crucis Lancers, joining the unit as a simple Astech in the last year of the Third Succession War. He was given the opportunity to pilot a 'Mech after the unit landed on Tikonov during the second wave of the Fourth Succession War. The courage and skill he showed during that operation garnered him a recommendation to Albion's OTS program, where he honed his leadership skills. Always a sharp strategist, Zibler makes the most of what he is given and is a demanding commander.

TACTICS

The Seventh Lancers take their motto of "Anything, Anywhere, Any Time" to heart. Though stripped of most of its conventional combat elements, the unit still fields a potent force of BattleMechs and armored vehicles and a complement of highly skilled warriors. The Seventh will not shy away from any mission it is given.

7th Crucis Lancers
Reinforced Regiment/Elite/Fanatical
CO: Leutnant-General Jasper Zibler
Aide: Leutnant-Colonel Ellen Isakov

Immediately following the Fourth Succession War, the Seventh was refitted with equipment manufactured in the Sarna March, lending it a noticeable Capellan look. Upon its assignment to the Lyran State, the Seventh attempted to trade much of that equipment for Lyran models. Though the unit fields only a handful of OmniMechs as a result, the Seventh is nevertheless considered the best-equipped Federated Suns unit in the Lyran Alliance.

7th Crucis Lancers Armor Brigade
Regiment/Elite/Reliable
CO: Leutnant-Colonel Sutton Scott Briggs
Aide: Kommandant Lila Dahar
20th Crucis Lancers ARR: Lut.-Col. Sutton Scott Briggs
7th Crucis Lancers AB: Kmd. Lila Dahar

The Twentieth Armored is rapidly finding itself overcommitted to the assignments the entire brigade once shared, which is cutting deeply into its weekly training regimen. At the same time, Leutnant-General Zibler has tasked Leutnant-Colonel Briggs and his unit with preparing Winter's Ninth Armored militia regiment for possible assimilation into the brigade.

405th Federation Interceptor Wing
Wing/Veteran/Fanatical
CO: Kommandant Sammy "The Slugger" Dahlke

The Slugger and his four squadrons were dispatched to Winter thirteen years ago for rest and refit and were promptly forgotten when the Clans appeared. They have been operating as a de facto element of the Seventh Lancers ever since.

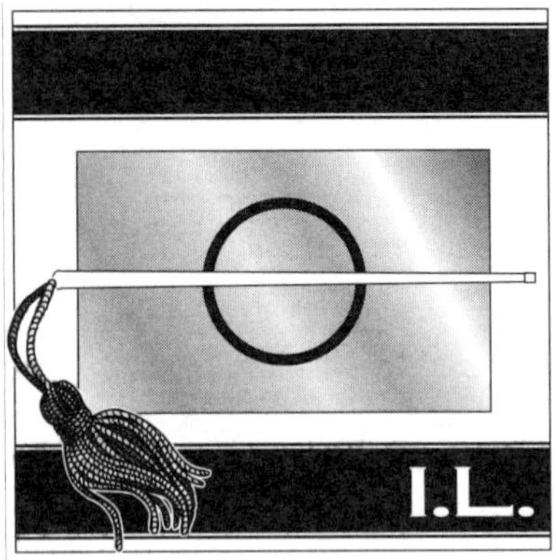

8TH CRUCIS LANCERS RCT: ISLAMABAD LANCERS

The Eighth Lancers have been assigned to the Draconis March for much of their history, and as a result they feel a deep resentment toward the Draconis Combine, a feeling strengthened after the unit's defeat during the War of 3039. Prince Hanse Davion and Field Marshal James Sandoval kept a tight leash on the Eighth, though it was cleared for a number of raids into Combine space after the War of 3039. Everything, of course, changed with the coming of the Clans.

The Eighth Lancers were one of the first units dispatched to help fortify the Lyran Commonwealth, removing the possibility of the unit taking advantage of the Combine's weakness to attack. The unit remained in the Lyran State for a number of years, continually repositioning itself in response to Falcon troop movements, until it was expelled following the Lyran secession.

The long years stationed along the Clan front have dulled the Eighth's hatred for the Combine, but many long-time Lancers still decry Prince Victor's close association with their traditional enemy and long for a return to the "old ways." Luckily, most of the unit's younger members neither share nor understand this feeling.

The unit uses crimson and silver highlights over the standard Lancers colors. The initials "I.L." are displayed below the Lancers insignia.

OFFICERS

Hauptmann General Jeremiah Marshall applied to the Robinson Battle Academy to escape his backwater homeworld of Tentavia. He immediately impressed Duke Aaron Sandoval, who closely followed Marshall's progress through the AFFS, occasionally nudging the man's career along the "right" path. That path finally led him to the helm of the Eighth Lancers, one of the few "outside" units loyal to the Sandovals. In his four months of command, Marshall has proved to be a skilled commander, though his temper has cowed a majority of his staff, a situation that could be disastrous for the unit.

TACTICS

While officially known as the Islamabad Lancers, the Eighth also has the unofficial nickname of "The Islamabad Sluggers," owing to the unit's decidedly heavyweight BattleMechs and combat vehicles. While the Eighth does not have the maneuverability of some of its sister units, its sheer firepower has often been enough to shatter even the most zealous defense.

8th Crucis Lancers
Reinforced Regiment/Regular/Reliable
CO: Hauptmann General Jeremiah Marshall
Aide: Kommandant Warner Alexander
AeroSpace Brigade: Lft. Gen. Hieremias Spaatz

The Eighth Lancers rely on the strength of their four battalions of heavy and assault BattleMechs, including the Fourth Battalion's complement of *Longbows* and *Salamanders*, to provide the hitting power for the entire RCT.

Along with the Fifteenth and Sixteenth Crucis Lancers Aero Wing, the Eight Crucis Lancers AeroSpace Brigade includes the Twenty-second Macintosh Heavy Interceptors.

8th Crucis Lancers Armor Brigade
4 Regiments/Green/Unreliable
CO: Leftenant General Frank Ritchie
Aide: Kommandant Shakti Anderson
22nd Crucis Lancers ARR: Lft. Gen. Penny Montgomery
23rd Crucis Lancers ARR: Lft. Gen. Munish Clark
24th Crucis Lancers ARR: Lft. Gen. Do Sy Truscott
144th Federation CR: Lft. Gen. Reva McCreery
8th Crucis Lancers AB: Kmd. Cassandra Bradley
1943rd F-C AB: Kmd. Nathan Devers

With two full artillery battalions to call upon, the Eighth RCT's armored formations often see only the remains of an enemy force.

8th Crucis Lancers Infantry Brigade
5 Regiments/Regular/Reliable
CO: Leftenant General Dorian Crerar
Aide: Leftenant General Gurion Dempsey
15th Crucis Lancers IR: Lft. Gen. Cerise Hodges
16th Crucis Lancers IR: Kmd. Phillipe Simpson
15th Crucis Lancers MIR: Lft. Gen. Chui Patton
16th Crucis Lancers MIR: Lft. Gen. Jim Patch
8th Crucis Lancers JIR: Lft. Gen. Guillaume de Lattre

Each of the Eighth's conventional infantry regiments has its own organic transportation and often fields large numbers of towed weapons.

DAVION BRIGADE OF GUARDS

THE PRINCE'S STRENGTH

The Davion Brigade of Guards, often referred to as the Royal Brigade, has been the backbone of the Federated Suns military for centuries. Once nothing more than the personal guard of the Davion family, the Royal Brigade gained prominence during Reynard Davion's reign as president of the Federated Suns. Since that time, the Brigade has grown to eight full Regimental Combat Teams and has been involved in every major military action the Federated Suns has fought.

The Brigade maintains a strong esprit de corps and an undying sense of devotion to the Davion family. Their first and foremost loyalty is to their Prince. In return, the Prince has consistently assigned them the best equipment and personnel the Federated Suns has had to offer and has often personally led one or more of the Brigade's units into battle. Until the ascension of Archon Katherine, the Brigade also provided personal security for the Prince and his family, as well as the Royal Castle and the Fox's Lair.

The Royal Brigade recruits the best candidates from the AFFC's schools and academies year after year. This, of course, ensures not only that the Brigade's level of experience remains high but also that the component units owe allegiance to no particular world or region—only to the Prince and their nation. The events of recent years have threatened to turn centuries of tradition on its head, however. Though no Royal Brigade unit has ever shown anything but loyalty to the Prince of the Federated Suns, none of the Brigade's commanders has sworn an oath of fealty to Archon Katherine, and several have openly spoken out against their nominal leader.

COMMANDING OFFICER

Marshal Bishop Sortek, a second cousin of long-time royal family advisor Major General Ardan Sortek (SLDF), has spent his entire life in the Royal Brigade. His parents, both officers in the Brigade, were killed in action during the Third Succession War, leaving him to grow up among the Brigade's families on New Avalon. Although guaranteed a position in the Albion Military Academy by virtue of his parentage, he nevertheless competed for and won a common position in the cadet corps. After graduation, Prince Hanse himself awarded Sortek a billet in the Heavy Guards. Proving himself an outstanding leader and strategist, he quickly rose through the ranks and took command of the First Guards at the age of thirty-four. When Field Marshal Jackson Davion stepped down from the command of the Guard to become Marshal of the Armies, the Brigade's commanders unanimously chose Sortek to take his place. Due to his long relationship with the Davion family, however, Marshal Sortek, along with his Brigade, has been the target of a number of Archon Katherine's bureaucratic assaults.

FORCES

Even in light of Archon Katherine's rise to power on New Avalon, the elements of the Royal Brigade want for nothing. Every RCT fields Inner Sphere–designed OmniMechs, and several also field a number of Clan-based 'Mech designs.

COLORS AND INSIGNIA

The 'Mechs and vehicles of the Royal Brigade commonly wear paint schemes based on the requirements of their current assignment, though for parade and ceremonial purposes they make use of a blue, white and red striped scheme, the colors of the Davion family. The Brigade's insignia is a variation on the Federated Suns crest, with each RCT adding its own modifications.

DAVION ASSAULT GUARDS RCT: THE CRUSHERS

Many an enemy has simply fled at the sight of the Davion Assault Guards bearing down on them. Fielding the heaviest BattleMechs and vehicles in the AFFC arsenal, the Assault Guards are indeed a fearsome sight and have lived up to their name time after time. Among the unit's half-dozen Limp Sword Flags are two won for victories over a Capellan Chancellor—the first for the death of Ilsa Liao in 2828 while the other was for cowing Chancellor Dainmar Liao into suing for peace after trapping him on St. Ives in the last years of the Second Succession War.

When the Clan hordes appeared, the Assault Guards remained in the Capellan March as a deterrent to Chancellor Romano Liao. The unit had its chance to face the Clans when it was assigned to Operation Bulldog. Unable to contest the Second Jaguar Regulars' drop onto McAlister, the Assault Guards later coordinated with the Combine's Fourteenth Alshain Avengers to crush the Tenth Jaguar Regulars on Jeronimo.

The Crushers' insignia shows the head of an *Atlas* set on a stylized version the Royal Brigade's crest.

OFFICERS

Marshal Stephan Cooper began his career in the cockpit of a Capellan March Militia *Valkyrie* more than three decades ago. He served briefly at the head of the Eighth Deneb Light Cavalry, where he distinguished himself and his unit during the War of '39. Impressed by the young general, Prince Hanse gave Cooper the Assault Guards when Marshal Winston Ashley and most of his command staff left to assist in forming the united AFFC. Highly skilled in light cavalry tactics, Cooper was forced to adapt to a radically different combat style, a feat he accomplished in a very short time.

TACTICS

The Assault Guards are the heaviest unit in the AFFC and by all accounts should also be the slowest. Unhindered by years of being told what assault-class 'Mechs and vehicles *couldn't* do, however, Marshal Cooper began to coax unheard-of maneuverability out of his unit just months after his appointment. The Guards rely heavily on echeloned fire, with fire teams able to shift support rapidly in any direction.

Davion Assault Guards
Reinforced Regiment/Veteran/Reliable
CO: Marshal Stephan Cooper
Aide: Colonel Susie Stokes
AeroSpace Brigade: Rear Adm. Laura Thao

While consisting primarily of assault-class machines, each of the Guards' three battalions fields an extra company of fast medium and heavy 'Mechs, capable of rapidly responding to any threat.

The Davion Assault Guards AeroSpace Brigade consists of the First Federation Attack, First Federation Interceptor and the 1201st New Avalon Bomber Wings.

Davion Assault Guards Armor Brigade
4 Regiments/Veteran/Fanatical
CO: Leftenant General Dwight Savignon
Aide: Leftenant Colonel Portia DuBois
2nd Federation HAR: Lft. Gen. Kent Frisk
3rd Federation AR: Col. Lisa Nicole Schon
43rd Saginaw ARR: Col. L. T. Niven
1009th Avalon AR: Col. Melissa Lane

The armored regiments of the Assault Guards were among the first AFFC line units to test the Alacorn and Challenger X in the field and have since been equipped with several full companies of each, supplementing their scores of Demolishers, Schrecks and SturnFeurs.

Davion Assault Guards Infantry Brigade
5 Regiments/Veteran/Reliable
CO: Major General Octavia Robedeau
Aide: Leftenant General Norman Ledesma
48th F-C BAB: Lft. Gen. Terias Sortek
8th Crucis LIR: Col. Hideaki Kaluzny-Hideoshi
23rd Federation MTR: Col. Los Virasith
870th Avalon Rifles: Col. Marah Nergaard
94th Remagen MRT: Col. Igor Nemirovsky
3290th Federation AB: Maj. Rishath "Seth" Kirkpatrick

Apart from fielding a reinforced battalion of Cavalier battle armor, the Guards' infantry brigade was also among the first in the AFFC to make extensive use of towed and crewed weapons.

DAVION HEAVY GUARDS RCT: STRENGTH OF THE DAVIONS

Since its formation in the mid-twenty-sixth century, the Davion Heavy Guards have been among the most active and loyal units in the Federated Suns. The unit became Prince Hanse Davion's favorite soon after he came to power, accompanying him on such actions as his raid on Halstead Station in 3014.

The RCT is permanently stationed on New Avalon but has always been available for temporary assignment to a key assault or task force. After serving in the Bulldog offensive, the unit returned briefly to New Avalon for resupply and refit. Two months later the Heavy Guards were reassigned to Galax, ostensibly to protect the shipyards of Federated-Boeing Interstellar. The unit's JumpShip squadron was immediately transferred back to the Naval Transport Division, however, which has been repositioning key units to similarly backwater assignments all over the Federated Suns for the past few months.

The Heavy Guards' insignia is a grinning fox set on a stylized version of the Royal Brigade's crest.

OFFICERS

Surprisingly, Marshal Ann Adelmana had no direct connection to the royal family before her assignment to the command of the Heavy Guard. In fact, she began her service in the Fifth Syrtis Fusiliers, but was transferred out after her superiors accused her of being disruptive—in this case, displaying a strong pro-Davion attitude. She soon found herself in the Heavy Guards, where she quickly worked her way up through the ranks. After Marshal Candent Sortek retired from active service, Prince Hanse gave Adelmana command of the unit, awarding the hauptmann general her Marshal's baton just two years later. Though she and her unit have been decorated a number of times since then, Archon Katherine appears to have abandoned them on Galax.

TACTICS

The Heavy Guards field a slightly lighter force than do the Assault Guards, but both units follow much the same combat doctrine. The Heavy Guards array their 'Mechs and vehicles into echelons, often mixed, which can mutually support each other while cavalry and scout elements reach out to cut off an enemy's escape.

Davion Heavy Guards
Reinforced Regiment/Elite/Fanatical
CO: Marshal Ann Adelmana
Aide: Leftenant Colonel Wendy Karner
AeroSpace Brigade: Rear Adm. Sen Wedde

The Heavy Guards lost several lances during Operation Bulldog and the preliminary Birddog raids that have yet to be replaced by the DQ. While Marshal Adelmana has secured a few replacements through other avenues of supply, she has had to rearrange her 'Mech regiment into three battalions each with an additional demi-company.

The 538th New Avalon Attack, Twenty-ninth Kestrel Attack and 374th Federation Fighter Wings deploy as the Davion Heavy Guards AeroSpace Brigade.

Davion Heavy Guards Armor Brigade
4 Regiments/Elite/Reliable
CO: Major General Jonathan Davion
Aide: Colonel DeLourdes Milland
14th Lexington ARR: Col. Khamal Shah
3rd Huron PR: Maj. Gen. Jonathan Davion
7th Lexington ACR: Col. Caroline Tully
Heavy Guards Auxiliary: Col. Donovan Kleif

Major General Jonathan Davion, a former professor of military sciences at the Sakhara Academy and a distant cousin of the royal family, reorganized the armor brigade after he transferred into the unit, re-forming the Seventh into a strict cavalry unit. He also built the 293rd Artillery into a full combat auxiliary regiment, fielding two additional battalions of heavy combat engineers.

Davion Heavy Guards Infantry Brigade
4 Regiments/Regular/Reliable
CO: Major General Harry Tuttle
Aide: Colonel Samson Lowry
1st F-C BAB: Lft. Gen. Quyen Nunez
80th Federation MI: Col. Bob Granger
4th Huron HIR: Col. Linda Iglar
2nd Necedah JI: Col. William Bareis
15th Kirklin MI: Col. Fred J. G. Joers
98th Kestrel CMR: Lft. Col. Atalanta Hartt

The Heavy Guards recovered a few dozen Clan Elemental suits after their action on Schuyler, assigning them to the Second Necedah Jump regiment.

DAVION LIGHT GUARDS RCT: SWIFT FOXES

While the rest of the Royal Brigade's units are designed to overwhelm an enemy with superior numbers and firepower, trading speed and mobility for additional weapons and armor, the Light Guards operate on the opposite principle. Primarily fielding light-class 'Mechs and armored vehicles, the unit was originally intended to carry out high-risk reconnaissance and lightning-raid missions. As the Succession Wars droned on and AFFS losses mounted, the Light Guards were assigned increasingly risky missions.

The unit saw a great deal of action during the Fourth Succession War, operating an experimental double-strength company—Delta Company—that eventually grew to form part of the now-deactivated First Kathil Uhlans, to great success. The Swift Foxes redeployed to the Lyran State expecting to deal with Clan raids into the remainder of the Tamar Pact, only to return to the Federated Suns a few years later following the Lyran secession.

The Light Guards' insignia is a fox's tail set against a stylized the Royal Brigade sunburst.

OFFICERS

After commanding the Light Guards for some thirty-five years, Field Marshal Jonathan Riffenberg finally retired in late 3060. Though he had been grooming Leftenant General Jessica Quarles to take his place, Archon Katherine instead promoted Hauptmann General Jon Buckvold. Promoted up through the ranks of the Lyran bureaucracy, Buckvold was moved into the Federated Suns State Command two years after the Truce of Tukayyid. As a vocal pro-Lyran, his career foundered following Archon Katherine's secession, only to be resuscitated when the Archon took the throne on New Avalon. While Buckvold is a competent officer, many of his men consider him little more than one of the Archon's toadies.

With Buckvold appointed to command the unit in her stead, Leftenant General Jessica Quarles' own career has stalled. Though unrecognized, she is the one element that has held the unit together, preventing an outright mutiny. On the other hand, she is rapidly losing patience with Buckvold and his small pro-Katherine contingent within the unit.

TACTICS

The Light Guards' component units have only their speed and maneuverability, combined with their skill and intrepidity, to rely on in battle. The Guards attempt to outflank and isolate portions of an enemy force, cutting their foes to ribbons and always remaining on the periphery of a battle.

Davion Light Guards
Reinforced Regiment/Veteran/Reliable
CO: Hauptmann General Jon Buckvold
Aide: Leftenant General Jessica Quarles
AeroSpace Brigade: Lft. Gen. Athena Davion-Ross

Though the Light Guards lost Delta Company to the Kathil Uhlans during the Fourth Succession War, Marshal Riffenberg re-formed the double-strength company soon after the war's end. This unit, with its primarily medium-class BattleMechs, is designed to take on a heavy or even assault 'Mech company.

The Davion Light Guards AeroSpace Brigade deploys two Wings: the 256th Crucis Recon and the Twenty-ninth Manassas Fighter.

Davion Light Guards Armor Brigade
4 Regiments/Veteran/Fanatical
CO: Hauptmann General Jacob Drathers
Aide: Leftenant General G. L. Froze
121st Federation SR: Kmd. Mazel Zardetto
12th Avalon LAR: Lft. Gen. Christine Lubek-Kokal
3rd Argyle TR: Lft. Gen. Vlado Chou
5th Mirach RR: Lft. Gen. F. James Weatherby
105th F-C AB: Kmd. Jesus Ramirez-Xolo

The 105th Artillery is equipped with the RCT's only heavy armor, fielding two full companies.

Davion Light Guards Infantry Brigade
5 Regiments/Regular/Reliable
CO: Hauptmann General Angela Voss
Aide: Kommandant Randy P. Hempe IV
19th MIR: Lft. Gen. Augustus Clarke
21st New Rhodes MR: Kmd. Timothy Heine
15th Mirach JI: Lft. Gen. Ashok Herman
7th Achernar JI: Lft. Gen. Aurelius Leger
69th Tybalt JI: Lft. Gen. Gus "Brick" Evans
343rd Federation JI: Lft. Gen. Richard Daley

Each of the four jump infantry regiments, including the battle-armor companies assigned to each, is equipped with its own organic transportation, allowing the entire infantry brigade to deploy or retreat quickly.

1ST DAVION GUARDS RCT: THE STRENGTH OF ALEXANDER

The First Davion Guards, as they exist today, date to the days following the Davion Civil War and the resulting battle against Dmitri Rostov. With the entire Brigade shattered in the wake of that devastating war, Prince Alexander rebuilt the Guards' units around the survivors. The First Guards were the first to be re-formed, staffed with those men and women who had shown him their undying loyalty and courage.

Since that time, the First Davion Guards have been among the few favorite units of almost every Prince. They were Prince Ian Davion's chosen, and though he died on Mallory's World while in command of the unit, he fell while defending his men. Since the time of Alexander Davion, every member of the RCT has sworn fealty to the Prince of the Federated Suns, with few notable exceptions; that none has done so for Archon Katherine speaks volumes.

The First Guards' insignia is a Greek helmet set atop the Royal Brigade sunburst.

OFFICERS

While Marshal Bishop Sortek is the nominal commander of the First Guards, it is Major General Wendy Adams who truly runs the unit. A seventh-generation member of the Royal Brigade, mother of two of the Brigade's officers and grandmother to another, Adams has known nothing but service to the royal family. She distinguished herself as a company commander in the Fourth Guards during the War of '39 and found herself in the First Guards as soon as Sortek could arrange a transfer. Since then, Sortek has been grooming her for command, all but giving it to her when he took on responsibility for the entire Royal Brigade. Like Marshal Sortek, she, along with her First Guards, is beginning to feel the subtle machinations of Archon Katherine.

TACTICS

The various elements of the First Guards train together extensively, allowing them to fight in unparalleled coordination. The unit refuses to prosecute a city fight at all costs, instead trying to minimize civilian casualties by finding a different battlefield.

1st Davion Guards
Reinforced Regiment/Elite/Fanatical
CO: Marshal Bishop Sortek
Aide: Major General Wendy Adams
AeroSpace Brigade: Rear Adm. Tabitha Steiner

Traditionally, one of the First Guards' four battalions has always remained on New Avalon to protect the Prince. With the assignment of the Third Robinson Rangers at the beginning of the year to provide royal security, the First can now field its full complement, including some ten lances of OmniMechs.

The First Davion Guards AeroSpace Brigade fields the First Davion Guards Air, Thirty-third Weldon Attack and Sixteenth Defiance Air Wings.

1st Davion Guards Armor Brigade
4 Regiments/Elite/Reliable
CO: Major General Terrell Chevalier
Aide: Leftenant Colonel Daru Klocke
1st Davion Guards AR: Col. Victor Montgomery
237th Defiance HA: Col. Dena Killson
370th Point Barrow Grenadiers: Col. Dan Hazard
18th Shubuta RR: Lft. Gen. Jaspar Robbins
1300th Federation AB: Major Jordan Reuchlen

Both the First Armored and the 237th Heavy Armor train extensively on the Avalonian continent of Rostock, giving them an edge over their brother units in mountainous regions.

1st Davion Guards Infantry Brigade
6 Regiments/Elite/Fanatical
CO: Major General Stacy Garver
Aide: Leftenant Colonel Travis Nemoure
1st Davion Royal Guards: Lft. Gen. Ana Felsner-Kipp
1st Davion Guards HI: Lft. Gen. Lucas Rojas
5th F-C BAB: Lft. Gen. James McPeak
352nd Crucis MI: Col. Erin Kirsanova
256th Avalon MI: Col. Zhang Mingjun
20th Minette JR: Col. Keith L. Knox

The First Davion Royal Guards, numbering a total of seven battalions, have guarded the Royal Compound and family for well over five centuries. Though that tradition recently ended, the First Royal continues to protect a number of the AFFC's highest-ranking personnel.

2ND DAVION GUARDS RCT: WRATH OF THE FREE

The Second Davion Guards field a unique force composition. While the majority of their component units operate light and medium 'Mechs and vehicles, a few key elements consist solely of heavy machines. While this gives the Second Guards a relatively light mix—in fact, only the Light Guards field a lighter force—this unique composition affords the unit additional flexibility.

The Second Guards returned to the Federated Suns shortly before the Lyran secession and took up station on Bromhead, alongside the Twenty-second Avalon Hussars. There both units refit following several difficult years along the Clan front; the two RCTs soon began to gravitate toward opposite sides of the growing political debate, however. The two commanders have wisely kept their units separated, and to date none of the petty internecine quarrels that have gripped many others within the AFFC have broken out on Bromhead.

A pair of broken shackles set against a stylized version of the Brigade's crest is the Second's insignia.

OFFICERS

Fleet Admiral William "Wet Willie" Kossacks is a maverick fighter pilot with no political ambitions and a tendency to speak his mind. Having earned his nickname after bailing out of his third plane and floating for days on Cylene's Tartarus Sea, he was awarded for gallantry and found his way into Sakhara's Officer Training Program. Fifteen years later, in preparation for taking command of the Second, he graduated from Albion's War College with honors, much to the puzzlement of several instructors. Though eligible for promotion years ago, he only accepted the Marshal's baton from Field Marshal Hasek last month to help stabilize matters within the Sirdar PDZ. He did, however, manage to wrangle a naval rank out of Hasek in recognition of his fighter-pilot origins.

TACTICS

Though significantly lighter than most of its potential adversaries, the Second Guards uses its heavy artillery and aerospace assets to more than even up any discrepancy. While this often forces the Second Guards to be fair-weather fighters, the mobility of their ground forces helps them evade their foe until they can bring their full force to bear.

2nd Davion Guards
Regiment/Veteran/Fanatical
CO: Leftenant General Gerald Mitchell
Aide: Leftenant Colonel Sheldon Doucette
AeroSpace Brigade: Flt. Adm. William Kossacks

The Second Guards' pride and joy are the eleven Clan 'Mechs, mostly mediums, they captured in action on Blackjack and Hot Springs. Though it took the unit's techs almost three years to scavenge enough parts to put them back together, these 'Mechs form the core of the regiment's two-company close-assault group.

The Second Davion Guards AeroSpace Brigade is an oversized unit, consisting of the Forty-seventh Federation Heavy Attack, Twenty-ninth Crucis Interceptor, Ninetieth Remagen Heavy Attack, 307th F-C Aero, Eighteenth F-C Recon and Seventh F-C Airlift Wings.

2nd Davion Guards Armor Brigade
3 Regiments/Elite/Fanatical
CO: Major General Bryce Staten
Aide: Colonel Belinda O'Rourke
2nd Davion Guards AR: Col. Bridget Griffith
50th Antietam HAR: Lft. Gen. Angelina Kartaly
25th Killarney AG: Col. Quyen Nzuyen
740th Avalon HAY: Col. Paragon Cannady

The 740th Artillery's four battalions are almost as mobile as the rest of the RCT, trained to stop only briefly for fire missions before moving on to a different position.

2nd Davion Guards Infantry Brigade
6 Regiments/Veteran/Fanatical
CO: Major General Phoebe Dreyfus
Aide: Major L. L. Ludlow
2nd Davion Guards MI: Col. James Beying Chen
25th Guards MIR: Col. Sax O'Bannon
29th Chirikof MI: Col. Dana Wigg
43rd Antietam MIR: Lft. Col. James Stock
140th Federation MI: Col. St. John Pleu
57th Trent JIR: Col. Lee Murph

While each of the brigade's regiments is mechanized, many companies are assigned only trucks.

3RD DAVION GUARDS RCT: DON'T TREAD ON ME

While not as celebrated as some of the AFFC's highest-profile units, the Third Guards are widely known, at least within the High Command, as one of the Federated Suns' workhorse RCTs. After emerging from conflagrations that would have consumed any other unit twice in this century alone, the Third could rightly be known as the Phoenix.

Following the unit's stellar performance during the Fourth Succession War, it was assigned to training duties within the Lyran State on the prominent world of Hesperus II. Unfortunately, the unit eventually fell prey to the politics of the Free Skye Movement and became entangled with the failed attempt of a number of senior Lyran officers to wrest control of Hesperus away from Archon Katherine following her secession. The Third retreated to the Federated Suns, where it was nearly deactivated but was instead given the opportunity to redeem itself in Operation Bulldog. The Third was assigned alongside ComStar's 208th Division to liberate Luzerne, but both units fell prey to the Jaguars' heavy naval defenses. Almost half of the assault force was wiped out or crippled in space. The remainder spent the next two months playing a game of cat-and-mouse with the Jaguars' two elite Clusters until the arrival of the Combine's Ryuken-Roko.

Today, the unit is operating at less than 40 percent strength following the actions on Hesperus and Luzerne. While very obviously out of favor with the pro-Katherine contingent in the AFFC, the Third has fostered a good relationship with ComStar, especially with the still-recuperating 208th Division.

The Third Guards' insignia is a phoenix, carrying a Davion sword in its claws.

OFFICERS

After coming into the Third Guards following their troubles on Hesperus, Leftenant General Margeurite McCaffee has developed a pragmatic outlook on life. Many of the Guards' officers developed a suicide complex on Luzerne in an effort to redeem themselves, and following the death of Marshal Jim Seymour, she emerged as the only officer able to pull the unit back together while still battling the Jaguars. Though promoted from Major to Leftenant General in near-record time, the prospects of her continued rise seem slim, at least while the current regime remains in power.

TACTICS

Since the days when Colonel Hanse Davion commanded the unit, the men and women of the Third RCT have frequently thrown caution to the wind. While often disastrous, as demonstrated by Luzerne, the sheer fire the Third shows makes them a deadly and feared opponent.

3rd Davion Guards
2 Battalions/Elite/Fanatical
CO: Leftenant General Margeurite McCaffee
Aide: Major Jena Davion-Bath
AeroSpace Brigade: Cdre. Kuch Jukita

Barely two battalions on paper, the Third's 'Mech complement fields only seventeen lances, many understrength.

The 134th Tsamma Air (Squadron) and Fifty-first Crucis Interceptor (Squadron) are all that is left of the Third Davion Guards AeroSpace Brigade.

3rd Davion Guards Armor Brigade
Regiment/Veteran/Reliable
CO: Colonel Vladimir Rostov
Aide: Demi Precentor III Chris Ludyan
22nd St. Robert AR: Col. Vladimir Rostov

The armor brigade has suffered the most in the past several years; of the two remaining regiments assigned to Luzerne, barely seven companies survived. After receiving a promotion to colonel, Major Vladimir Rostov took on the task of rebuilding the Third's once-heavy armor complement.

Assisting Rostov is Demi-Precentor III Chris Ludyan, also a veteran of Luzerne from the 208th. Though technically only a communications liaison, a position used to circumvent the High Command's refusal to allow an officer exchange, Ludyan is proving a valuable asset in training the shattered brigade.

3rd Davion Guards Infantry Brigade
2 Regiments/Veteran/Reliable
CO: Colonel Chien-Fen Ktorides
Aide: Captain Tellus Nemitz
45th Cerulean Mountain: Col. Giovanni Gervasi
35th Federated Suns MR: Colonel Alexander Horvat
3rd Davion Guards CA: Demi-Ptr. IV Tallulah Cloudwalker

Another 208th veteran and ComStar "specialist," Demi-Precentor Cloudwalker unofficially commands the Third's Auxiliary, a hodgepodge battalion of artillery and engineering platoons.

4TH DAVION GUARDS RCT: DRAGON'S BANE

Though traditionally assigned to the Draconis March, the Fourth Guards have faced off against the Jade Falcons along their OZ for the past decade. Like several other units, such as the Eighth Crucis Lancers, the Fourth Guards were reassigned during the Clan invasion not only to bolster the Lyran defenses but also to prevent any unauthorized strikes into the Combine.

The members of the Fourth Guards easily transferred their deep hatred of the Combine onto the Clans. From its station within the Pandora Theater, the Guards can strike at dozens of Falcon- and Wolf-held systems and have been fighting both with unmatched fervor for years. This alone has won the unit some amount of favor within the LAAF. On the other hand, some elements within the LAAF, including the Fifth Alliance Guards, with whom the unit shares garrison duties, apparently consider the Fifth a potential threat, if rumors filtering out of the Alliance can be believed.

The Fourth Guards' insignia shows a hand holding up a dragon's head.

OFFICERS

Though dedicated only to her unit and to defending the home soil of her current master, Alberta Orsina, and many within her command, can't help but feel slighted by the LAAF. Though once an AFFC marshal, as the commander of an RCT the LAAF recognizes her only as a leutnant-general. In addition, several of her officers and senior NCOs received similarly inconsistent "promotions" within the LAAF. In response, a majority of the Fourth's men and women have taken to wearing traditional Federated Suns insignia on their uniforms. Additionally, many have had specially designed dress uniforms made, replacing the standard Steiner-blue jacket with a Davion-green version.

TACTICS

With a complement of quick light- and medium-class vehicles, the Fourth has made a career of attacking from the most unusual sector, be it from a turbulent sea, the middle of a burning forest or down a nearly sheer cliff.

4th Davion Guards
Regiment/Elite/Fanatical
CO: Marshal Alberta Orsina
Aide: Leftenant Colonel Korath Adams
AeroSpace Brigade: Rear Adm. Pallas Czaban

In the past ten years, the Fourth has pulled off seven successful raids on Wolf-held worlds, and many more against the Jade Falcons. Though highly successful, the Fourth has been ordered to cease further operations against Clan Wolf. At the very least, the Fourth has gained nineteen Clan 'Mechs, from which it fields a specially trained company designed to look and fight just like a Clan Trinary. This company, the Dopplegangers, has on several occasions successfully shifted the blame for a raid from the Fourth Guards to a rival Clan unit.

The Fourth Davion Guards AeroSpace Brigade fields the Eighty-third and Eighty-fourth F-C Fighter Wings.

4th Davion Guards Armor Brigade
4 Regiments/Elite/Reliable
CO: Leftenant General Hohiro Ma
Aide: Hauptmann-Kommandant Billy Lutz
7th Zebebelgeubi LAR: Lft. Col. Francis Daebel
19th Bolan ARR: Col. Hetty Rutt-Steiner
15th Chesterton ARR: Col. Z. Alan Maloney
30th Federation LA: Col. Dewey Simms
4th Guards AB: Maj. Jeb Richelieu

Both the Fifteenth and the Thirtieth almost exclusively field hovercraft units. The two remaining regiments, while predominantly made up of hovercraft, also field two companies each of heavier armor.

4th Davion Guards Infantry Brigade
5 Regiments/Regular/Reliable
CO: Leftenant General Estes Malevannyy
Aide: Colonel Hak-Joong Kim
478th Avalon Rifles: Col. Kyle Elms
18th Manassas MI: Col. Pam Kunin
177th F-C MI: Col. Alexander J. Beyer V
178th F-C MI: Lut.-Col. Eugenia Bello
3rd Saginaw JI: Lft. Col. Zachary Rybacki

The Third, 177th and 178th have all been equipped with a full battalion each of Cavalier battle armor, while the Third has also recently managed to acquire a few Lyran battle-armor suits.

5TH DAVION GUARDS RCT: THE BENT SWORD BRIGADE

Though one of the Prince's own personal units, the Fifth Guards owe much of their history and lineage to the Capellan March. Positioned within that region many decades ago to ensure the loyalty of the Duke of the Capellan March, the Fifth has since come up against the Capellan Confederation Armed Forces (CCAF) on a number of different occasions, winning the Limp Sword Flag four times.

Its most trying hours were spent not on the Capellan front, however, but in battle with the Clans. Stationed on Bessarabia at the start of the Clan War, the Fifth learned to counter the Clans' overwhelming power, but not before taking severe losses and being forced from two worlds by the Clans. The Fifth remained on the Clan frontier for a few years following the Truce of Tukayyid before rotating back into the Suns' Capellan March, where they finally re-formed their crushed conventional units. They remained on Monhegan for a few years before the Eighth Donegal took their place, freeing them to take up station on the shared world of Spica in response to the Confederation's invasion of the St. Ives Compact. Following the last cease-fire, however, Archon Katherine ordered the Fifth to Alcyone, placing them two jumps away from the war and leaving the defense of Spica in question.

The Fifth Guards paint their equipment in the standard colors of the Brigade of Guards. Its insignia is a parody of the Capellan standard set on a stylized version of the Brigade's crest.

OFFICERS

With the Fifth bereft of its senior leadership following its action on La Grave against Clan Jade Falcon in late 3051, newly promoted Hauptmann General Linda Archer took nominal command of the decimated unit after its retreat to Mkuranga. Though it took years to re-form the RCT, as surviving elements had been scattered across nearly a dozen planets and the death of Hauptmann General Galloway was not confirmed for almost three years, Archer rallied her troops and soon had the unit back in fighting condition. As a result, she was awarded her Marshal's baton in early 3057.

TACTICS

Once considered a generalized unit, the Fifth learned to use the cover offered by the hedgerows and forests of La Grave's Gaullic Highland to their best effect, on both the offense and defense.

5th Davion Guards
Regiment/Regular/Fanatical
CO: Marshal Linda Archer
Aide: Colonel Tasset Rinnaldi
AeroSpace Brigade: Rear Adm. Nech MacAllister

Though the Fifth experienced two crushing defeats at the hands of the Clans, in which they were reduced to less than 50 percent strength, they managed to escape Bessarabia and La Grave with some salvageable equipment. In addition to their two companies of standard OmniMechs, the Fifth fields a full lance of Clan heavy and assault OmniMechs.

The Fifth Davion Guards AeroSpace Brigade fields the Fortieth Avalon Ravens and 294th Attack Wings.

5th Davion Guards Armor Brigade
3 Regiments/Veteran/Fanatical
CO: Major General Connor K'Tenga
Aide: Leftenant Colonel Fred Korleon
197th New Avalon PR: Lft. Gen. Dana Han-Pehn
43rd Albion Grenadiers: Col. Seth Roman Varnay
225th RR: Col. T. J. Bathe
2748th Federation AB: Lft. Col. Pan Ti DeGalley

Much like the rest of the Royal Brigade, the Fifth has been assigned several companies' worth of new vehicle designs, including a platoon of Kuritan Tokugawas.

5th Davion Guards Infantry Brigade
4 Regiments/Regular/Reliable
CO: Major General Claude Raines
Aide: Colonel Marcus Tovovich
23rd F-C BAB: Lft. Gen. Rolf von Gath
449th McHenry Rifles: Lft. Gen. Glenn Hutchissen
98th Galax JR: Col. Sabin Fang
199th MR: Col. Yvonne Courrinas

Following its action against the Clans, the Fifth's infantry brigade is still short by one full regiment, though Marshal Archer considers the recent assignment of a full Cavalier battalion more than adequate replacement.

DENEB LIGHT CAVALRY

STRENGTH THROUGH TRADITION

Like the Avalon Hussars, the origins of the Deneb Light Cavalry Regimental Combat Teams lie in the breakup of the SLDF. Long before the Reunification War and the subsequent conflict with Stephan Amaris and his Rim Worlds Republic, the SLDF High Command saw the need for quick-response units that could rapidly deploy to any League world. With its Marine and Independent regiments ill-equipped and ill-trained to deal with these assignments, the SLDF soon developed unique formations to take them on. Within a few years, these new Rapid Deployment Mixed Arm Forces (RDMAF), officially considered Special Operations units and thus not found on any Regular Army TO&E, were deployed in every Combat Region.

Of course, the RDMAFs were an open secret and were soon immortalized in the holos as the Deneb Light Cavalry, after their HQ world of Deneb Kaitos. After General Kerensky's Exodus, only one of the RDMAFs remained in the Inner Sphere—the Fourth Deneb Light Cavalry.

As soon as the Fourth and its homeworld were accepted into the Federated Suns, the AFFS began to form a number of additional Light Cavalry units, using the Fourth to train them. While this largely kept the Deneb Light Cavalry out of the First Succession War, the units more than proved themselves in the Second and Third Succession Wars and in this century.

COMMANDING OFFICER

Marshal Andrew Terlecki Jr. comes from a long line of loyal and capable DLC officers. He ascended to the command of the Fifth DLC just as his father, Marshal Andrew Terlecki Sr., took command of the entire DLC in the mid-3030s. As with the rest of the AFFC, the DLC was thrown into a state of confusion by the Clan invasion. Then-Hauptmann General Terlecki Jr. and his Fifth Cavalry were immediately transferred to the Lyran front. Though involved in a few operations before the Truce of Tukayyid, most of the Fifth's actions occurred afterward, including its final mission to rescue several hundred alleged POWs on Sudeten. While technically successful, Terlecki returned with only a fraction of his command, due to an intelligence analysis that underestimated the world's defenders by 50 percent.

The Fifth DLC was subsequently deactivated and its survivors transferred to other Cavalry units. Terlecki, on the other hand, received a promotion. His father was on the verge of retiring after more than forty-five years in the AFFC, and the senior Terlecki had named his son as his choice of successor. Promoted to marshal and given command of the Addicks PDZ, Andrew Jr. has led the DLC since. Like his father before him, Marshal Terlecki maintains his primary headquarters on Deneb Kaitos in deference to DLC tradition.

FORCES

The Deneb Light Cavalry is organized slightly differently from the AFFC's "standard" RCT. Around the central BattleMech regiment are two hovercraft regiments, two jump infantry/battle-armor regiments and anywhere from three to six aerospace fighter wings. Additionally, each Light Cavalry RCT operates an independent ranger battalion and a combat engineer unit. As the DLC is devoted to highly mobile warfare, none of its RCTs permanently field any artillery assets, instead relying on their own fighters and the artillery of another friendly unit to provide fire support.

The DLC was once a Special Operations unit, fielding exotic equipment such as *Exterminator* BattleMechs and Nighthawk infantry suits, though centuries of heavy combat have stripped it of all its Star League-era equipment. The units remain true to their original intent and still field a light to medium force. 'Mechs such as the *Stealth* and the *Spector* are popular, as are battle-armor suits like the Infiltrator Mk II and the Cavalier.

COLORS AND INSIGNIA

The subordinate regiments of the DLC prefer the SLDF's standard light tan paint scheme, even when circumstances would tend to dictate otherwise. The Cavalry's insignia is a black rectangle surrounding the bust of a white horse, which in turn is set upon a gold box. Each unit's numerical designation is written in roman numerals inside the box.

4TH DENEB LIGHT CAVALRY RCT: EVELINE'S ORIGINALS

Today's Fourth Cavalry is the direct descendent of General Eveline McDennis' original Fourth Deneb Light Cavalry; in fact, the officers of the Fourth have, whenever possible, recruited from the families of McDennis' Fourth. As a result, the Fourth DLC is one of the most tradition-bound units in the AFFC, devoted to the people of their homeland more than to any one leader.

Assigned to the Lyran State for the past sixteen years, the Fourth DLC was one of the half-dozen units assigned to patrol duty along the Clan front after the Truce of Tukayyid. Since taking up permanent garrison again on Dalkeith, the Fourth has championed a new cause—helping their fellow man. Conditions on a number of nearby worlds have deteriorated since the Alliance's secession, and the men and women of the Fourth have pledged to assist in any way they can.

The Fourth uses the standard DLC colors and insignia but adds a single silver general's star to its insignia in tribute to General McDennis.

OFFICERS

Leutnant-General Fran Vidal is a native Terran and one of the few surviving members of the Fifth DLC. With no immediate billets open in the Fourth following the Fifth's deactivation, she attended the Nagelring's War Command College, where she learned to lead an entire RCT in battle, skills that came in handy when she returned to the Fourth as its XO. She soon moved up to command the entire unit when Hauptmann General Harrison left the service for personal reasons. Vidal is not as hands-on as many commanders, instead preferring to see the big picture before embarking on a plan of action.

TACTICS

Like many of the AFFC's other light units, the Fourth Cavalry uses its speed to outmaneuver its opponents. Under the leadership of Leutnant-General Vidal, the Fourth's entire complement is constantly updated with the current positions of both friendly and hostile forces, allowing the unit's component elements to work as a team even when separated by hundreds of kilometers.

4th Deneb Light Cavalry
Regiment/Regular/Reliable
CO: Colonel Andrew Terlecki III
Aide: Kommandant Gareth Evers
AeroSpace Brigade: Lut.-Kdre. Pablo Escobedo

The son of Marshal Andrew Terlecki Jr., Colonel Terlecki is likewise making a name for himself. He has spearheaded a number of efforts to help Dalkeith's people, and as a result his regiment has been lauded by the Interstellar Red Cross on a number of occasions. Though his 'Mechs were designed for war, Terlecki has proved time and again that they can also be used to for peacetime construction.

The prototype for the rest of the DLC, the Fourth Deneb Light Cavalry AeroSpace Brigade still retains its six original Wings: 3512th Aerospace, 644th Independent, 645th Independent, 646th Independent, 2193rd Royal Attack and 2204th Air.

4th Deneb Light Cavalry Armor Brigade
2 Regiments/Veteran/Reliable
CO: Colonel Santiago Nunez
Aide: Hauptmann-Kommandant A. J. Rust-Deetjen
4898th Royal CR: Col. Santiago Nunez
714th LIR: Lut.-Col. Del Brodzller

Using its historic ties with the Lyran Royal Guard, the 4898th Royal has been assigned nearly two companies of Plainsman hovercraft over the past few years.

4th Deneb Light Cavalry Infantry Brigade
2 Regiments/Regular/Unreliable
CO: Leutnant-General Tracy Mokhtaruddin
Aide: Kommandant Barry Seles
223rd JR: Lut.-Col. Clarise Arn
224th Royal JR: Col. Sarah Colletto

On a recent assignment to Symington, the infantry brigade, supported by one of Colonel Terlecki's BattleMech battalions, encountered a terrible scene of destruction. A group of Symington natives had hunted down and wiped out an entire colony of Buddhist monks who had emigrated from the Combine. The Fourth's men and women were instrumental in bringing the guilty to justice, but they have lost faith in the Alliance and its justice system.

8TH DENEB LIGHT CAVALRY RCT: THE GREEN LEOPARDS

The second of the DLC units created by the AFFS after the dissolution of the Star League, the Eighth Cavalry saw extensive action against the Capellan Confederation during the Second and Third Succession Wars and has suffered greatly, especially in the third assault on Tikonov in 2834.

Following the creation of the unified AFFC, the Eighth was transferred to the Lyran State Command. Assigned along the Lyran border with the Combine, and inside the Skye March, the Eighth was literally besieged from all sides. In several instances prior to 3050, members of the Eighth found themselves under attack by DCMS forces while anti-Davion protesters sabotaged their supplies and equipment. With no other help immediately forthcoming, the Eighth turned to the Free Skye Movement to assist them. Even after the unit's relocation to the Melissia Theater, the Eighth DLC maintained its contact with Duke Robert Kelswa-Steiner and his movement.

Though the Eighth has been involved in a number of combat situations over the past two decades, it has received few replacement 'Mechs and vehicles through official AFFC or LAAF channels. Intelligence reports indicate that Duke Kelswa-Steiner has provided a number of new vehicles, apparently to further ensure the Eighth's loyalty.

The Eighth Cavalry uses a variation on the standard DLC paint scheme, substituting green drab for the SLDF tan. Likewise, the Eighth colors the horse on its insignia green instead of white.

OFFICERS

A native of Kittery, Leutnant-General Lisa Aileen Bortman's parents moved to Zebebelgenubi while she was still a toddler. She spent most of her formative years there before her mother's assignment in the Lyran Commonwealth ended and her family returned to the Federated Suns, but in her heart she knew she was a "child of Skye." She entered the AFFS as soon as she could and eventually earned a transfer to the Sanglamore. Bortman bounced from assignment to assignment until she landed in the DLC and found herself in the Eighth Cavalry, apparently due to the machinations of Duke Ryan Steiner. Though her unit is loyal to the Cavalry and its traditions, she still considers herself a child of Skye.

TACTICS

The men and women of the Eighth Cavalry are accomplished guerrilla fighters, a tactical style forced on them by a number of near-disastrous campaigns. The unit prefers hit-and-run tactics that make the most use of the RCT's elements but do not put them all in jeopardy at once.

8th Deneb Light Cavalry
Regiment/Veteran/Reliable
CO: Leutnant-General Lisa Aileen Bortman
Aide: Kommandant Uriah Jervais
AeroSpace Brigade: Lut.-Kdre. Bruce Dewey

The majority of the Eighth's 'Mech complement is more than eighty years old, and a full quarter of the unit is often down for repairs at any one time. Rather than parceling out her new BattleMechs throughout the unit, Leutnant-General Bortman has kept them for her command company.

The Eighth Deneb Light Cavalry AeroSpace Brigade is composed of the Twenty-second Deneb Air, the Third Deneb Independent and the Fourth Deneb Independent Wings. The Fourth is composed solely of conventional VTOL aircraft.

8th Deneb Light Cavalry Armor Brigade
2 Regiments/Elite/Reliable
CO: Colonel Jean-Henri de Jeruc
Aide: Hauptmann-Kommandant Wilson Dornacher
3rd Deneb LA: Lut.-Col. Nichole Tertadian
4th Deneb LA: Col. Armin Brodzller

The Eighth's two armored regiments are in much the same shape as its 'Mech regiment, though the relative availability of vehicle parts ensures both units operate at near-full strength.

8th Deneb Light Cavalry Infantry Brigade
2 Regiments/Regular/Fanatical
CO: Colonel Jessup Robinette
Aide: Kommandant Desmond Poulakos
75th Deneb Kaitos JI: Col. Ignazio Fulcher
80th Deneb Kaitos JI: Col. Dona Hibaki

10TH DENEB LIGHT CAVALRY RCT: THE CHEETAHS

Having been created from the remains of the Ninth Deneb Light Cavalry in the early years of the Third Succession War, the Tenth Deneb struggled to find form and break out of the loop of inexperience leading to losses, which in turn led to inexperienced recruits. The RCT's leadership and troops were tenacious but could never quite break the vicious cycle. That ended with the Fourth Succession War.

The RCT was deployed to Glenmora, where it fought alongside Wolf's Dragoons Alpha Regiment in their bitter conflict with the Draconis Combine. The fighting was brutal, costing many lives and much materiel. These losses were concentrated in the Light Cavalry's armor forces, with only twenty hovertanks remaining from the two armor regiments. The Dragoons were grateful for the Light Cav's efforts on their behalf, and Jaime Wolf said he he would be proud to fight anywhere alongside the Light Cavalry's warriors. When the war ended, the Tenth was rebuilt, but at a pace dictated by Marshal Carl Sanders, allowing the RCT to retain its hard-won veteran status.

In addition to the standard white horse insignia of the Light Cavalry, the Tenth bear the wolf's head motif of the Wolf's Dragoons, a tradition started on Glenmora by the grateful mercenaries and continued with the full approval of Jaime and Maeve Wolf.

OFFICERS

During the Fourth Succession War, Hauptmann General Jameston Gaston commanded a hovertank company in the Tenth and saw first-hand the carnage of war against the Draconis Combine. This left him with an abiding hatred of the Combine and the Kurita family in particular. Ex-Prince Victor's public association with the Kurita family and his rumored dalliance with Omi Kurita have placed the general squarely in Archon-Princess Katherine's camp. His views are shared by many in the RCT, who view Victor's alliance with the Kuritan enemy a betrayal of everything they have stood for, tantamount to treason.

TACTICS

The Tenth favors mobile warfare, making good use of their light 'Mechs and hovertanks, but is also adept at ambushing enemy troops. The soldiers receive extensive training in fieldcraft and can disappear into the terrain, reappearing only to harass the enemy.

10th Deneb Light Cavalry
Regiment/Veteran/Reliable
CO: Hauptmann General Jameston Gaston
Aide: Leftenant General Al Rowse
AeroSpace Brigade: Lft. Gen. Basil Collier

Like all Light Cavalry regiments, the Tenth are almost entirely composed of light 'Mechs, though the past decade has seen an influx of fast medium and in some cases heavy 'Mechs using Star League-era technology. Even when on garrison duty, such as the present posting on New Avalon, the unit's combat skills are maintained by exercises and competitions between the companies. The company judged the best is chosen to represent the Light Cavalry in the Outreach Cup, an annual friendly competition held between the regiment and Wolf's Dragoons on Outreach. Over the past thirty years, the Light Cav have won the competition eight times and the Dragoons twelve. Four contests have been judged draws and six cancelled because of war or combat deployments.

Though on paper the Ninth, Eleventh and Twelfth Independent Wings of the Tenth Deneb Light Cavalry AeroSpace Brigade are attached to the 'Mech regiment, in practice they operate independently.

10th Deneb Armor Brigade
2 Regiments/Regular/Reliable
CO: Leftenant General Callista Lamouroux
Aide: Kommandant Guy Agento
19th Deneb LA: Lft. Gen. Catherine "Cat" Deely
22nd Deneb LA: Lft. Gen. Hugo Gaunce

A rolling program over the past decade has upgraded 90 percent of the hovercraft used by the Armor Brigade with Star League technology. Many of these have been field upgrades, but the Light Cavalry have also received new designs like the Fulcrum and the updated Pegasus. The unit also has a platoon of Bandit hovercraft, obtained from the Wolf's Dragoons.

10th Deneb Infantry Brigade
2 Regiment/Veteran/Reliable
CO: Leftenant General Hubert Kvaener
Aide: Kommandant Elizabeth Stanford-Tuck
102nd Deneb Kaitos JI: Lft. Gen. Theo Osterkamp
103rd Deneb Kaitos JI: Lft. Gen. Jessica Duart

The addition of two companies of Cavalier battle-armor suits has given the infantry brigade a level of firepower rarely seen in jump infantry units. The brigade has yet to use them in combat, but the suits have performed well in exercises and General Kvaener is hoping to expand the armored force to company or battalion size.

12TH DENEB LIGHT CAVALRY: THE BLUR-SNAKES

The Twelfth DLC came into being at the height of the Third Succession War, when the AFFS High Command realized they needed a force of quick-response "free" 'Mech regiments to complement the larger RCTs. After spending a century and a half along the Combine border, the Twelfth redeployed to the Lyran Commonwealth as soon as the Clans appeared. During the war years, the Twelfth was held mostly in reserve. It was only after the Truce of Tukayyid that the High Command released the Twelfth for action against the Clans.

Forces from both sides of the OZ conducted dozens of raids into the other's territory during the first year. During that time, the AFFC High Command learned that the Falcons had a base of operations on Sudeten and that, in addition to millions of tons of stockpiled supplies, they were keeping hundreds of POWs and political prisoners on the planet. Amid a flurry of smaller raids aimed to keep the Falcons off-balance, the AFFC dispatched the Fifth and Twelfth DLC, along with the 81st F-C and 403rd Lyran Independent Aero Regiments, to Sudeten.

Unfortunately, the AFFC's intelligence directorates sorely misjudged the Falcons' garrison force. More than a full Galaxy was stationed on Sudeten with an *Aegis*-class cruiser in orbit. The Cavalry force appeared at a pirate point scant hours away from the world. While the aero regiments and supporting assault ships faced off against the WarShip, the Cavalry units landed. The Twelfth dropped on top of the Falcon logistics depot—and the Eighth Falcon Regulars. Though armored elements of the Fifth RCT diverted to assist the Twelfth, the arrival of the Twelfth Falcon Regulars, on temporary assignment to Sudeten, sealed the Twelfth DLC's fate. It escaped with less than a third of its complement and without having significantly damaged a single warehouse. The Fifth likewise retreated with almost 80 percent losses—though with several hundred POWs in tow. A mere twenty-seven of nearly 120 fighters survived the battle with the *Jade Talon.*

Expelled following the Lyran secession, the Twelfth Cavalry slowly rebuilt its shattered command, assisted in part by Field Marshal James Sandoval. The Twelfth's soldiers, while devoted to the Cavalry and their nation, are extremely wary of Archon Katherine and her Lyran Alliance. At the same time, they have little confidence in the AFFC High Command, whom they blame for the Sudeten "mistake."

The Twelfth Cavalry uses standard DLC colors and insignia but colors the horse blood red in memory of the men and women sacrificed on Sudeten.

OFFICERS

Frustrated by the lack of support for him and his unit, Leftenant General Jeffrey Neece finally stepped down from command of the Twelfth after almost fourteen years at its helm. Field Marshal Sandoval promoted Ariel Zibler, his XO and a survivor of the Fifth DLC, in his place. Hailing from a long line of AFFC officers—in fact, six of her cousins and siblings hold the rank of Leftenant General or better in the fractured AFFC—Zibler is, at least outwardly, an excellent officer. Since her unit's ordeal on the Clan front, however, she has become increasingly belligerent and restless. Unchecked, she could become unpredictable.

TACTICS

The unit's preferred deployment is a low-level combat drop directly on their objective. Though this tactic turned deadly on Sudeten, the unit's MechWarriors use the surprise, along with the exhaust from their DropShips, to confuse and scatter their enemy.

12th Deneb Light Cavalry Regiment/Regular/Reliable
CO: Leftenant General Ariel Zibler

Almost half of the Twelfth's MechWarriors are survivors of Sudeten, many originally hailing from the Fifth, while the rest transferred in through the DLC's headquarters on Deneb Kaitos. Nearly all are natives of the Draconis March.

12th Deneb Combat Auxiliary Battalion/Veteran/Unreliable
CO: Kommandant Lex Cartegena

The two armored and three infantry companies of the Auxiliary are all that remain of the Fifth's conventional forces.

3498th Federation Aero Wing Wing/Green/Reliable
CO: Kommandant Stanislov Mintaka

Originally a part of the Eighty-first F-C Aero Regiment, the 3498th and its fifteen fighters remain with the Twelfth Cavalry.

15TH DENEB LIGHT CAVALRY RCT: THE WHIP BEARS

Originally a "free" 'Mech regiment, the Fifteenth Deneb Light Cavalry was upgraded to RCT status in 3027 as part of Hanse Davion's preparations for Operation Rat, the invasion of the Capellan Confederation. Based on Verlo (though deployed to Kittery during the conflict), the Fifteenth helped contain the CCAF, guarding against a counterstrike into the Capellan March while poised to strike into the heart of the Sian Commonality should the need arise. Eventually the Whip Bears deployed into the Confederation, but not as invaders. Instead, they were part of the Federated Suns contingent invited into the St. Ives Commonality by Candace Liao to ensure her nascent realm's survival in the first months after it seceded from the Confederation. The Fifteenth faced several CCAF raids, but it was the determination of the St. Ives people to remain independent of the Federated Suns that eventually prompted the RCT to withdraw.

Having been based in the Sirdar PDZ for many decades, the Light Cavalry's move to Demeter in 3059 caused many problems for the unit. Since it was deemed a temporary deployment to deal with Capellan activity in the so-called Tikonov Reaches, families and dependents were not relocated, causing considerable resentment in the unit. Subsidized travel permits, introduced in 3060, have done much to alleviate this discontent, but the suspension of leave in August 3062 as tensions increased in the wake of the Solaris crisis has placed considerable strain on the unit's morale. Seeking an outlet for their frustration, the cavalry's troops have singled out the "enemy" Fifth Crucis Lancers, and clashes between personnel have occurred.

The Fifteenth uses the standard Deneb insignia but prefer a khaki rather than tan paint scheme.

OFFICERS

The recent promotion of Michael Lipstein to marshal has done much to ensure the Fifteenth's loyalty to Archon Katherine, but the move has infuriated Olaf Richardson of the Fifth Crucis Lancers and the commander of the Valexa PDZ, Piper Burullo. Though Lipstein is an excellent commander and administrator who is respected throughout the AFFC, the pro-Victor Richardson and Burullo regard the promotion as an attempt to undermine support for Prince Victor on Demeter, placing Lipstein—as the senior officer—in command of the planet's defense. Lipstein has argued that his promotion provided an essential chain of command on the planet, given the upsurge in Capellan-sponsored terrorism throughout the Tikonov Reaches.

TACTICS

The Fifteenth's fourth battalion of 'Mechs, principally heavy designs like the *Caesar* and *Falconer*, allows the RCT to use a broader range of tactics than the other Deneb units. Normally these 'Mechs are used en masse to crush enemy positions, but they have also been deployed in lance-sized elements to reinforce lighter companies.

15th Deneb Light Cavalry
Reinforced Regiment/Regular/Reliable
CO: Marshal Michael Lipstein
Aide: Leftenant General Nicole Haswell
Air Wing: Lft. Gen. Ian Beske

The Fifteenth Deneb's 'Mech regiment has been sidelined to restrict the operations of the Capellan Zhanzeng de Guang terrorists on Demeter. Instead, they have staged a number of training exercises, both internally and against the Fifth Crucis Lancers. These have provided a way for the warriors to let off steam, though their feud with the Lancers has led to considerable "accidental" damage to both units.

Unusually for an RCT, the Fifteenth Deneb Light Cavalry has only a single Wing, which operates independently.

15th Deneb Armor Brigade
2 Regiments/Regular/Reliable
CO: Leftenant General Kirpal Bath
Aide: Kommandant Eva de la Rua
24th Deneb LA: Lft. Gen. Kirpal Bath
11th Deneb RR: Lft. Gen. Michel Saunders

The Eleventh Recon Regiment makes heavy use of ultralight vehicles such as the Savannah Master and the Ferret VTOL.

15th Deneb Infantry Brigade
2 Regiments/Regular/Questionable
CO: Leftenant General Burim Jashari
Aide: Kommandant Sweeny Robins
11th Verlo IR: Lft. Gen. Herbert Austin
12th Verlo IR: Lft. Gen. Pam Barton

Having been recruited from the RCT's homeworld of Verlo, the infantry brigade is the focus for much of the discontent in the Fifteenth DLC.

THE FEDCOM CORPS

STRENGTH THROUGH UNITY

One of the earliest manifestations of the Federated Commonwealth alliance was increased cooperation between the Armed Forces of the Federated Suns (AFFS) and the Lyran Commonwealth Armed Forces (LCAF). The ultimate goal of this process was the amalgamation of the AFFS and LCAF (which would not take place until the 3040s), but in the bitter fighting of the Fourth Succession War the two nations agreed the alliance should take on tangible form in the shape of a series of military units.

Thus were born the Federated Commonwealth RCTs, deployed throughout the new combined realm as a symbol of unity. They recruited the best and brightest soldiers from the two realms' academies and received the best equipment in the alliance.

However, of the twelve original FedCom RCTs, only six remain. Five were absorbed by the nascent Lyran Alliance and renamed Alliance Guards, while the other was destroyed in combat. Additionally, the decision of several FedCom units to withdraw from the Sarna March rather than fight the Liao and Marik invaders has done considerable damage to the entire FedCom Corps. What was once seen as a force for unity in the Federated Commonwealth is now perceived as a politicized force seeking to advance its own position.

COMMANDING OFFICER

Marshal Alistair Venger has commanded the Federated Commonwealth Corps since 3037 when the previous CO, James Dassert Sandoval, became commander of the Draconis March. He is one of the longest-serving senior officers in the AFFC, with a formidable reputation. However, he has seen his once-great command denuded by the tribulations of the past dozen years, undermining his position and forcing him to take desperate measures. Though initially standing aloof from the factional politics that have consumed the realm, Venger ultimately sided with the Archon-Princess, believing she offers the best hope of reunifying the alliance the corps was created to serve.

FORCES

The FedCom RCTs make use of the best equipment in the AFFC, including the *JagerMech III* and *Nightstar* 'Mechs, the Challenger heavy tank, and even the Cavalier battle armor. However, though its PR says otherwise, the corps is far from the pinnacle of FedCom military prowess. Though the corps' warriors are individually very good, their inexperience in working as a team has hampered the RCT's battlefield performance.

COLORS AND INSIGNIA

The insignia of the Federated Commonwealth RCTs is a hybrid of that used by the Federated Suns and the Lyran Commonwealth; the backdrop is the Davion Sunburst, over which is the mailed fist of House Steiner, and in front of that the Davion sword.

The parade colors of the Federated Commonwealth Corps are navy blue with gold banding on the arms and legs. Gold is also used on the upper torso of 'Mechs and the turrets of armor units to highlight panels, resulting in a checkerboard pattern.

1ST FEDCOM RCT: GUARDIANS OF A NEW ORDER

Prince Hanse Davion and his AFFS were presented with a gift from Melissa Steiner and the Lyran Commonwealth on the eve of their wedding: a full BattleMech regiment. Prince Davion immediately staffed this new regiment with recent graduates from the CMS and Albion, as well as from the Nagelring on Tharkad, and stationed it in the Draconis March under the command of Marshal James Sandoval. There, the green warriors of the First F-C learned firsthand the horrors of war, but they emerged a united and skilled unit.

The combined Marik-Liao Offensive of '57 caught the RCT, assigned to the Sarna March, unawares, forcing the unit to fall back to Nanking. From there, the First launched a number of punitive raids against several Capellan-controlled worlds within the Chaos March, though that lasted only until the High Command cut the unit's supply allotment. Since then, the members of the First F-C nave grown restless about their inability to avenge their failure. The situation is further exacerbated by members of the Nanking SMM RCT, who blame the First RCT for losing control of the Sarna March and for its historic ties to the Davion family.

The insignia for the First RCT is a dragon clutching a sword, flanked by two lightning bolts.

OFFICERS

Marshal Ally Swanson took command of the dispirited RCT just months after its retreat to Nanking. After spending the majority of her 20-year career as a problem solver for the High Command, she was given the task of rebuilding the First into a capable fighting unit. With the RCT's morale at an all-time low, Marshal Swanson resorted to rather draconian measures to breathe life back into the unit. Though despised by most of her soldiers, the marshal has at least succeeded in returning the First to fighting form.

TACTICS

For the past several years, the elements of the First have rarely fought in formations larger than a battalion, though its movements have been carefully orchestrated by Marshal Swanson so that its battalions can quickly swap positions, even during the thick of battle, making the RCT far tougher than it appears.

1st FedCom Regiment
Regiment/Veteran/Reliable
CO: Leftenant General Dixon Zibler
Aide: Kommandant Beau Gentry
AeroSpace Brigade: Lft. Gen. Demos McCarron

Official reports indicate two of Leftenant General Zibler's medium 'Mech companies were down for repairs for all of 3061 and half of 3062. When they resurfaced, each was equipped with new BattleMechs, predominantly designs manufactured by GM. That fact, along with secondhand reports of one of the First RCT's DropShips in orbit around the planet Genoa, seems to indicate elements of the First F-C unofficially took part in the recent operation sponsored by GM and Johnston Industries to liberate the worlds of Acamar and Genoa from Capellan-friendly governments.

The First FedCom AeroSpace Brigade fields two Wings.

1st FedCom Armor Brigade
3 Regiments/Regular/Unreliable
CO: Hauptmann General Simeon Petrovich
Aide: Leftenant General LaSalle Shoshanna
1st FedCom RR: Lft. Gen. Karen D'Erico
1st FedCom HA: Kmd. Zurich Johann Meitzler
1st FedCom AC: Lft. Gen. Billy O'Connell

More than a few of the brigade's members hail from worlds in the middle of the Chaos March, leading to quite a bit of friction within the unit and animosity toward the AFFC and its High Command.

1st FedCom Infantry Brigade
5 Regiments/Regular/Reliable
CO: Hauptmann General Carmen Azurria
Aide: Kommandant Karl Yuan
1st FedCom IR: Lft. Gen. Grayson Applegarth
2nd FedCom IR: Lft. Gen. Skyler Pree
1st FedCom MI: Lft. Gen. Burian Thi
2nd FedCom MI: Kmd. Sandy Roland
1st FedCom JI: Lft. Gen. Patrice Young

Two battalions of infantry, one each from the First Mechanized and First Jump, dropped out of sight and returned to active duty at the same time as the First's two 'Mech companies.

3RD FEDCOM RCT: THE PENITANTS

Formed in 3034, the Third Federated Commonwealth RCT (abbreviated to the Third FedCom) recruited from throughout the Lyran Commonwealth and the Federated Suns, seeking to build loyalty to the Federated Commonwealth alliance as a whole. Nonetheless, the unit built a strong relationship with the people of Sarna, which served as their base of operations for more than two decades, swearing to defend them against all threats. When Free Worlds League forces raided the world in 3049, the Third worked closely with the local militia and the Sarna Training Battalion to drive off the attackers, sustaining severe damage in the process.

In 3056, with the increasing threat of rebellion in the Skye March, the RCT relocated to New Earth, where they would be better able to deal with any secessionist activity. The Skye revolt fizzled out, but the unit remained on New Earth until the summer of 3057. Departing for home on 22 July, the Third FedCom was at New Aragon when their adopted people needed them most, during the Marik-Liao Invasion of the Sarna March. They could do little more than watch as the Sarna March collapsed under rival claims into the Chaos March. Though not directly occupied, Sarna fell under the influence of the Capellan Confederation. The RCT had failed in its promise.

Many in the unit blamed Victor Steiner-Davion for this—his actions had triggered the invasion—and argued in favor of accepting Katherine's call to come home. Others felt that to do so would be to abrogate their responsibility to the people of the former Sarna March, and this view grudgingly became accepted by the majority of the RCT's warriors.

The Third eschews the traditional FedCom livery and paints its 'Mechs and vehicles jet black. As a reminder of their failure, few members of the RCT wear the Federated Commonwealth logo, substituting a stylized map of Sarna, painted in white on a red rectangle.

OFFICERS

Charles Finnigan took command of the Third in 3058, when Hauptmann General Gregory Sykes retired rather than serve under Prince Davion. Though a moderate within the Third, Finnigan is a staunch opponent of Prince Victor and is thus often at odds with his superiors in the Valexa PDZ. He has walked a fine line between duty and insubordination, citing "administrative problems" for failing to comply quickly or at all with unpopular orders. Particularly notable was his refusal to cooperate with a Transport and Resupply Command audit of the RCT's stocks of ammunition and parts, though his official position was that faulty stock management computers made the investigation impossible.

TACTICS

The Third favors defensive tactics, making use of its heavy 'Mechs and vehicles as well as field emplacements and fortifications built by its engineering battalion.

3rd FedCom Regiment
Regiment/Regular/Reliable
CO: Hauptmann General Charles Finnigan
Aide: Kommandant Pat Rutherford
AeroSpace Brigade: Lft. Gen. Fred Perry

The Third FedCom 'Mech Regiment consists of one medium and two heavy battalions. Almost all the 'Mechs use some form of advanced technology, though most of this comes from field refits rather than factory-built models. Since Katherine's rise to power, the proportion of new 'Mechs in the unit has increased considerably, reflecting the Archon's policy of rewarding loyalty. This new equipment includes two lances of Inner Sphere OmniMechs, including the FS9-O *Firestarter* and the BJ2-O *Blackjack*.

The Third FedCom AeroSpace Brigade consists of two Wings.

3rd FedCom Armor Brigade
3 Regiments/Regular/Reliable
CO: Leftenant General Shaw Millikan
Aide: Kommandant George McMahon
3rd FedCom RR: Lft. Gen. Sara Robbins
3rd FedCom HA: Lft. Gen. Esteban Norton
3rd FedCom AC: Lft. Gen. Mitchell Fellows

The Third FedCom's armor brigade also contains the RCT's artillery and air defense detachments. The latter includes a company of Partisan tanks, linked using the C^3 system to provide a formidable "umbrella" against enemy aerospace fighters.

3rd FedCom Infantry Brigade
5 Regiments/Regular/Reliable
CO: Leftenant General Jessica Mao
Aide: Kommandant Bernhard Biesterfeld
5th FedCom IR: Lft. Gen. Jessica Mao
6th FedCom IR: Lft. Gen. Edward Dowlais
5th FedCom MI: Lft. Gen. Ronald Knox
6th FedCom MI: Lft. Gen. Lavinia Strutt
3rd FedCom JI: Lft. Gen. George Baldwin

The veteran jump infantry regiment has received several platoons of the new Infiltrator Mk II battle armor suits, in addition to its standard Cavalier complement, though General Mao has done her best to conceal their existence from outsiders.

5TH FEDCOM RCT: THE BLACK RATS

The Fifth has stuggled to meet the expectations of the Federated Commonwealth RCTs as "the best soldiers and best equipment both states can produce." A series of accidents have befallen the unit, steadily sapping morale and establishing a self-perpetuating cycle: the RCT's troops believe bad things will happen to them, so they do.

The first disaster to befall the unit occurred in 3037, shortly after the unit became operational. While on a routine flight transporting troops to their training area, one of the Fifth's DropShips suffered engine failure, crashing and killing almost a third of the MechWarriors. Scarcely six weeks after this, and before the losses could be made good, the unit suffered further losses when Marik troops attacked Tsitang. The raiders were driven off, but this did little to alleviate the physical and psychological damage the RCT suffered.

Their preparations for the War of '39 were careful and deliberate, planning for every contingency. The unit's commanders had high expectations, and it performed well in the initial stages of the drive into Dieron, seizing Halstead Station after a brief confrontation. Then the Combine counterattacked with then-unheard-of Star League technology, shattering the RCT. The unit had to withdraw, leaving behind much of its equipment.

Though the RCT was rebuilt after the Halstead Station debacle, the new equipment and personnel were not of the same quality as originally used to form the unit. This reinforced the view of the Fifth as "losers," a mind-set they have never recovered from.

The Fifth uses the traditional blue and gold livery of the Federated Commonwealth RCTs, but it is often defaced. Other units have taken a perverse pleasure in infiltrating the RCT's facilities and painting a yellow streak down the spine of each 'Mech or on the turrets of vehicles. The regimental insignia depicts a black rat on a white disc.

OFFICERS

James White has commanded the Fifth since 3047, taking charge in the wake of the previous commander's suicide. A competent leader and administrator, White led the RCT to one of its few victories, smashing a Zhanzeng de Guang-led rebellion in 3055. However, White is known to be addicted to Taleridol, a prescribed narcotic whose side effects include insomnia and paranoia. The High Command watched the general closely but have refrained from replacing him, knowing that to do so would shatter what little self-esteem the unit still has. Instead, they have encouraged Annette Leyland, the unit's XO, to take over its day-to-day operations. Unfortunately, this has not gone down well with White, who believes Leyland is plotting to usurp him.

TACTICS

According to their AFFC status report, the Fifth FedCom RCT is composed of generalists, lacking any specialization or special skills. Unofficially, the RCT has a reputation for withdrawing if given the opportunity or fighting by any means necessary if cornered. It is this trait that earned the RCT the nickname of the "Black Rats," in which they have perversely taken a great deal of pride.

5th FedCom Regiment
Regiment/Green/Reliable
CO: Hauptmann General James White
Aide: Leftenant General Annette Leyland
AeroSpace Brigade: Lft. Gen. Harry "Rockhopper" Peterson

The Fifth is a mix of light and medium 'Mechs, principally older models but including a handful of modern designs acquired through the canny dealings of Marika Satala, the regiment's unscrupulous quartermaster. Though officially comprising three battalions, the regiment can only field around eighty 'Mechs.

The Fifth FedCom AeroSpace Brigade fields two Wings.

5th FedCom Armor Regiment
Regiment/Green/Reliable
CO: Leftenant General Klaus von Ludendorf
Aide: Kommandant Ladislav Potemkin

Rebuilt after Halstead Station, the Fifth's armor brigade was one of the few parts of the RCT that met AFFC proficiency targets. But the unit suffered a steady deterioration of its fighting abilities as troops and equipment were transferred out to make good losses in other units. In 3050, this trickle became a flood: over the space of eighteen months, the Fifth lost two whole regiments to units damaged in the war against the Clans.

5th FedCom Infantry Brigade
5 Regiments/Green/Reliable
CO: Leftenant General Dan Luger
Aide: Kommandant Joely Rouse
9th FedCom IR: Lft. Gen. Dan Luger
10th FedCom IR: Lft. Gen. Charisse Turner
9th FedCom MI: Lft. Gen. Giles Rowsell
10th FedCom MI: Lft. Gen. Viv Batchelder
5th FedCom JI: Lft. Gen. Elspeth Burnside

6TH FEDCOM RCT: THE TIGER BAITERS

Though barely a year old, the Sixth FedCom RCT performed admirably in its first combat role, helping envelop and reduce the Kuritan positions on Ancha in the War of '39. Though eventually forced to withdraw, they did so in good order, forcing the Combine reinforcements to pay dearly for their victory. The regiment sat out the Clan War on Small World, guarding against any Combine attempt to exploit the crisis, where they staged a number of exercises on the planet's tundra, experience that stood them in good stead when they relocated to the frigid world of Dalkeith on the Clan front in 3055.

When the Lyran Alliance seceded from the Federated Commonwealth, the Sixth withdrew from its cantonment on Dalkeith, rejecting Archon Katherine's offer of a place in the LAAF, and returned to the Terran Corridor. Unable to take up residence on the now-independent Small World, the RCT settled on Addicks, where it has built a strong relationship with the local population and the Addicks DMM.

Though most of the Sixth uses the blue and gold color scheme of the FedCom RCTs, Baker Battalion and the Sixth FedCom Light Armor, which usually take the role of "enemy" forces in the unit's exercises, often use a black and yellow tiger-stripe paint scheme. The RCT's insignia continues this motif, showing a stylized tiger's head.

OFFICERS

When Katherine Steiner-Davion took command of the Federated Commonwealth in late 3060, Hauptmann General Isbella Rahn chose to retire. Her XO, James McConnel, was promoted to command of the RCT. The new Hauptmann General is suave and sophisticated, with a calm, polite manner and an ability to see the strengths and weaknesses of those around him. The few people who have mistaken his quiet demeanor for timidity have quickly been disabused of the notion; McConnel brooks no challenges to his authority and believes in sharp, decisive action when diplomacy fails. The general is smart enough to see the clouds of war gathering on the horizon and has worked hard to sway Nancy Bannson of the Addicks DMM from her position of neutrality. He believes that once the Federated Commonwealth and Lyran Alliance descend into civil war—which he regards as a certainty—neither side will look kindly on neutral forces.

TACTICS

Over the past two decades the Sixth has gained a reputation for bringing enemy forces to battle, luring them out of fortifications or onto terrain of the RCT's choosing. Feigned retreats, radio interceptions and superior mobility are central to this "tiger baiting" skill, keeping the enemy off balance and where the Sixth wants them.

6th FedCom Regiment
Regiment/Regular/Reliable
CO: Hauptmann General James McConnel
Aide: Leftenant General Marco Le Tissier
AeroSpace Brigade: Lft. Gen. Gavin Stamp

Light and medium 'Mechs dominate the Sixth, many of which use advanced technology. The regiment's few heavy 'Mechs are gathered into two assault companies that are deployed as needed. Intelligence gathering plays a major role in the regiment's operations, so the unit's scout 'Mechs make extensive use of active probes, and designs like the *Stealth* and *Firestarter* are common in scout lances. These operate independently from the main regiment, often sneaking behind enemy lines to gather valuable data. Such missions require incredible skill and courage, and the recon pilots are the best troops in the RCT, usually of veteran or elite status.

The Sixth FedCom AeroSpace Brigade fields two Wings.

6th FedCom Armor Brigade
3 Regiments/Regular/Reliable
CO: Leftenant General Loui Camara
Aide: Kommandant Robert Owen
6th FedCom LA: Lft. Gen. Halmar Baardson
6th FedCom HA: Lft. Gen. John Fox
22nd Addicks AR: Lft. Gen. Ruth Clegg

Many vehicle crews opted not to follow the Sixth in its exodus from Dalkeith, and the remaining lances were consolidated into two regiments. Fortunately, the RCT's good relations with the people of Addicks have allowed them to work in conjunction with the planetary militia, in effect absorbing the Twenty-second Addicks Armor Regiment.

6th FedCom Infantry Brigade
5 Regiments/Regular/Reliable
CO: Leftenant General Barry Booker
Aide: Kommandant Louise Easton
11th FedCom IR: Lft. Gen. Marco Bakke
12th FedCom IR: Lft. Gen. Quinn Marcelino
11th FedCom MI: Lft. Gen. Tore Foley
12th FedCom MI: Lft. Gen. Joan Reddy
6th FedCom JI: Lft. Gen. Powell Armand

7TH FEDCOM RCT: SPECTRAL MENACE

The Seventh entered service on schedule in March 3039, less than a month before the first wave of AFFS troops poured into Combine space. While the High Command rated the Seventh as a full-strength RCT, the unit entered service at less than 30 percent authorized strength, with the remainder existing only as a phantom paper trail to cloak the massive buildup in the Draconis March from Combine intelligence analysts. It took the High Command almost four years to build the Seventh up to its full authorized strength.

Assigned to the defense of the St. Ives Compact for most of its life, the Seventh RCT developed a close relationship with that nation's people. The unit naturally took offense at the stationing of foreign troops within the Compact by then-First Lord Liao, even under the auspices of SLDF peacekeepers and in the form of a supposedly allied force. The members of the Seventh took the presence of the Fifth Lyran Regulars on Nashuar personally, seeing the Fifth not only as an invading force but also as the embodiment of the problems that were besetting the Federated Commonwealth.

Combat between the Seventh RCT and the Fifth Regulars commonly degenerated into all-out fights, such as the massive battle in the Khingan Foothills, in which few soldiers from either side walked away. The Seventh calmed down a bit following the Fifth's retreat from the world but continued to defend Nashuar vigorously until the cease-fire on the planet was called. The unit eventually pulled offworld to Spica, where it remains, serving as the world's primary defending force as it licks its wounds.

The Seventh's insignia is a phantom warrior brandishing a sword.

OFFICERS

Serving as executive officer of the armor brigade prior to the start of hostilities in the Compact, then-Colonel Torri Hughes was thrust into command of the entire RCT after more than a year of constant fighting, in addition to the Thuggee nerve-gas attacks, robbed the Seventh of most of its senior officers. Though not the most imaginative leader in the AFFC, Hughes is a strong leader who is still coming to grips with her new rank. Her confidence as the Seventh's commanding officer is growing daily.

TACTICS

The Seventh has never favored subtlety in battle, which has been even more evident since the demise of most of its command staff. The RCT simply attempts to flood the field of battle with its overwhelming strength.

7th FedCom Regiment
2 Battalions/Regular/Reliable
CO: Leftenant Colonel Pat Williams
Aide: Major Lee Trebelhorne
AeroSpace Brigade: Major Andrew "Jojo" Means

The Seventh's actual strength falls below four provisional companies when not counting the 'Mechs still in need of major repairs, though it can field two additional short companies of heavily damaged but marginally functional machines.

The Seventh FedCom AeroSpace Brigade suffered relatively few casualties and is made up of two operational Wings.

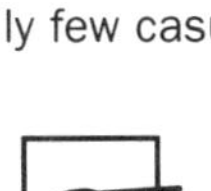

7th FedCom Armor Brigade
2 Regiments/Regular/Reliable
CO: Leftenant General Torri Hughes
Aide: Captain Frederick Bally
7th FedCom HA: Lft. Gen. Torri Hughes
7th FedCom AC: Lft. Col. Jeremiah Bellamy

The brigade's two regiments field barely fourteen functional companies between them.

7th FedCom Infantry Brigade
3 Regiments/Regular/Reliable
CO: Leftenant General Gus Streebach
Aide: Major Han Lo
1st St. Ives Legionnaires: Col. Thomas Overton
14th FedCom MI: Maj. Maia Poole
7th FedCom JI: Lft. Col. Gina Torres

The First St. Ives Legionnaires formed at the beginning of the Confederation-Compact war and attached itself to the Seventh during the fight on Nashuar. With the Confederation controlling the entire St. Ives thumb, the regiment chose to remain in the seriously understrength RCT.

8TH FEDCOM RCT: THE KATZBALGER

The formation of the Eighth FedCom RCT, originally scheduled for 3039, was delayed until 3041 because of the need to make good the losses in other units during the War of '39. Though generally considered a failure, the war allowed a new generation of FedCom troops to gain combat experience, so the officers assigned to the RCT had more current combat experience than many of those assigned to earlier FedCom units, experience they passed on to the new recruits assigned to the Eighth.

The RCT quickly gained a reputation for excellence and was granted regular status by the AFFC, rare for an untested unit. This status was vindicated by their defense of Second Try during the Marik-Liao Invasion, in which they inflicted considerable damage before being forced to withdraw.

Since withdrawing from Second Try in 3057, the Eighth FedCom RCT has played a major role in defending the vital industrial world of Kathil against the resurgent Capellan Confederation. However, with the formation of the Kathil CMM and the return of the First Capellan Dragoons at the Duke of Kathil's request, the ruler of the Capellan March, Duke George Hasek, has deemed the RCT redundant and has just issued orders for the Eighth to take up a posting on Halloran V.

The new insignia of the Eighth FedCom RCT shows a *Huron Warrior* BattleMech transfixed on a broadsword.

OFFICERS

Mitchell Weintraub has commanded the Eighth since its creation, having formerly served with the Eighth Donegal Guards in the Fourth Succession War and the War of '39. A Katherine loyalist, the general has little respect for Duke Hasek, regarding him as little more than a shadow of his illustrious father and more akin to his grandfather. Indeed, within the command staff of the RCT, Weintraub has referred to the duke as "the traitor's blood run true," citing his resistance to Archon Katherine as a throwback to his grandfather, Michael Hasek-Davion, who resisted Hanse Davion's rule and sought to betray his liege lord to the Capellan Confederation. The ejection of the Eighth FedCom from Kathil is, according to General Weintraub, the latest step in a conspiracy to take the Capellan March away from the Federated Commonwealth. He believes that Prince Victor is using his links to the Hasek family to shatter the Federated Commonwealth rather than allow his sister to rule. That is not something Weintraub will allow to happen.

TACTICS

The Eighth favor open field engagements where their heavy 'Mechs and armor can be used to best effect. They excel in assault operations, successfully dislodging more experienced units from fortified positions.

8th FedCom Regiment
Regiment/Regular/Reliable
CO: Hauptmann General Mitchell Weintraub
Aide: Leftenant General Karen Fallon
AeroSpace Brigade: Lft. Gen. Jocasta Carlton

Heavy and assault 'Mechs dominate the Eighth, divided equally among the nine companies. However, only half of these 'Mechs use upgraded technology, though this rises to almost three-quarters among the lighter elements. The general has ordered all members of the Eighth to alert status, and elements of the 'Mech regiment have been deployed to guard the RCT's base. These 'Mechs have been authorized to use force if threatened by Duke Hasek's forces, though Weintraub does not believe the duke will risk a confrontation.

The Eighth FedCom AeroSpace Brigade fields three Wings.

8th FedCom Armor Brigade
3 Regiments/Regular/Reliable
CO: Leftenant General Michelle Dettori
Aide: Kommandant Kim Fortune
8th FedCom HA: Lft. Gen. Michelle Dettori
9th FedCom HA: Lft. Gen. Garry Duffield
8th FedCom AC: Lft. Gen. Natalie Pollard

The heavy Eighth Armored Brigade uses some of the most advanced tanks in the AFFC, including the Challenger X and the Typhoon. Most of these are production models, but two of the Typhoons use the tracked drive system only found on the prototype model.

8th FedCom Infantry Brigade
5 Regiments/Regular/Reliable
CO: Leftenant General Nicky Little
Aide: Kommandant Donald Mernagh
11th FedCom IR: Lft. Gen. Casey Siddons
12th FedCom IR: Lft. Gen. Norton Polli
11th FedCom MI: Lft. Gen. Charles Price
12th FedCom MI: Lft. Gen. Evantha Lowe
6th FedCom JI: Lft. Gen. Yoshitomi Tendo

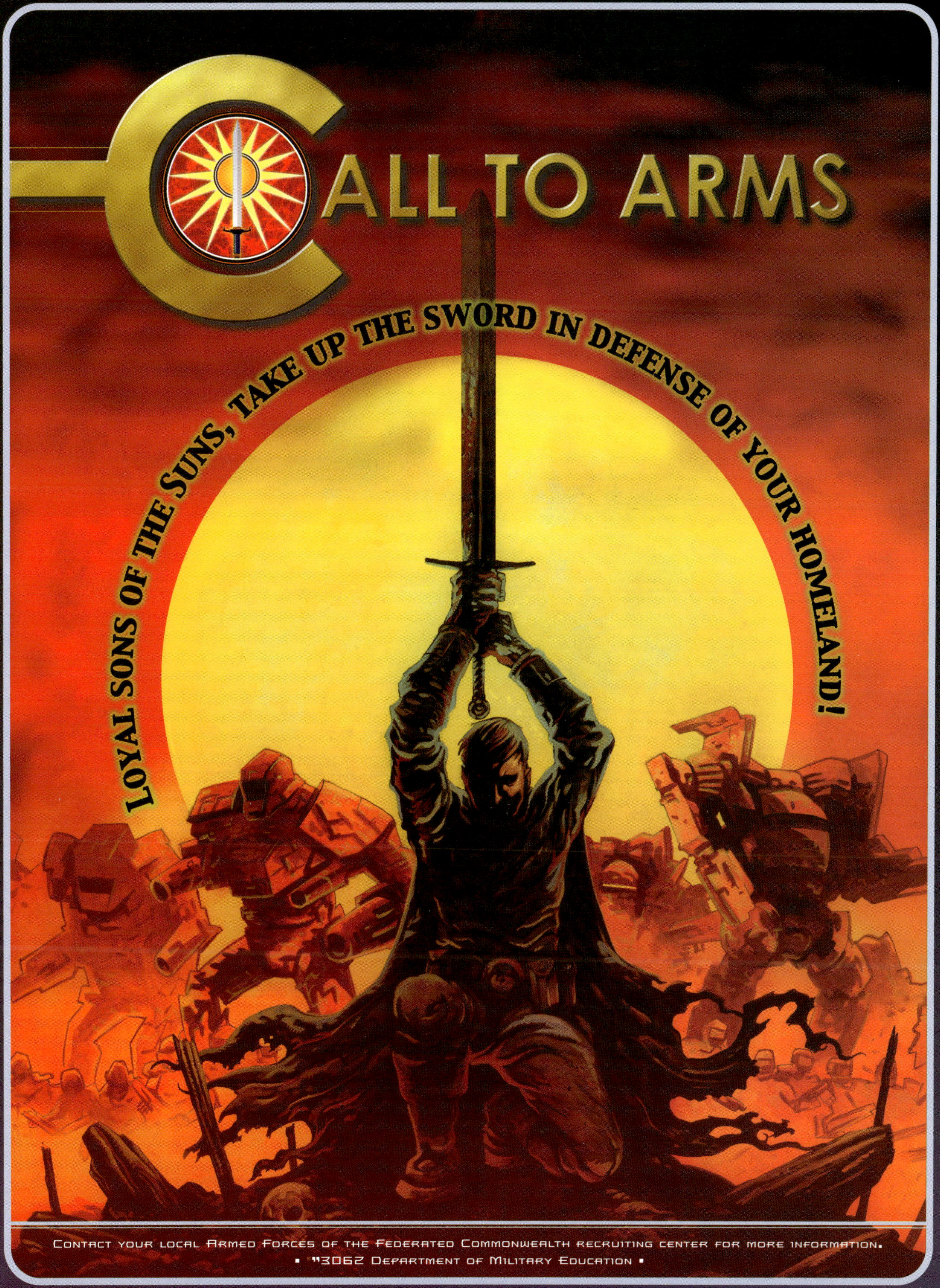
CALL TO ARMS
LOYAL SONS OF THE SUNS, TAKE UP THE SWORD IN DEFENSE OF YOUR HOMELAND!
CONTACT YOUR LOCAL ARMED FORCES OF THE FEDERATED COMMONWEALTH RECRUITING CENTER FOR MORE INFORMATION.
• ©3062 DEPARTMENT OF MILITARY EDUCATION •

Federated Suns

An Avalon Hussars MechWarrior, with comprehensive checklist in hand, prepares to test a prototype neurohelmet for the NA15.

Ranking Insignias

Recruit

Private

Private, 1st Class

Corporal

Sergeant

Sergeant-Major

Command Sergeant-Major

Cadet/Subaltern (Cadet)

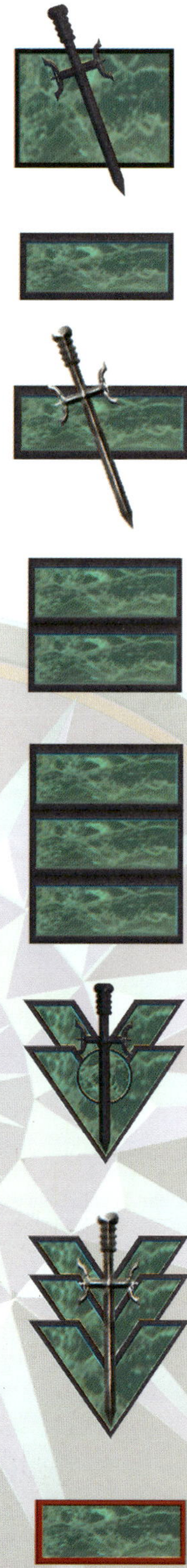

The Sunburst Medal

Federated Suns

Ranking Insignias

The Medal Excalibur

A Davion Heavy Guards aerospace pilot prepares to test a variant of the standard AFFC flight suit for Federated-Boeing Interstellar.

Federated Suns

Syrtis
Medal of Honor

This Tenth Deneb Light Cavalry infantryman carries one of the perks of being stationed on New Avalon—a non-standard issue pulse laser rifle.

Federated Suns

Knights of the Federated Suns

A First Crucis Lancers RCT officer wears the latest version of the AFFC uniform, but demonstrates his loyalties by wearing only the House Davion logo.

FEDERATED SUNS

1st Crucis Lancers

• JagerMech III •

1st Robinson Rangers

• Centurion •

4th Deneb Light Cavalry

• Hatchetman •

1st Davion Guards

• Enforcer III •

Federated Suns

17th Avalon Hussars

• JagerMech III •

5th Syrtis Fusiliers

• Centurion •

1st NAIS Cadet Cadre

• Hatchetman •

1st FedCom RCT

• Enforcer III •

FEDERATED SUNS

An overlarge integral jump pack and a retractable parafoil provide the Mk. 11 with superior high-altitude insertion capability.

Taking the original Infiltrator's sensor-baffling capabilities a step further, the Mk. 11 uses active electronic countermeasures.

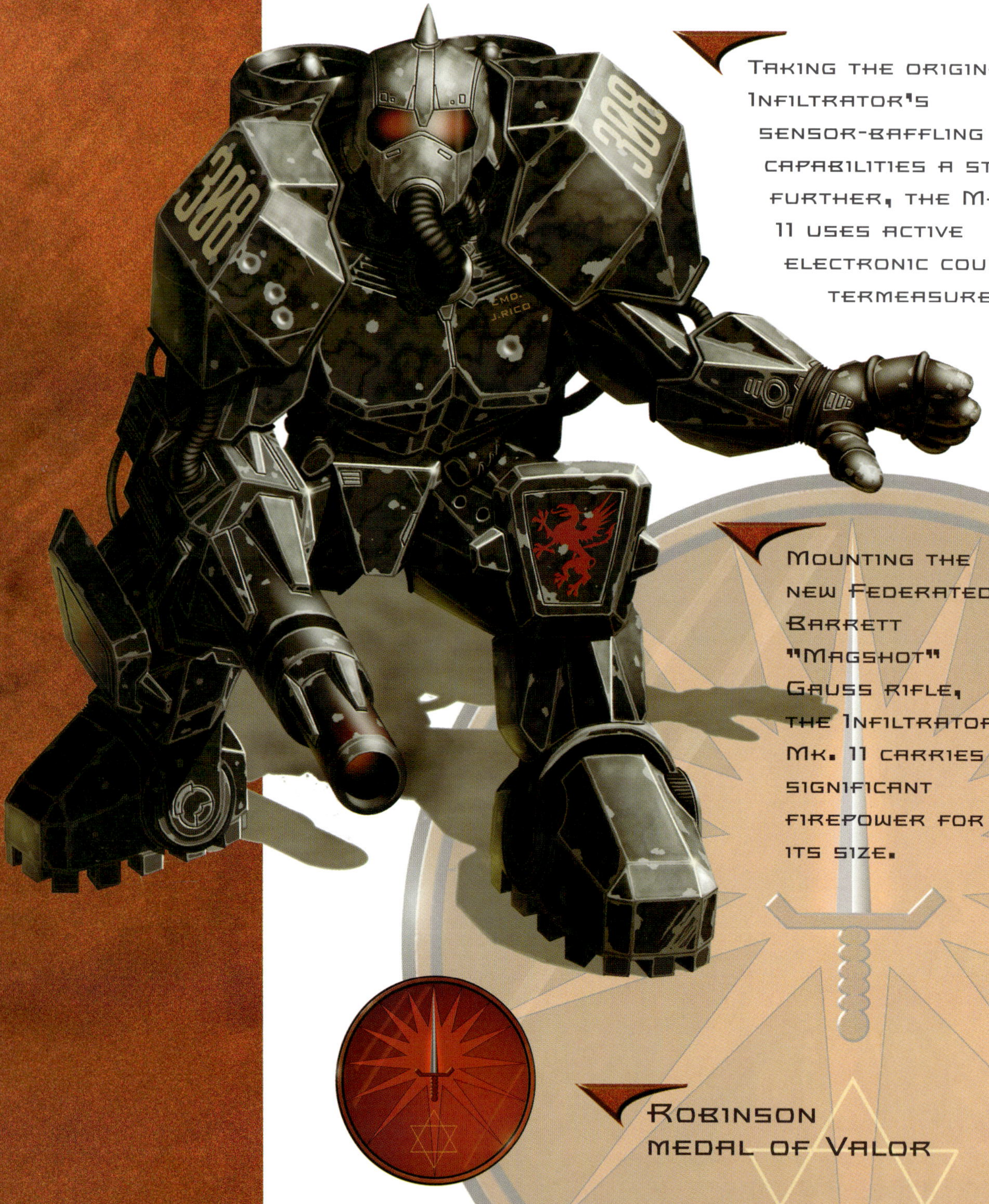

Mounting the new Federated-Barrett "Magshot" Gauss rifle, the Infiltrator Mk. 11 carries significant firepower for its size.

Robinson Medal of Valor

NEW IVAARSEN CHASSEURS

PIKEMEN

As with the First Capellan Dragoons, the two regiments of the New Ivaarsen Chasseurs hold dual positions as the family guard to Duke Leto Stephenson of New Ivaarsen and as elements of the Armed Forces of the Federated Commonwealth. Less than a jump from the current Combine border, the world of New Ivaarsen has been the object of conquest dozens of times throughout the centuries of hostility between the Combine and the Federated Suns. Following the Suns' final recapture of New Ivaarsen almost two centuries ago, the world's new duke began to organize his own private guard.

As New Ivaarsen's farming and manufacturing industries recovered and began to grow, the Stephenson family's wealth likewise increased, prompting the duke to expand his private guard. By the beginning of this century, the New Ivaarsen Chasseurs had grown to two full BattleMech regiments, in addition to supporting elements. Though the Second was almost destroyed by the Jade Falcons during the initial months of the Clan invasion, Prince Victor ordered the regiment rebuilt in 3060 in recognition of the Chasseurs' many years of devotion to the Federated Suns.

The two Chasseurs units are technically a part of the AFFC, though the Stephenson family was accorded a number of privileges in deference to their years of fanatical loyalty to the Federated Suns. At least one of the Chasseur regiments has remained on New Ivaarsen since their attachment to the AFFS, and the Duke has the right to quash any Chasseur assignment he objects to (unless ordered by the *rightful* Prince, a technicality Duke Leto has used at least once since Archon Katherine's assumption of the throne on New Avalon). Additionally, the Chasseurs are under the direct authority of the High Command, though for practical purposes Chasseur officers routinely coordinate with the commanders on Le Blanc and Robinson.

COMMANDING OFFICER

Major General Kristi Johnson took command of the Chasseurs in 3056, after Hauptmann General Ophelia Stephenson's retirement. Though not related to the Stephensons, Johnson's family has a long history with the Chasseurs, and in fact she led the First Regiment for almost ten years. While she holds a deep-seated resentment toward the Draconis Combine for pillaging her homeworld years ago, as do most of her soldiers, she is well aware of her duty and of the debt they owe Prince Victor.

FORCES

The Chasseurs are the only units in the AFFC composed solely of personnel from a single world, from lowest astech to commanding general. All Chasseurs swore an oath of fealty to the duke prior to entering New Ivaarsen's Military Training Academy and again before leaving to attend any required AFFC technical or specialty training.

The regiments are equipped with a wide variety of 'Mechs and vehicles, owing mainly to the re-formation of the Second Chasseurs. While the First Chasseurs have received little new equipment—and before its eradication, so did the Second—that was by choice. Many still pilot 'Mechs original to their unit and prefer to continue doing so.

COLORS AND INSIGNIA

Both Chasseur regiments commonly use a dark blue, red and gold paint scheme, even while operating in the field. The Chasseur insignia is a golden-winged sword set against a blue triangle.

1ST NEW IVAARSEN CHASSEURS: THE WINGS

The First Chasseurs have been one of the staples of the Draconis March, right alongside the Robinson Rangers, for the past 150 years. Critical to the defense of their homeworld and, as New Ivaarsen sits directly between Robinson and the Draconis Combine, the entire Draconis March, the First Regiment has proved a tough opponent to the dragon.

Temporarily detached to the Capellan border during the Fourth Succession War along with the Second Regiment, the First Chasseurs returned to the Draconis front and found themselves battling for their lives against the DCMS counterattack in 3039. Bloodied, the First made a number of retaliatory strikes over the next few years, including one highly publicized raid on Prosperina. The damage sustained on these raids began to add up, however, giving the Chasseurs no choice but to cease the attacks.

As soon as Duke Stephenson learned of his Second Chasseurs' destruction at the hands of Clan Jade Falcon in the early months of 3050, he authorized the First's reassignment to the Clan front. Though unable to take part, the Chasseurs supported a number of raids into the Falcon OZ, including the missions to Somerset and Sudeten that freed several hundred POWs, or "bondsmen" in Clan parlance, including a few of the Second's men. With little hope of freeing any more, the First began the long trip home in early 3054.

Since the re-formation of the Second Chasseurs, the First Regiment has been actively training with the new unit on New Ivaarsen and on Robinson's CTC. Though unofficially blacklisted by Archon Katherine because of the unit's fanatical loyalty to her brother, the First remains at top form, supplied by its duke's deep coffers and by Field Marshal Sandoval's HQ on the nearby world of Robinson.

The First Regiment uses the standard Chasseur colors but adds the roman numeral I to the brigade insignia.

OFFICERS

Leftenant General Sal Cole, nephew to Duke Leto Stephenson by way of his wife, Lissette, has been the master of the First Regiment since Major General Johnson moved up to command the entire demi-brigade. On the surface, Cole seems a foppish dilettante, appointed to command solely because of his relationship to the duke. Underneath the façade, however, is a cold and calculating officer with ambitions to be more than just an officer within the Chasseurs. Devoted solely to himself, he sees his support of Victor Steiner-Davion as the best chance to jump-start his career.

TACTICS

The First Regiment prefers to fight in open-field engagements where its strong aerospace forces can readily support its actions.

1st New Ivaarsen Chasseurs
Regiment/Elite/Fanatical
CO: Leftenant General Sal Cole
Aide: Major Lassiter Batren

Almost half of the regiment's BattleMechs are original to the unit, many handed down within the same family for generations. Combat losses and mechanical failures have accounted for the remaining half over the course of some 150 years. In the past two decades, the DQ has offered upgrades or outright replacements for each of the unit's 'Mechs, and while few pilots have turned down the upgrades, most have refused to give up their 'Mechs for newer designs. Fewer than two dozen of the regiment's 'Mechs are less than ten years old.

1st Chasseurs Air Group
Wing/Elite/Fanatical
CO: Rear Admiral Elliott Stephenson
Aide: Major Gillian Han
1st New Ivaarsen Air Wing: Maj. Justin Foxx
2nd New Ivaarsen Air Wing: Maj. Geoff Wallace
3rd New Ivaarsen Attack Wing: Lft. Gen. Julie Smith

The First Chasseurs have always fielded a strong aerospace force, believing correctly that almost any battle begins long before an invading force reaches the ground. The First and Second Wings are accomplished dogfighters and have proved themselves capable of taking on a force twice their size. The Third specializes in air-to-ground and anti-ship attacks, delivering tons of ordnance in precise bombing runs.

2ND NEW IVAARSEN CHASSEURS: THE SWORDS

Assigned along the coreward reaches of the Lyran Commonwealth shortly before the Clans made their appearance, the Second Chasseurs fell before the Jade Falcon onslaught. The unit experienced almost 60 percent losses on Anywhere before withdrawing to Somerset, where it finally fell alongside the Eighth Arcturan Guard and the Somerset Training Battalion. Though a few members escaped the Falcon OZ and a number were later rescued from reeducation camps, for all practical purposes the Second Chasseurs were defunct.

Duke Stephenson and his First Chasseurs did not forget the sacrifices of the second regiment, however. Neither did Prince Victor. In 3060 he authorized the re-formation of the unit, giving Duke Stephenson and Major General Johnson full access to the DQ and their pick of academy graduates. Remaining true to Chasseur traditions, they recruited only New Ivaarsen natives.

Within two years the new Second Chasseurs were operational, if still green. Major General Johnson moved a number of experienced officers and NCOs from the First Regiment to fill out command billets, but by and large the unit is made up of fresh graduates from the Federated Suns' many academies and training units. The regiment has been training almost constantly since it came together, however, and was recently upgraded to regular status by the High Command. The unit's unfettered access to the DQ ended after Archon Katherine's ascent to power, but, like the First Regiment, the Second will likely have few supply problems.

The Second's insignia is similar to that of the First, only with a roman numeral II set atop the sword. The sword's hilt is white, signifying the sacrifice of the original Second Regiment.

OFFICERS

Leftenant General Winona Stephenson-Voss is Duke Leto's youngest cousin and a sixteen-year veteran of the AFFC. Though a MechWarrior by profession, she served in a number of support roles in the Draconis March at the behest of her husband, one of Field Marshal Sandoval's many ministers. The two divorced just a few years ago when the call went out to help rebuild the Second, however, and she and her three children returned to New Ivaarsen. Initially a part of Hauptmann General Johnson's administrative staff, Stephenson-Voss proved a shrewd leader, and Johnson tapped her to command the temporary Second New Ivaarsen MechWarrior Training Group. Once the regiment was fully operational, Johnson turned command over to the new leftenant general. While she had never commanded anything larger than a lance in battle, Stephenson-Voss is a born analyst and quickly adapted her talents to the battlefield.

TACTICS

The Second is still training and has yet to display any marked tactical tendencies. On the other hand, the strong logical methodologies of Stephenson-Voss are quite noticeable and have served the unit well on the training fields.

2nd New Ivaarsen Chasseurs
Regiment/Regular/Fanatical
CO: Leftenant General Winona Stephenson-Voss
Aide: Leftenant Colonel Julian Poole

One hundred twelve brand-new BattleMechs, fresh off the lines on Kathil, Talon, New Avalon and Robinson, compose the new Second Regiment. Considered a heavy unit, 'Mechs like the *Enforcer III*, the *JagerMech III* and the *Nightstar*, all staples of the AFFC, are heavily represented in its three battalions.

2nd Chasseurs Air Wing
Wing/Green/Fanatical
CO: Light Commodore Anya Velacek
Aide: Captain Rain Wedi

As with their ground-bound brethren, the pilots of the Second Air Wing all fly brand-new aerospace fighters. A full squadron each of *Thunderbirds* and *Stukas* provide the wing its punch while *Eagles* and *Hellcats* round out its complement.

ROBINSON RANGERS

HONOR THROUGH VENGEANCE

Originally a volunteer brigade that participated in the defense of Robinson, that incarnation of the Rangers was wiped out after the Draconis Combine captured the world during the First Succession War. After retaking the world in the early twenty-ninth century, Prince Peter Davion authorized his AFFS to raise two BattleMech regiments to commemorate those heroes, with the members recruited from the world's population and trained at the Duke of Robinson's expense.

The Rangers have fought valiantly against the Draconis Combine in five major wars, spurred on not only by the Federated Suns' long-standing bad blood with the Combine but also by the loss of Robinson early in their history. The Rangers have vowed to repay the Combine for the damage done to their home at that time.

Two prime qualities define the Robinson Rangers. The first is the hatred of the Draconis Combine, though that hatred has cooled significantly. The second, and far more important, is an undying loyalty to the Duke of Robinson. Since their inception, the Rangers' regiments have acted almost as the private guard of Sandovals, who have always rewarded the Rangers' loyal service handsomely.

Some 225 years after the AFFS raised its two regiments from the bevy of volunteers on Robinson, Archon Katherine ordered the creation of a third unit within the command, named simply the Third Robinson Rangers. While undoubtedly motivated by political aims, this move gave Field Marshal Duke James Sandoval and his homeworld a definite boost in strength and standing. On the other hand, a relatively small percentage of the Third claims any ties to Robinson, and in fact the company owes more to the Archon and her supporters than to the Sandovals, making it a somewhat hollow gesture.

COMMANDING OFFICER

As the Duke of Robinson, Field Marshal James Sandoval serves as the titular commander of the Rangers. His position as commander of the Draconis March prevents him from overseeing the day-to-day needs of the Rangers, however, responsibilities he delegates to Major General Mai Fortuna and Leftenant General Travis Vehrson. Not a MechWarrior, Vehrson is an accountant by training who transferred into the command three years ago; he is a fourth-generation Ranger. Vehrson is a talented number-cruncher who not only ensures the Rangers remain fully operational despite cutbacks from New Avalon but also heads a major staff section within the Draconis March HQ that does the same for the entire region.

FORCES

Until the creation of the third regiment, every member of the Rangers was recruited from Robinson's populace and from graduates of the Robinson Battle Academy. In his capacity as Duke of Robinson, Field Marshal Sandoval personally approves each transfer into the unit. The Third Rangers is still somewhat outside his purview, however, at least as long as the rest of the RCT is being formed around the 'Mech regiment.

Unlike the Capellan and Crucis Marches, which a majority of the Federated Suns' largest war industries call home, the Draconis March is not blessed by a surplus of 'Mech production facilities. As a result, the Rangers' regiments are not equipped with the best equipment, at least by the high standards of the Davion Guards. On the other hand, with the extensive fortunes of the Sandoval family to draw on, the Rangers are by no means ill-equipped.

COLORS AND INSIGNIA

The Robinson Rangers commonly use a ceremonial paint scheme of deep red and black. The unit insignia is a single red arrow set against a gold shield.

1ST ROBINSON RANGERS: THE VOLS

The First Robinson Rangers have come a long way since their inception. At first nothing more than a band of rogue MechWarriors looking to avenge the injustices the Draconis Combine visited upon their homeworld, the First quickly shaped into a truly professional unit. Even after Robinson fell to Coordinator Miyogi's Marathon in 2858, the surviving elements of the regiment restrained themselves from launching a suicidal assault to recapture their home.

The First Rangers saw more than their fair share of combat along the Draconis front throughout the Second and Third Succession Wars and the War of 3039. They were transferred to the Lyran Commonwealth along with many other units when the Draconis March was denuded of its defenders early in the Clan war. Stationed on Kooken's Pleasure Pit for a number of years, the Rangers learned the potency of the Clan war machine at the hands of the Jade Falcons and the Steel Vipers.

Several years later, the Rangers were again tasked to defend a world against a Clan menace. This time they were stationed deep within the Smoke Jaguar OZ, alongside some of their most hated enemies. Invited by Prince Victor to be part of the multinational Task Force Bulldog, the Rangers found themselves sharing the planet Wolcott with four elite Combine and two Capellan regiments, as well as one of Katherine's most loyal units. Though numerous incidents reportedly broke out between FedSuns and Combine personnel on Wolcott, to their credit, no charges were ever levied against the Rangers. The Rangers traveled to Clan space and back with Prince Victor, returning as heroes to their homeworld.

Since Archon Katherine's takeover of the Federated Suns, the Rangers' members are wary. While the Archon has gone out of her way to appease Duke Sandoval and the Rangers, her actions ring hollow to many of them, to whom she is simply another manipulative Lyran.

The First Rangers uses the standard Rangers colors and insignia.

OFFICERS

Cousin to Field Marshal Duke James Sandoval, Mai Fortuna is one of Robinson's most loyal and capable officers. However, she and her cousin have had numerous heated disagreements on policy in the Draconis March over the years. Major General Fortuna knows her place in the grand scheme though, and publicly she has always been her cousin's greatest supporter, especially within the halls of the Robinson Battle Academy.

Captain Jethro Stephenson-Sandoval is another relative of Duke Sandoval, only one with far less personal motivation and tact. Demoted twice for insubordination in the past five years, his only redeeming factor is his unmistakable talent as a MechWarrior. In an effort to rein him in while putting his talents to use, General Fortuna has given him command of her headquarters company.

TACTICS

Owing to the nature of its assignments—often in defense of Robinson itself—the First Rangers regiment has developed a profound inclination for open field fighting. By arranging its forces into several echelons, it can rapidly fall back or advance fresh units for a renewed attack. Whenever possible, the unit attempts to build temporary field works, which it uses quite effectively on the defensive.

1st Robinson Rangers
Regiment/Veteran/Fanatical
CO: Major General Mai Fortuna
Aide: Major Frederico Temms

The First Rangers field a full range of 'Mechs, from the *Devastators* and *Victors* common in the first battalion to the numerous *Sentry*, *Stealth* and *Watchman*-class 'Mechs in the remaining two battalions. Though a few newer-tech 'Mechs are assigned to the unit, the majority fields older and refitted equipment.

1st Robinson Air Wing
Reinforced Wing/Regular/Reliable
CO: Major Perdition Feff
Aide: Leftenant Luis Rogas-Tassett

The First Air Wing just received eight new *Stuka* fighters to replace several outdated and increasingly difficult-to-repair airframes, courtesy of Duke Sandoval, making it one of the heaviest and deadliest fighter units in the Draconis March.

2ND ROBINSON RANGERS: THE PATRIOTS

Built around the few freedom fighters who survived the Combine purges on Robinson during the First Succession War, the Second Rangers have always considered themselves the most loyal to Robinson. At the same time, the Second has also been far more unpredictable than the First, leading the AFFS, and later the AFFC, to keep it on a tight leash.

Assigned to Breed as the Fourth Succession War began, the Second Rangers nearly lost the world to the Eleventh Benjamin Regulars. Though the regiment finally pushed the DCMS troops off the world, the Combine made another, more successful, effort less than a decade later during Gunji no Kanrei Kurita's counteroffensive. A third of the regiment refused to abandon Breed in 3039, instead carrying out a guerrilla campaign while the rest of the unit regrouped with AFFS reinforcements meant for the thrust toward Benjamin. The Rangers eventually carried the day, but only after a tremendous blow to their honor and an accompanying loss in manpower.

The Rangers spent much of the next two decades in the former Sarna and Skye Marches. Hamstrung by the loss of its JumpShips, the Rangers were just scant jumps away from the Chaos March when the Marik-Liao Offensive hit. Likewise, when DCMS troops took up station within the Lyons Thumb, only a lack of available funds prevented the unit from hiring private ships for a retaliatory strike against the Combine peacekeepers.

Still frustrated in their need for revenge against the Combine, the men and women of the Second Robinson Rangers barely notice the second-rate treatment they are receiving from the LAAF. On the other hand, they also secretly long for a return to their home. The LAAF finally assigned the regiment a JumpShip to reward its loyal service to the Lyran Alliance. Since then, rumors indicate the unit is slowly pulling up stakes and preparing to leave Phecda.

The regiment uses the standard Rangers colors but modifies the insignia by adding a second head to the arrow.

OFFICERS

As soon as he took command of the unit in late 3056, Colonel Theodor Mikul changed the entire tone of the regiment. The Rangers' former commander, Major General Cluey Montserrat, is an outspoken, tactless officer with a singular desire to transform everything he touches into his own vision of the AFFC. While those qualities are now being put to good use as head of the Filtvelt Military Academy, they were often cause for embarrassment for the unit and the LAAF. Mikul, on the other hand, is a quiet and unassuming commander who exudes an air of confidence and ambition. At the age of thirty-one, he is one of the youngest commanders of a 'Mech regiment in the LAAF or the AFFC. He has almost certainly plotted out the future of his unit; where that future leads is still in question, however.

TACTICS

The Second Rangers prefer close-range combat, often meaning they will take a battle into the midst of a city or rocky terrain. The regiment often practices coordinated battles in such arenas, though its members have historically eschewed any preset strategies and often any honor when it comes to fighting the DCMS.

2nd Robinson Rangers
Regiment/Regular/Reliable
CO: Colonel Theodor Mikul
Aide: Hauptmann Kommandant Lissa Katsunaki

The Second Rangers consist of primarily medium-class BattleMechs. As with the First regiment, the Sentry and Watchman, both built on Robinson, are well-represented in the unit. Ironically, the Second has managed to piece together a full company of C^3-equipped designs built in the Combine for the Lyran Alliance. After experimenting extensively with those 'Mechs, the men and women of the Second are well-versed in how to defeat the C^3 system in battle.

3RD ROBINSON RANGERS: DEATH OR GLORY BOYS

Formed in 3061, the Third Robinson Rangers are the newest line unit in the AFFC. Their creation, on the orders of Archon-Princess Katherine Steiner-Davion, has captured the imagination of the Draconis March, whose people regard the new unit as recognition of their importance to the Federated Commonwealth and the injustices done them by the Kurita-sympathetic Prince Victor. This shift in public opinion has done much to bring militia units stationed along the combine border into Archon Katherine's camp.

In contrast to the existing Ranger regiments, units from throughout the Draconis March contributed personnel and equipment, allowing the new unit to enter service a scant four months after the Archon announced her plan. Training and exercises continue, but Katherine has already recognized the unit's abilities and pedigree by assigning them to the prestigious New Avalon Garrison. They have principally been used for parades and guard duty, though Leftenant General McBride has requested the regiment be released for independent operations against the troubled Draconis Combine. The Archon has refused, stating it would not be wise to commit the unit to such an engagement before it is upgraded to RCT status (scheduled for 3064) and while the New Avalon garrison is understrength, following the relocation of the Davion Heavy Guards to Galax.

The Third uses a triple-head red arrow on the gold shield motif of the Robinson Rangers.

OFFICERS

Jason McBride previously served as XO of the Robinson DMM, and before that served as a captain in the First Rangers. His staunch Draconis March background—he was born on Breed and attended the Battle Academy on Robinson—has combined with his trivid-star good looks to make him the darling of the Draconis March press. Less charitable individuals in the AFFC have claimed McBride's looks complement the Rangers perfectly: a good-looking public-relations exercise without any real substance. However, a succession of creditable performances in exercises against the Tenth Deneb Light Cavalry and the First Davion Guards has demonstrated the falsehood of these claims.

Aristide Sandoval-Lee, a distant cousin of the dukes of the Draconis March, is credited with the regiment's surprisingly fast formation. A formidable administrator and politician, he is regarded by many as the "power behind the throne" in the Third Rangers, reporting to Duke Sandoval on Robinson as well as to the AFFC.

TACTICS

As a new unit, the Third Rangers have yet to develop any signature tactics, concentrating instead on improving their general fighting skills and unit integration. Their duties have allowed them to develop their drill and parade skills to a high degree, something General McBride is looking to exploit by establishing a series of preplanned battle maneuvers.

3rd Robinson Rangers
Regiment/Veteran/Reliable
CO: Leftenant General Jason McBride
Aide: Kommandant Aristide Sandoval-Lee

The Third Rangers comprises two medium battalions and one heavy battalion. All three use a diverse mix of BattleMechs, ranging from hundred-year-old Jenners to new-built JagerMech IIIs. Alpha Battalion contains the unit's only OmniMechs, a pair of Blackjacks.

3rd Robinson Guard
Reinforced Regiment/Regular/Fanatical
CO: Leftenant General Heinrich Gruber
Aide: Kommandant Jarman Koch

The Third Guard came together in nearly record time, with volunteers hailing from the Federated Commonwealth's best units. In recognition of that great accomplishment and the sacrifice of the March's people, Archon Katherine gave the Third Guard responsibility for protecting the Royal Compound and the Fox's Den.

3rd Robinson Air Wing
Wing/Regular/Reliable
CO: Leftenant General Sarah Churchill

The Rangers' Air Wing is a mix of light and medium fighters. The light fighters provide reconnaissance and air defense while the medium fighters are also configured for use in a ground-support role.

SYRTIS FUSILIERS

PRIDE AND PREJUDICE

Like the Royal Brigade and the Robinson Rangers, the Syrtis Fusiliers were once the private army of one of the Federated Suns' most powerful families, in this case the Hasek family of New Syrtis. Following Prince Alexander's reformations in the wake of the Davion Civil War, during which the Fusiliers had backed the Varnay faction and had suffered considerably for it, the surviving elements of the Fusiliers were absorbed into the AFFS.

Though a part of the Federated Suns military and led by officers at least nominally appointed by New Avalon, the Fusiliers have remained fiercely loyal to the Hasek family throughout their history. This loyalty nearly cost the Fusiliers everything when it ensnared them in Duke Michael Hasek-Davion's machinations against Prince Hanse in the early part of this century; indeed, it brought the Fifth Fusiliers down during the Fourth Succession War. Since then, however, the Fusiliers have moved beyond the petty rivalries Duke Michael infused in them, due in no small part to his son, Marshal of the Armies Duke Morgan Hasek-Davion.

Today, many within the High Command are again questioning where the Fusiliers' loyalties lie. Few in the Fusiliers' command structure have publicly taken a position in the debate over the true leader of the Federated Commonwealth. However, the Fusiliers exhibit a vehement devotion to Duke George Hasek, who, while not outwardly a proponent of Prince Victor, could hardly be considered a supporter of Archon Katherine.

COMMANDING OFFICER

Field Marshal George Hasek is the titular leader of the Syrtis Fusiliers by virtue of his dual position as Duke of New Syrtis and Duke of the Capellan March, though General Nathaniel Hasek maintains operational control. Once under the sway of Duke Michael, Nathaniel was nearly cashiered from service following the Fourth Succession War until his cousin Duke Morgan took him under his wing. Saved from an ignominious end, Nathaniel proved himself enough by 3047 that Prince Hanse personally elevated him to the command of the Fifth Fusiliers. Nine years later, he earned the mastery of the entire Fusiliers Brigade.

FORCES

The Fusiliers remain on the High Command's second logistics tier, even in light of their decades of loyal service. While this often means the Fusiliers cannot draw the most modern equipment from the Department of the Quartermaster, they have other means of supply. The Capellan March is home to a number of the Federated Suns' largest military industries, giving the Fusiliers a supply chain that bypasses the AFFC entirely. That, in addition to Duke Hasek's ability to reroute supplies within his own March, ensures the Fusiliers are kept at full authorized strength.

COLORS AND INSIGNIA

The Syrtis Fusiliers are one of a growing number of organizations that use the old Federated Sunburst and Sword rather than the FedCom insignia. Likewise, the Fusiliers' personnel have recently begun to wear the AFFS uniform and ranks. Their insignia is a variation on the Hasek family's crest. The unit uses a standard dark green paint scheme on its vehicles.

5TH SYRTIS FUSILIERS RCT: THE EVIL EYE

The Fifth Fusiliers is a proud unit that has shown the Duke of New Syrtis nothing but unlimited loyalty. While that trait was often a boon for the Federated Suns, it also nearly proved to be the unit's undoing. Following the murder of Duke Michael Hasek-Davion by Chancellor Maximillian Liao during the Fourth Succession War, the soldiers of the Fifth clamored to avenge their fallen duke. When Prince Hanse finally released the RCT against Sarna, the Fifth struck out ahead of its supporting task force, hoping to win the world alone in the name of its duke. Instead, Sarna's defenders, led by four elite regiments of McCarron's Armored Cavalry, all but wiped them out. After its few survivors were absorbed into the Kathil Uhlans and the Davion Heavy Guards, the Fifth existed in name only.

Following the war, Duke Morgan Hasek-Davion invested his family's fortunes into rebuilding the unit. Under the direction of Colonel Solomon Hasek, the new Fifth Fusiliers were recruited from Capellan March natives spread across the AFFS. Rather than build a cult of personality around himself, however, Duke Hasek-Davion tried to raise a unit loyal first to the Federated Suns. With the Capellan March united so strongly behind him and against the Confederation, that effort was only partially successful. Following Duke Hasek-Davion's death and Archon Katherine's rise to power, the Fifth lost the few direct allegiances it still held toward the Prince, instead placing its full confidence in its new duke, Field Marshal George Hasek.

The RCT's insignia is the All-Seeing Eye with a blood-red pupil.

OFFICERS

Major General Nathaniel Hasek recruited a young officer directly from Sakhara fifteen years ago. A distant relative by marriage, Serena Thompson-Hasek showed great promise even though her family were predominantly farmers and general laborers. Of course, the recommendation of Marshal Solomon Hasek, whose youngest son she had married, carried some weight. She has since surpassed every expectation and quickly risen to the rank of colonel. Mindful of the unit's headstrong history, she is determined to keep a firm hand on her 'Mech regiment.

TACTICS

This incarnation of the Fifth is far more reserved than its predecessor. The RCT uses massed long-range suppression fire to throw the enemy into confusion while cavalry and infantry units advance under an umbrella of missiles and autocannon shells to break apart the enemy formation.

5th Syrtis Fusiliers
Regiment/Regular/Reliable
CO: Colonel Serena Thompson-Hasek
Aide: Major Fortino Agassi
4 Wings: Rear Adm. Lew Froelich

This regiment was built entirely with BattleMechs constructed in the Capellan and Sarna Marches, making it a slightly heavier than standard unit. While most of its 'Mechs are fully upgraded, the regiment has received few new units in the past several years.

5th Syrtis Fusiliers Armor Brigade
2 Regiments/Green/Fanatical
CO: Major General Jing-Quo Lucifer
Aide: Major Sarah-Brynn Smith
65th New Syrtis ARR: Col. Caanan Doyle
3rd Jaipur ARR: Lft. Col. Theresa McGinty
2933rd Federation AB: Maj. Bobby Tapp

Both armored regiments field dozens of missile platforms, capable of raining down withering long-range barrages.

5th Syrtis Fusiliers Infantry Brigade
5 Regiments/Green/Reliable
CO: Leftenant General Ho Tingh
Aide: Major Al Pressman
584th Syrtis IR: Col. Beria Nikolayivich
20th Hobbs MI: Lft. Col. Eustachia Tobin
6th Talon MRT: Col. Sherry Bork
Royal Diefenbaker MDR: Col. Benton Frasier
245th Syrtis JR: Lft. Gen. Terrence LaCasse

Each regiment has several platoons of specially equipped spotters trained to direct artillery and missile barrages.

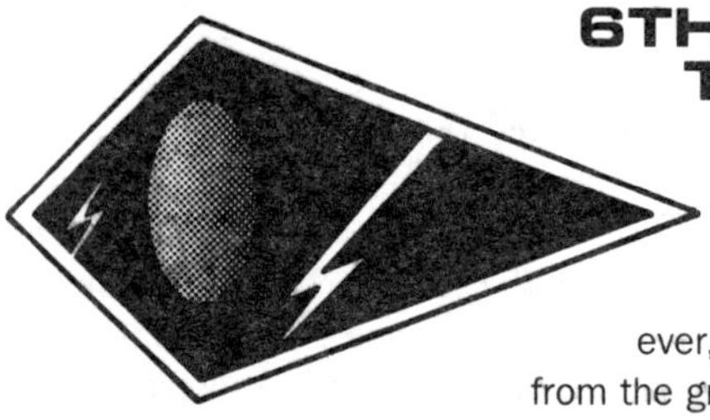

6TH SYRTIS FUSILIERS RCT: THE DUKE'S OWN

Traditionally the personal guard of the duke of the Capellan March, the Sixth was reassigned following the death of Duke Michael during the Fourth Succession War. With many of its officers cashiered out of service or brought up on charges of sedition, the Sixth Fusiliers lost much of its potency. Rather than let it fade away, however, Duke Morgan Hasek-Davion temporarily took direct command of the Sixth and rebuilt it from the ground up.

By the time of the Clan invasion, the Sixth was back up to its original strength. Assigned to defensive duties in the Sarna March, the Sixth never had the chance to face the Clans. During the Marik-Liao Invasion, however, the Fusiliers experienced more than their share of combat fighting the Confederation's assault on the Sarna March. With elements of the RCT spread across almost half a dozen worlds, and little hope of receiving any reinforcements, the Fusiliers could do little more than make the Confederation pay for every centimeter of ground it took. Although they experienced heavy losses, they left a number of Capellan units in far worse shape.

Since its withdrawal from the former Sarna March, the Sixth has been assigned to Novaya Zemlya. There, they have not only rebuilt their losses but have also been actively training other AFFC units in countering Capellan combat models.

The insignia for the Sixth is that of the Hasek family, the All-Seeing Eye.

OFFICERS

Major General Richard Silver assumed command of the Sixth Syrtis Fusiliers on 24 June 3053 amid the confusion of a terrorist bombing that cost the unit fourteen of its most senior officers. Rather than fall prey to the same emotions that brought on the Kentares Massacre, Silver left the investigation to others. This cool-headedness later allowed him to coordinate the Sixth's fighting withdrawal from the Sarna March, earning him the Syrtis Medal of Honor. A self-professed apolitical, Silver has sworn to defend the people of the Federated Suns as well as his duke and former subordinate, Field Marshal George Hasek.

TACTICS

The Sixth Fusiliers specialize in defensive and garrison operations, which allowed them to take a heavy toll on the Capellan military, and especially Kamakura's Hussars, during Operation Guerrero.

6th Syrtis Fusiliers
Reinforced Regiment/Elite/Reliable
CO: Major General Richard Silver
Aide: Colonel Hercules Senn
2 Wings: Rear Adm. Samson Bin Hurach

The Fusiliers' BattleMech regiment consists mainly of heavy and assault-class BattleMechs, further emphasizing its mission as a defensive unit. Early in its history, the Duke of New Syrtis expanded the unit beyond standard regimental size, and though the Fusiliers were transferred from their posting on New Syrtis some three decades ago, two of its 'Mech companies are still assigned as his honor guard.

6th Syrtis Fusiliers Armor Brigade
3 Regiments/Veteran/Reliable
CO: Major General Randy Hasek-Bills
Aide: Colonel Andrea Poemo
14th Syrtis Grenadiers: Lft. Gen. Stephen Duran
18th Verlo ARR: Lft. Gen. Doyle Rollins-Vuchovich
99th Syrtis Armor: Col. Jared El-Khaled
184th Federation Auxiliary: Lft. Col. Darius Nex

The 184th Auxiliary includes not only the RCT's artillery assets but also two battalions of combat engineers.

6th Syrtis Fusiliers Infantry Brigade
6 Regiments/Regular/Reliable
CO: Major General Gillian Singh
Aide: Colonel Jorge De Jesus Ramirez
1st Syrtis Royal IR: Lft. Gen. Shane Lucas Entebbe
113th Syrtis Rifles: Lft. Gen. Perry Ellens
22nd Syrtis Mountaineers: Col. Laramie Pough
55th Jaipur MIR: Col. Antonio Temmeni
249th Airborne Regiment: Col. Zermey Koch

The First Syrtis Royal Infantry Regiment is a nominal element of the Sixth Fusiliers but is permanently assigned, along with the Sixth's only battle-armor battalion, to New Syrtis as Duke Hasek's personal guard.

8TH SYRTIS FUSILIERS RCT: THE SNOW COBRAS

The Eighth Fusiliers have seen little change over the past several centuries, surviving the machinations of Duke Michael and the sweeping changes of his son, Duke Morgan, relatively intact. Like the remaining two Fusiliers units, the Eighth is fiercely loyal to the Hasek family and to the people of the Capellan March. Likewise, with two Limp Sword Flags proudly displayed on the unit standard, the Eighth has few fans, but many sworn enemies, in the Capellan Confederation.

The Eighth was one of the first AFFC units to report any significant problems with the citizenry of their assigned world following Archon Katherine's ascent to power on New Avalon. As tensions on Lothair increased, the Eighth was given instructions to keep the peace at all costs, something many of the unit's personnel were loath to do. When groups like the Citizens for Davion Purity and Action Democratic began violently clashing, new troops highly trained in anti-riot actions arrived to bolster the police forces, threatening to devolve into civil war. Rather than be forced into action against Lothair's people, the Eighth's Major General Palu removed Lothair's governor and declared martial law. She then used her Fusiliers to disarm *every* faction not willing to stand down and recognize her power, leading to a number of quite brief encounters with local police and anti-terrorist units. Though this action concluded five months ago, Lothair remains under martial law until Duke Hasek appoints a new governor.

The unit's insignia is a white cobra.

OFFICERS

Major General Deborah Palu is a native of Cumberland; because of that and her life of service in the Fusiliers, she considers all the people of the Capellan March her own. She made few allies in the High Command after her actions on Lothair. While the Department of Military Justice quickly swooped on her, a few words from Field Marshal Hasek quashed any charges.

TACTICS

The Eighth Fusiliers practice highly mobile combined-arms warfare, using their fastest hovercraft and VTOL aircraft to constantly harry and probe an enemy force while the rest of the light RCT moves into a killing position.

8th Syrtis Fusiliers
Regiment/Regular/Reliable
CO: Leftenant General Entropy Richards
Aide: Major Lorena Felton
1 Wing: Lgt. Cdre. Jordana Meloch

After sparring with the Confederation for many centuries, the Eighth fields quite a few Capellan-designed BattleMechs, including two brand-new *Men Shens*.

8th Syrtis Fusiliers Armor Brigade
3 Regiments/Elite/Fanatical
CO: Major General Bruce Durnne
Aide: Leftenant Colonel Iakobos Malenkov
8th Ogilvie LA: Col. Hachi Lee-Kwane
88th Lyons CR: Col. Ingvar Haakensen
100th Talon ARR: Lft. Col. Darlene Shouth
259th Syrtis AB: Capt. Dusty Bondo

The Fusiliers' armor brigade is made up predominantly of hovercraft and other light and fast-moving armored vehicles. The Eighty-eighth Cavalry is unique in that of its five battalions, three are exclusively equipped with combat VTOLs, many built specially for the unit.

8th Syrtis Fusiliers Infantry Brigade
4 Regiments/Regular/Reliable
CO: Major General Rezi LaFayette
Aide: Leftenant Colonel Paul Farchione
384th Federation MIR: Col. Meda Brown
390th Federation MIR: Col. William James Rohm
7th Alcyone MRT: Lft. Col. Veronica Hasek-Romanov
189th Syrtis JR: Col. Rafaella Delacruz

Though technically assigned as combat support elements, the transport units of the Fusiliers' infantry regiments field AFVs like the Maxim and even a number of Bandits, and have often been given combat assignments normally delegated to a recon or cavalry unit.

AFFC MARCH MILITIAS

Long before the days of the Star League, AFFS leaders realized they could not continue to defend the vast expanses of the Federated Suns with just their Regular Army units. Even when the thousands of independent militias and home guard defense forces were taken into account, the AFFS generals knew that if the Capellan Confederation or the Draconis Combine ever launched a coordinated campaign, the Suns would almost certainly fall. In an effort to solidify the defense of the Federated Suns, the AFFS formed the March Militias.

The AFFS knew it was the uncoordinated efforts of defenders scattered across dozens of light-years that contributed most to the downfall of a nation's defense. As it created the Polymorphous Defense Zones, it brought together the strongest militia units in each PDZ into a united military command. By forming the March Militias in this way, the AFFS not only ensured the most vigorous possible defense of its borders but also gave these new units a strong tie with the many other independent militias in the region.

Under the direct authority of their March's commander, these Militia RCTs are nominally assigned solely to defensive duties, with subordinate commands often spread out across their PDZ's most important worlds. These commands constantly alternate assignments between worlds to ensure that no world feels it has been left undefended and that the March Militia maintains a solid relationship with the independent militias in its jurisdiction. As the commander of a PDZ can take command of the planetary militias in his or her region in times of emergency, a close relationship between the March Militias and each world's militias is paramount.

Though it is not unheard of, the majority of March Militias do not see action outside their assigned PDZ. This is well-known by the Federated Suns' enemies and was used to much success during the last three major wars fought by the Federated Suns against its neighbors, when Prince Hanse ordered elements of the Capellan and Draconis March Militias into action on the front lines. Though virtually bereft of JumpShip support, the March Militias are capable fighting units, a fact that has surprised many an enemy commander.

FORCES

Each March Militia is predominantly made up of AFFC personnel native to the region and is almost evenly split between untested recruits and long-time veterans who are looking to finish their careers close to home. As a result, while the High Command gives each of the Militia RCTs a blanket status rating of "green" or "regular," most can put up a much stronger defense than their official AFFC rating may indicate. Because these Militias are home to so many combat veterans, the High Command has unofficially used them as proving grounds for new recruits and recent academy graduates not considered fit for service in a Regular Army unit.

The Department of the Quartermaster (DQ) is supplying the March Militias with as many new and refitted 'Mechs and vehicles as it can. Because of the losses experienced by the entire AFFC in the military campaigns over the past decade, it is years behind schedule in re-equipping the Militias. Though the Militias field little new equipment, the majority are equipped with designs refitted with upgraded technologies.

CAPELLAN MARCH MILITIA

The Capellan March has been rife with confusion in the past several years. Traditionally owing ultimate fealty to their duke, the men and women of the Capellan March's Militias have found themselves torn between conflicting loyalties. Field Marshal George Hasek is their duke, but they have begun to receive far more support from the AFFC since Archon Katherine's rise to power on New Avalon.

On the other hand, the Militias are resolute in their hatred of the Capellan Confederation, a feeling made stronger since the invasion of the Sarna March and the St. Ives Compact. Since the dissolution of the Sarna March, the Achernar and Nanking SMMs have been assigned to the Capellan March Regional Command, further muddying the waters of loyalty in the Militia.

COMMANDING OFFICER

Marshal Christopher Tice took charge of the CMM on New Syrtis in 3058, amid the resignations of a number of senior AFFC officers. Holding the dual position of CMM commander and director of operations for the entire Capellan March, Tice is well aware of the difficulties brewing in his units. Though unable to change the conflicting loyalties, he has been trying to reassign potential "problem" units to worlds where they would pose little threat to the security of the Federated Suns.

COLORS AND INSIGNIA

All CMM units wear the insignia of the Federated Suns and the Hasek family, in addition to that of the Capellan March. The March insignia is a torch set upon a red shield, over which is placed a ribbon inscribed with the words "Duty. Honor. Loyalty." Each Militia RCT further adds the initials of its PDZ to the shield. The units commonly wear dark green and red colors.

ALCYONE CAPELLAN MARCH MILITIA

The members of the Alcyone CMM have earned a reputation as hard fighters who will do whatever is needed to defend the worlds under their protection. Historically the militia has stood aloof from politics, knowing that to become involved would threaten their defenses against the Capellan Confederation. However, their unwillingness to take sides has drawn the ire of other units in the region, particularly from line units like the Twentieth Avalon Hussars. Defending themselves against such criticism has slowly drawn them into the camp of the Archon-Princess, whose policies, the CMM believes, stand the best chance of maintaining peace along the Capellan border.

OFFICERS

General Dan Kendall is torn between his loyalty to the Federated Commonwealth and his duty to defend the PDZ. He knows that to maintain the unit in fighting trim he must work with the Archon, but he also knows that taking sides in the dispute will bring additional hostile forces to bear. He has no illusions of the CMM's ability to withstand a concerted attack by Victor loyalists in the unit or by an outside agency like the Avalon Hussars. He has strived to build bridges and reduce tensions, protesting that any conflict in the AFFC would only aid the Federated Commonwealth's enemies. However, with his pleas falling on deaf ears and time running out, there may be only one option remaining, and the Alcyone Militia has never shied from a fight.

TACTICS

The Militia favors defensive operations and makes extensive use of minefields and prepared positions. Artillery also plays a major role in their battle plans, with preplotted barrages breaking up enemy assaults and FASCAM minefields used to channel hostile movement.

Alcyone Capellan March Militia
Regiment/Regular/Reliable
CO: Leftenant General Dan Kendall

Alcyone CMM Infantry Brigade
5 Regiments/Regular/Reliable
CO: Leftenant General Shannon Sarksyan

Alcyone CMM Armor Brigade
2 Regiments/Regular/Reliable
CO: Leftenant General Jo Revill

Alcyone CMM Aerospace Wing
Wing/Regular/Reliable
CO: Leftenant General Ilir Bollat

AFFC MARCH MILITIAS

KATHIL CAPELLAN MARCH MILITIA

Stripped of much of its component armor and infantry support during the first waves of the Fourth Succession War, and soon thereafter losing its 'Mech and aerospace forces to the newly formed First Kathil Uhlans, the Kathil CMM was decommissioned at the end of that war. With the replacement of wartime losses and the refitting of the Regular Army with lostech equipment a priority, the High Command simply assigned additional defenders to the Kathil PDZ rather than re-form the RCT.

In light of the action in the nearby Chaos March, Field Marshal Duke George Hasek began to rebuild the destroyed Militia on the QT in the past several years. Quietly diverting supplies and funding from other units in his March and recruiting heavily from the warriors cashiered from service by Archon Katherine's machinations, Duke Hasek reformed the Militia before anyone in the High Command realized what he was doing. Officially commissioned just a few weeks ago, the RCT has already taken up its posting, though the Eighth F-C RCT, which it was to relieve, has yet to reposition itself.

OFFICERS

Major General Donald Sampreis is one of Duke Hasek's closest allies and his former chief of staff. A veteran of the Clan War, General Sampreis is an outstanding administrator but only an average strategist, and as such is only temporarily leading the unit until a new commander can be found.

TACTICS

The RCT has had little time to drill together and has so far mastered only the most basic combat tactics.

Kathil Capellan March Militia
Regiment/Regular/Reliable
CO: Major General Donald Sampreis

Kathil CMM Infantry Brigade
4 Regiments/Green/Reliable
CO: Major General Lars-Erik Gennady

Kathil CMM Armor Brigade
2 Regiments/Green/Reliable
CO: Leftenant General Chandra Lo

Kathil CMM Aerospace Wing
Wing/Green/Reliable
CO: Major Jerald Northrup

NEW SYRTIS CAPELLAN MARCH MILITIA

In addition to being charged with the security of the historic seat of power for the Capellan March and the Hasek family, this unit has responsibility for the Werke system, home of one of the Federated Suns' largest military suppliers, Kallon Industries. To better defend these two key systems, the Duke of New Syrtis long ago authorized the creation of several additional units in the RCT, tasked with the defense of the planet Talon, where some third of the unit is permanently stationed. Because the RCT is responsible for such prominent worlds, the Dukes of New Syrtis have always ensured the unit is supplied with better-than-average equipment and some of the most loyal recruits from the Warrior's Hall.

OFFICERS

Recently promoted to the rank of Hauptmann General, Tia Caruthers is a twenty-four-year veteran of the AFFC and a dedicated supporter of the Hasek family. She has turned down a number of promotions since taking command of the CMM in 3048, preferring to remain on New Syrtis. In addition to her duties as commander of the Militia, she serves as a professor of political science at the Warrior's Hall.

TACTICS

The component regiments of the RCT are all intimately acquainted with the terrain of the worlds under their charge and have set up prepared defenses capable of shredding an enemy force on each.

New Syrtis Capellan March Militia
Reinforced Regiment/Regular/Reliable
CO: Major General Tia Caruthers

New Syrtis CMM Infantry Brigade
6 Regiments/Green/Reliable
CO: Leftenant General Erin Rodgers

New Syrtis CMM Armor Brigade
4 Regiments/Green/Reliable
CO: Leftenant General Lyle Tomaine

New Syrtis CMM Aerospace Brigade
3 Wings/Green/Reliable
CO: Leftenant General Thomas Milkovich

RIDGEBROOK CAPELLAN MARCH MILITIA

The relationship between the Federated Commonwealth and the Taurian Concordat has never been good, and the recent Taurian alliance with the Capellan Confederation has done little to help. As such, the Ridgebrook CMM spend much of their time monitoring the border, as much to prevent actions that could be construed as provocative by their Periphery neighbors as to defend against an attack. As such, the militia has a formidable intelligence network and work closely with corporate and private security forces, with joint exercises commonplace.

OFFICERS

Like all officers in the Ridgebrook CMM, General Seth Miller is politically astute. Though the relationship with the Concordat is foremost in his mind, he also has to work closely with civil authorities throughout the region. Though ultimate responsibility for such missions lies with the commander of the PDZ, Marshal Gil Carson prefers to leave them in Miller's hands. The general is thus away from Ridgebrook on a regular basis, and his aide, Emeline Jones, handles many of the day-to-day operations.

TACTICS

The Ridgebrook CMM favors defensive engagements and has a number of preplanned strategies. Usually, only a few commands are needed to instigate a strategy, allowing the unit to respond with an eerie lack of radio communication. However, when faced with situations not covered by their scenarios, the CMM's response is often rushed and inappropriate, leading to heavy losses.

Ridgebrook Capellan March Militia
Regiment/Green/Reliable
CO: Leftenant General Seth Miller

Ridgebrook CMM Infantry Brigade
5 Regiments/Green/Reliable
CO: Leftenant General Emma Lazenby

Ridgebrook CMM Armor Brigade
2 Regiments/Green/Reliable
CO: Leftenant General Emeline Jones

Ridgebrook CMM Aerospace Wing
Wing/Green/Reliable
CO: Leftenant General Frank Jenkins

SIRDAR CAPELLAN MARCH MILITIA

With their area of responsibility including the largest salient into Capellan territory, the Sirdar CMM has not lacked for combat. Over the past century, the militia has fought CCAF forces on sixty-two occasions, ranging from small lance-on-lance engagements to pitched battles. The unit believes prevention is better than a cure and has taken part in numerous tactical assaults against Liao positions. The unit has been rebuilt five times but always gets up for more, earning them a formidable reputation in the PDZ.

OFFICERS

Russel Ito is widely regarded as a glory hound, his actions taken as much to improve his standing as to safeguard the people of the Sirdar PDZ. His arrogance has done little to disprove this image, but his Gold Sunburst testifies to his skill and bravery. He inspires confidence in those under his command and has a fanatical following. Indeed, Sirdar CMM troops have been known to counter snide remarks directed at their commander with violence, prompting the deployment of double the number of MPs normally found in a March Militia.

TACTICS

The Sirdar CMM is principally a defensive unit, holding what others take and maintaining order among the civilian population. Historically, the Militia's 'Mech regiment operates independently of the main force, undertaking spoiling raids against Capellan positions, but in the détente following the Clan invasion, such missions have been rare.

Sirdar Capellan March Militia
Regiment/Regular/Reliable
CO: Leftenant General Russel Ito

Sirdar CMM Infantry Brigade
5 Regiments/Regular/Questionable
CO: Leftenant General Allen Harris

Sirdar CMM Armor Brigade
2 Regiments/Regular/Reliable
CO: Leftenant General Burford Priory

Sirdar CMM Aerospace Wing
Wing/Regular/Reliable
CO: Leftenant General Kay Jonze

VALEXA CAPELLAN MARCH MILITIA

Following the Fourth Succession War, the Valexa PDZ suddenly found itself in the interior of the Federated Suns. With little chance of a major Confederation advance toward the region, the High Command cut the PDZ's training and supply budget to the absolute minimum. Over the years, skills honed by decades of opposing Capellan raids atrophied, and when the Marik-Liao Offensive of '57 tore through the Sarna March, the Valexa CMM could do little more than hope the Capellan forces wouldn't cross into their region. Many officers in the unit placed the blame for the invasion squarely on Archon Katherine, who refused support to the Sarna March and sundered the nation.

OFFICERS

Leftenant General Sarah Delittle was a promising officer who got on the bad side of an AFFC marshal and found herself "promoted" out of the Crucis Lancers and into the Valexa CMM. Surprisingly, especially to herself, she found a family in the RCT and remained with the unit even after the court-martial of her nemesis, Marshal Aldrich King. She has gone to great lengths to retrain the RCT following the '57 offensive, and her efforts have been rewarded with the Syrtis Medal of Honor and a "regular" rating for her unit.

TACTICS

The RCT relies on its strong infantry forces to slow an enemy while its remaining units form up for a single, powerful advance.

Valexa Capellan March Militia
Regiment/Regular/Reliable
CO: Leftenant General Sarah Delittle

Valexa CMM Armor Regiment
Regiment/Regular/Reliable
CO: Leftenant Colonel Dora Carlisle

Valexa CMM Infantry Brigade
6 Regiments/Regular/Reliable
CO: Leftenant General MacKenzie Bolan

Valexa CMM Aerospace Brigade
2 Wings/Regular/Reliable
CO: Commodore Stephane de Argall

WARREN CAPELLAN MARCH MILITIA

The Warren CMM suffered considerable casualties in the border skirmishes that followed the Landmark Incident, when a Federated Commonwealth freighter misjumped into the Taurian Concordat, setting off Thomas Calderon's paranoia. Even though the crisis eventually wound down, the Warren CMM, like other units stationed along the border, remained on high alert until Thomas was replaced in 3055. However, with allegations rife that the Federated Commonwealth was behind the assassination of Protector Jeffrey Calderon in 3061, the CMM has stepped up its readiness in anticipation of renewed raids. These preparations include a number of cross-theater intelligence exchanges with the Islamabad Crucis March Militia.

OFFICERS

Martin Masar is a career soldier with little inclination for politics. His forthright manner has earned him many enemies, with whom he deals in the same manner as battlefield foes: directly and without compromise. Several planetary rulers who have sought to undermine the general's position have found their fiefs the site of unscheduled Militia exercises.

Rumors are circulating that Gi Verda, commander of the CMM's infantry brigade, is a member of Citizens for Davion Purity. The CDP is a radical and violent group opposed to the Steiner-Davion alliance and the Federated Commonwealth; it has strong support along the Periphery border. If this is the case, Verda is one of their highest-placed and most powerful operatives.

TACTICS

As might be expected, given their commander's up-front style, the Warren CMM favors uncomplicated open-field engagements that rely on brute force rather than finesse.

Warren Capellan March Militia
Regiment/Green/Reliable
CO: Leftenant General Martin Masar

Warren CMM Armor Brigade
2 Regiments/Green/Reliable
CO: Leftenant General Viktoria Spenser

Warren CMM Infantry Brigade
5 Regiments/Green/Questionable
CO: Leftenant General Gi Verda

Warren CMM Aerospace Wing
Wing/Green/Reliable
CO: Leftenant General Tim Crisp

AFFC MARCH MILITIAS

ACHERNAR SARNA MARCH MILITIA

The Achernar PDZ is composed of worlds that, prior to the Fourth Succession War, were claimed by both the Federated Suns and the Capellan Confederation. As soon as Chancellor Liao made clear his intentions during the '57 offensive, the Achernar SMM, along with the region's few other AFFC units, went on the highest possible alert, awaiting the Capellan invasion forces. The invasion stalled long before it reached the PDZ's borders, however. With the entire PDZ on alert, pro-Capellan rebel and terrorist activities were immediately quashed. The region did experience a strong resurgence in pro-Tikonov feelings, however, which left little room for loyalty to the AFFC.

OFFICERS

Hauptmann General Walter Flostet and his commanding officer, Marshal Kleindienst, have a deep and long-lasting friendship. Both hail from New Hessen and followed Colonel Pavel Ridzik when he chose alliance with the Federated Suns during the Fourth Succession War. Though confronted with Federated Suns loyalists in their own units, both officers are apparently trying to gain supremacy in the Achernar PDZ, perhaps in hopes of re-forming a Tikonov state.

TACTICS

After decades of putting down pro-Capellan demonstrations and terrorist bands, the Militia is adept at rooting an entrenched enemy from even the most heavily populated cities.

Achernar Sarna March Militia
2 Battalions/Green/Reliable
CO: Leftenant General Bruce Haas

Achernar SMM Infantry Brigade
5 Regiments/Green/Reliable
CO: Leftenant General Leah Gennadiy

Achernar SMM Armor Brigade
2 Regiments/Green/Reliable
CO: Leftenant General Marko Luczenko

Achernar SMM Aerospace Wing
Wing/Green/Reliable
CO: Kommandant Ho-Li Krulov

NANKING SARNA MARCH MILITIA

Situated in the Federated Suns' precarious salient into the Chaos March, Nanking is a world rife with trouble. With the Capellan Confederation slowly absorbing world after world in the Chaos March and supporting terrorist and rebel organizations throughout the region, the Militia has its hands full maintaining order on its own homeworld. The assignment of the First F-C to Nanking has only made matters worse. The members of the Militia already blame the AFFC Regular Army and Prince Victor for the situation in the Chaos March, and while the First F-C continues to covertly support resistance movements on nearby worlds, thousands of the region's citizens die every day.

OFFICERS

Leftenant General Fritz Tull willingly took the assignment to the Nanking SMM to remain close to his family on Elgin. With his unit bereft of JumpShips, he was unable to defend his homeworld and eventually learned his entire family had been wiped out in the continuing fighting. Tull is a man consumed by anger at his former Prince. He is a public proponent of Archon Katherine and her changes, but those who know him believe he is simply using that as a means to gain better support for his unit.

TACTICS

Many of the Militia's personnel are relatively new recruits, leaving the RCT with little battlefield cohesion. Tull recruits members who have a strong sense of family and who would willingly die in the defense of their home.

Nanking Sarna March Militia
Regiment/Green/Reliable
CO: Leftenant General Fritz Tull

Nanking SMM Infantry Brigade
2 Regiments/Green/Reliable
CO: Leftenant General Jontel Freeman

Nanking SMM Armor Brigade
2 Regiments/Green/Reliable
CO: Leftenant General Min Huong

Nanking SMM Aerospace Wing
Wing/Green/Reliable
CO: Hauptmann Billy Preston

CRUCIS MARCH MILITIA

The members of the Crucis March Militia have always shown an intense loyalty to their Prince, at least outwardly. After Archon Katherine secured her position on New Avalon, she bought the loyalty of the Crucis March Militia by giving them a one-step increase on the DQ's priority list. While assigned to protect the heart of the Federated Suns, the CMM has also rarely seen action in the past century, and as a result has been among the last to receive replacement equipment and men.

COMMANDING OFFICER

Hauptmann General Maurice Konrad is the new commander of the CMM, appointed by Field Marshal Gallagher soon after his own appointment by the Archon. Konrad is not a combat officer but a lawyer by profession who spent most of his career in the High Command bureaucracy. A long-time friend of Gallagher, Konrad is considered a political appointee with no practical skills in running a combat unit.

COLORS AND INSIGNIA

The CMM's insignia is sword placed over a red and white shield. Each unit adds the initials of their Combat Region to the shield. The CMM uses a blue, white and red paint scheme.

ANJIN MUERTO CRUCIS MARCH MILITIA

Perhaps the biggest challenge faced by the Anjin Muerto Militia is the major influx of colonists to the world of Okefenokee following the discovery of major petrochemical deposits there. The natives of the world do not look kindly on the interlopers and have been waging a guerrilla war against the newcomers and corporate concerns for almost forty years. Two full regiments of the militia's infantry are based on the world as peacekeepers, but they have been caught between the two factions. On top of this, the planet's newfound wealth has made it an inviting target for pirates operating from the Periphery.

OFFICERS

Despite the stringent military requirements of the Anjin Muerto Combat Region, Earl Renquin is a political appointment, made by the Archon-Princess contrary to the recommendation of Suzane Lipstein, commander of the Combat Region. Unlike Renquin or her brother Michael (commander of the Fifteenth Deneb Light Cavalry), Lipstein is a staunch supporter of Prince Victor, and Renquin's promotion is likely an attempt to undermine her influence. Fortunately, the CMM's strong tradition of independent operations has allowed them to function with minimal disruption, though as he has become more confident, Renquin has begun to meddle with deployments. Whether this will be beneficial remains to be seen.

TACTICS

The Anjin Muerto CMM is proficient in a broad range of operations, but it excels at small-unit tactics and combat in difficult terrain. Though the unit's reliance on hovercraft and jump-capable 'Mechs limits the firepower it can bring to bear, the maneuverability advantage allows the CMM to strike where it is least expected.

Anjin Muerto Crucis March Militia
Regiment/Regular/Reliable
CO: Leftenant General Earl Renquin

Anjin Muerto CMM Armor Brigade
2 Regiments/Regular/Reliable
CO: Leftenant General Quentin Rathcaven

Anjin Muerto CMM Infantry Brigade
5 Regiments/Veteran/Reliable
CO: Leftenant General Marta Andrews

Anjin Muerto CMM Aerospace Wing
Wing/Regular/Reliable
CO: Leftenant General Stella Dee

AFFC MARCH MILITIAS

ISLAMABAD CRUCIS MARCH MILITIA

One of the oldest units in the AFFC, the Islamabad CMM began life as the Messengers of Shiva, the combat arm of the United Hindu Collective (UHC). When the collective joined the Federated Suns in 2540, the Messengers were absorbed into the AFFS. Their traditions have been eroded, but the thirty-first century has seen a rise in nationalism in the semi-independent collective, which has in turn prompted similar changes in the CMM. Chief among these is a resurgence in caste-based politics that threatens to undermine the unit's organization.

OFFICERS

Born on New Syrtis, Carrie Zetso has faced many problems since taking command of the Militia. As an outsider, she is regarded with suspicion, though her willingness to respect local traditions—she spent several months learning Urdu—has done much to allay this. Ironically, Zetso's family history is entwined with that of the UHC. Back in the Davion Civil War of the 2530s, one of her ancestors, a supporter of Cassandra Varnay, led troops into the collective in an effort to outflank enemy forces. This force was crushed and the survivors imprisoned on the UHC capital. Zetso has kept this secret from her troops, but the Duke of Panpour, Pravin Singh, has discovered the secret and is using it to blackmail her.

TACTICS

The proud warriors of the Islamabad CMM are excellent fighters, but they lack the discipline to bond as an effective fighting force. They perform well in lance- or company-sized engagements, but larger gatherings quickly degenerate into chaos. However, the resurgence of interest in the Messengers of Shiva has improved their discipline and coordination, as many hope to one day live up to their potential.

Islamabad Crucis March Militia
2 Battalions/Green/Reliable
CO: Leftenant General Carrie Zetso

Islamabad CMM Armor Brigade
2 Regiments/Green/Reliable
CO: Leftenant General Mick Gyat

Islamabad CMM Infantry Brigade
5 Regiments/Green/Reliable
CO: Leftenant General Lark Neville

Islamabad CMM Aerospace Wing
Wing/Green/Reliable
CO: Leftenant General Solange Avery

KESTREL CRUCIS MARCH MILITIA

The units assigned to the Kestrel Combat Region were called to action a number of times throughout the Second and Third Succession Wars, as the Combine pushed deep into the Suns, virtually placing the Region on the front lines. Though the Militia has not been called into active service in more than a century, the unit retains many of the trophies of war it took throughout the years, including one fully functional Star League-era *Thug* once piloted by Coordinator Jinjiro Kurita and the battle standard for the First Sword of Light, taken in the unsuccessful battle for Robinson in 2854.

OFFICERS

Leftenant General Vivian Petty is a career officer whose career stalled when she took command of the Kestrel CMM. Long past recriminations and resentment, Petty is now content with her position. Since her headquarters was a virtual museum of AFFS history when she took charge of the unit, she has turned it into just that, earning a doctoral degree in history and sociology and eventually becoming the Crucis March's official historian.

TACTICS

Students of military history, the members of the Kestrel CMM seem like seers on the battlefield, often predicting an enemy's moves long before the foe has even considered them.

Valexa Capellan March Militia
Regiment/Regular/Reliable
CO: Leftenant General Garman Doucette

Valexa CMM Armor Brigade
2 Regiments/Regular/Reliable
CO: Leftenant General Joachim Brandeis

Valexa CMM Infantry Brigade
4 Regiments/Regular/Reliable
CO: Leftenant General Lucinda Sanches

Valexa CMM Aerospace Wing
Wing/Green/Reliable
CO: Hauptmann Rich "Blower" Blauer

MALAGROTTA CRUCIS MARCH MILITIA

Site of the battle that triggered the Reunification War between the Star League and the Periphery states, Malagrotta has long been a source of conflict between the Federated Commonwealth and the Taurian Concordat. Despite the distance involved, Taurian raiders have targeted the world more often than any other border world, with the valuable titanium mines of Grex-Dex on the moon Fontana the principal target. Initially private security forces managed the defense of the world, but as the Succession Wars raged and Taurian attacks became more common, the Malagrotta CMM was created to counter them.

OFFICERS

A graduate of the NAIS College of Military Sciences class of '44, Andrew Grundy has had a meteoric rise to success. He served as a leftenant with the Twenty-sixth Lyran Guards during the Clan invasion, facing the Wolves on Kobe and Thun and rising to field command of an ad-hoc battalion during the battle in the Logan Delta. After the war, his brevet-rank of Kommandant was confirmed, and he remained with the Twenty-sixth as they rebuilt. However, his mix of political, administrative and tactical skills made him a standout officer, and he was marked for promotion. No posts of suitable rank were available in the Lyran State Command, so Grundy was offered the Malagrotta Militia, which he accepted in 3060.

TACTICS

Often required to fight on Fontana, the Malagrotta CMM has learned to maneuver and fight in a gravity only one-third that of Terra.

Malagrotta Crucis March Militia
Regiment/Green/Questionable
CO: Leftenant General Andrew Grundy

Malagrotta CMM Armor Brigade
2 Regiments/Green/Questionable
CO: Leftenant General Jess Hoskins

Malagrotta CMM Infantry Brigade
5 Regiments/Green/Questionable
CO: Leftenant General Mervyn Wahwhani

Malagrotta CMM Aerospace Wing
Wing/Regular/Reliable
CO: Leftenant General Hannah Goodheart

MARLETTE CRUCIS MARCH MILITIA

Although the threat to the Archemar Combat Region has declined since the Fourth Succession War shattered the Capellan Confederation's Tikonov Commonality, the Marlette CMM remains a potent force. The Militia undertakes frequent exercises, both internally and against other units. The CMM frequently spar against the Goshen War College Training Battalion, providing the "enemy" force for the cadets' graduation exercise. However, with tensions in the region increasing, thanks to Capellan agitators and the widening political rift in the AFFC, the Marlette CMM has declined to participate in this year's exercise and is instead preparing for war.

OFFICERS

Despite his reputation for eccentricity, Dennis Waxton is an excellent tactician and an inspirational commander. Fond of quoting poetry, he leads by example, commanding from his upgraded *Cyclops* BattleMech. The High Command has requested that the general be more circumspect, but Waxton shrugs this off as the worry of jealous "desk warriors."

TACTICS

The Marlette CMM is aggressive, favoring offensive strategies even when cast as the defender in an engagement. Enemy forces commonly find themselves the target of tactical assaults and spoiling attacks, designed to keep them off balance while the main force of the Militia maneuvers into position. These raids are carried out by fire companies that operate independently from the main battalions and are equipped with fast and well-armed 'Mechs like the *Centurion* and the *Enforcer.*

Marlette Crucis March Militia
Reinforced Regiment/Regular/Reliable
CO: Leftenant General Dennis Waxton

Marlette CMM Armor Brigade
2 Regiments/Regular/Reliable
CO: Leftenant General Charlotte Cobham

Marlette CMM Infantry Brigade
5 Regiments/Regular/Questionable
CO: Leftenant General Nick Massarik

Marlette CMM Aerospace Wing
Wing/Regular/Reliable
CO: Leftenant General Jen Harvey-Kelly

NEW AVALON CRUCIS MARCH MILITIA

Though deployed to New Aragon in the Fourth Succession War, the New Avalon CMM is more at home on the parade ground than the battlefield. Their usual duties are ceremonial, guarding key locations in Avalon City and the Royal Palace. The militia also deploys a combined arms battalion at the NAIS to discourage a repeat of the 3029 Death Commando raid. Discipline in the militia is strict, and troops are expected to behave appropriately at all times, in or out of uniform.

OFFICERS

Prim and proper, with a reputation for meticulous planning, Russel Payne is the ideal commander for a ceremonial unit like the New Avalon Militia. The grandson of Christopher Payne, who commanded the Davion Brigade of Guards during the reign of Ian Davion, Payne is also the Duke of Augusta and a permanent member of the royal court. Though lacking the tactical brilliance of his forefathers, the general is a fair tactician and, as the Militia is unlikely to see combat, sees little reason to change.

TACTICS

The New Avalon CMM is highly skilled at close-quarters fighting. They make regular use of the urban combat arena at the NAIS College of Military Sciences, with particular emphasis on precision so as not to damage the buildings they are defending.

New Avalon Crucis March Militia
Regiment/Regular/Reliable
CO: Leftenant General Russel Payne

New Avalon CMM Infantry Brigade
5 Regiments/Regular/Reliable
CO: Leftenant General Devon Hyams

New Avalon CMM Armor Brigade
2 Regiments/Regular/Reliable
CO: Leftenant General Hank Swengly

New Avalon CMM Aerospace Wing
Wing/Regular/Reliable
CO: Leftenant General Polly Caplan

REMAGEN CRUCIS MARCH MILITIA

Once known as the Remagen Combat Region, the Nunivak CR was renamed as part of the internal reorganization·that followed the creation of the Federated Commonwealth. However, the region's Militia retained its old name, fighting off several attempts to rename it during the 3040s. The most recent, in 3045, almost resulted in a mutiny and persuaded the AFFC to leave the unit alone.

OFFICERS

Hans Sheller has just celebrated his tenth year in command of the militia and his twenty-fifth in service to the Davion military. Outwardly cold and severe, Sheller's friends know him as a caring officer who distances himself from others as a buffer against the day he must send them to their death in battle. Though widely respected as a tactician, Sheller lacks the confidence to be an effective front-line commander, and it was for this reason that he transferred from the Crucis Lancers to the Militia in 3048.

Lucy Davion, one of the Victoria Davions, has occasionally been cited as a potential claimant for the New Avalon throne. Though she dismisses such speculation, opponents of the Federated Commonwealth alliance point to Lucy's pure Davion heritage and her military experience, something both Archon Katherine and her sister Yvonne lack.

TACTICS

The Remagen CMM excels at night operations, using darkness to hide its maneuvers from enemy recon assets. Such operations are normally carried out in lance- or company-sized groups to minimize the chance of detection. Sabotage is another favored tactic of the Militia, with several platoons of infantry trained to infiltrate enemy lines and destroy supply dumps, communication facilities and command posts.

Remagen Crucis March Militia
Regiment/Regular/Reliable
CO: Leftenant General Hans Scheller

Remagen CMM Infantry Brigade
5 Regiments/Regular/Reliable
CO: Leftenant General Mika Scarletta

Remagen CMM Armor Brigade
2 Regiments/Regular/Reliable
CO: Leftenant General Lucy Davion

Remagen CMM Aerospace Wing
Wing/Regular/Reliable
CO: Leftenant General Jaime Francis

TSAMMA CRUCIS MARCH MILITIA

Despite its location far from any hostile borders, the Tsamma CMM has not been short of combat. Between 3034 and 3037, the unit was involved in a bloody conflict on St. Robert, fighting to restore the duke after religious extremists staged a coup. The Barnet Accords brought peace to the world, but a full regiment remained until 3042. Most of the next decade was taken up hunting pirates, but the assassination in 3055 of Antonio Garibaldi, Duke of Parma, prompted a deployment to Gambier. The Dukes of Parma were also major landholders on the neighboring world, and the CMM deployed there as a precaution against the spreading conflicts among the antagonistic Parmian nobles.

OFFICERS

Michael Buckley is a former gunnery instructor at the Albion Military Academy. In 3042, after six years on New Avalon, he chose to return to active duty and became XO of the Tsamma CMM. He was promoted to command of the RCT only three weeks before the Gambier deployment and had to master his new role under difficult conditions. It is a testament to his skill that there were no major incidents during the yearlong operation.

TACTICS

The Tsamma CMM specializes in non-lethal combat, interceding in disputes and using their equipment to keep hostile factions apart. They show admirable restraint in the face of verbal and physical abuse, regarding any occasion on which they have to resort to violence as a failure. However, as several units discovered on Gambier to their regret, the militia's preference for negotiation should not be mistaken for weakness. When forced to fight, they are efficient and deadly.

Tsamma Crucis March Militia
Regiment/Regular/Reliable
CO: Leftenant General Michael Buckley

Tsamma CMM Infantry Brigade
5 Regiments/Regular/Reliable
CO: Leftenant General Byre Gilarist

Tsamma CMM Armor Brigade
2 Regiments/Regular/Reliable
CO: Leftenant General Liam Clemmens

Tsamma CMM Aerospace Wing
Wing/Green/Reliable
CO: Leftenant General Arsene Rostov

DRACONIS MARCH MILITIA

Like the Capellan and Crucis March units, the Draconis March Militia is experiencing a shake-up in loyalties, though to a far lesser extent. Field Marshal James Sandoval has kept a tight rein on his region, allowing no one outside his command to make any decisions for his units. While the slight increase in supplies to the Federated Suns March Militias has bought Katherine some perfunctory loyalty in the Draconis March, the rank and file support their field marshal. Frustration at the continuing detente with the Draconis Combine is beginning to show in the Militia's performance, however.

COMMANDING OFFICER

After working her way through the ranks of the Robinson Rangers, Hauptmann General Katheryn Sandoval-Ito took charge of the Draconis Militia seven years ago. The younger sister of Field Marshal James Sandoval, Katheryn was once considered the rogue of the family. Though maturity has drained most of her impetuous tendencies, she sates her restlessness by traveling the length and breadth of the Draconis March, mercilessly inspecting her troops.

COLORS AND INSIGNIA

The DMM RCTs paint their equipment a deep red. The March insignia is a crown on a vertical white bar, set against a red shield. Each unit adds the initials of its Combat Region to the shield.

ADDICKS DRACONIS MARCH MILITIA

Assigned to one of the most important, and most contested, regions of space, the Addicks DMM is one of the better-equipped Militias in the AFFC. With a number of planets in the so-called Terran Corridor to defend, the Militia is also one of the largest. Still bereft of any JumpShips, the unit could do little to help when the Marik-Liao Offensive crushed the Sarna March. The unit maintained control of its own PDZ, however, yet it technically lost three worlds to the Chaos March—each of which maintains close relations with the Federated Suns. Since the conclusion of the AFFC's latest military endeavors, the men and women of the Militia have requested numerous times to take up station on Caph and Small World again. Even with the Sixth F-C RCT stationed on Addicks, the High Command has refused each time.

OFFICERS

Leftenant General Nancy Bannson is a patriot and a pragmatist. Directing the PDZ's secondary headquarters on Addicks, as Marshal Terlecki commands the region from his HQ on Deneb Kaitos, Bannson is privy to reports and projections that paint a dismal future for the Federated Suns. Unconvinced that either Archon Katherine or Prince Victor could pull the nation back together, she has secretly been preparing her unit to reassemble on Addicks by pre-positioning a number of commercial and military JumpShips at strategic locations in the region.

TACTICS

Normally arrayed throughout the PDZ, the Militia commonly fights small-unit defensive and guerrilla campaigns, slowing an enemy force long enough to allow reinforcements to arrive.

Addicks Draconis March Militia
Regiment/Regular/Reliable
CO: Leftenant General Rupert Pesch

Addicks DMM Infantry Brigade
5 Regiments/Regular/Reliable
CO: Leftenant General Donal Pauolos

Addicks DMM Armor Brigade
3 Regiments/Regular/Reliable
CO: Leftenant General Nancy Bannson

Addicks DMM Aerospace Wing
Wing/Green/Reliable
CO: Kommandant Durina Soto

BREMOND DRACONIS MARCH MILITIA

Once regarded as a problem unit, the Bremond DMM rose to prominence under the command of Mary Tallman (now the marshal commanding the Bremond PDZ). Tallman rooted out corruption and lawlessness in the unit and instilled a level of professionalism and competence that would do justice to an elite line unit, let alone a March Militia. The militia's abilities stood it in good stead when it was deployed to Galtor in 3025, taking part in the last great battle of the Third Succession War.

OFFICERS

Syraman Simpreeni was hand-picked by Tallman to succeed her as head of the DMM when Tallman finally accepted a promotion in 3042. Sympreeni shares many of her characteristics—charisma, drive and intelligence—and has worked hard to maintain the unit's high standards. Unlike many senior officers in the Draconis March, the general is not consumed by hatred of the Draconis Combine, though he doesn't believe the peace with the Kurita family can continue indefinitely.

TACTICS

Like all March Militias, the Bremond DMM operates best in a defensive posture. However, their experience on Galtor demonstrated that militia units need not restrict themselves to such operations. Thanks to the preparation of Tallman and Simpreeni, the Bremond DMM is classed as a quick-response force and assigned its own JumpShips and DropShips.

Bremond Draconis March Militia
Regiment/Regular/Reliable
CO: Leftenant General Syraman Simpreeni

Bremond DMM Infantry Brigade
5 Regiments/Regular/Reliable
CO: Leftenant General Ruby Chan

Bremond DMM Armor Brigade
2 Regiments/Regular/Reliable
CO: Leftenant General Markus Cleaver

Bremond DMM Air Wing
Reinforced Wing/Regular/Reliable
CO: Leftenant General Simpson Mann

BRYCELAND DRACONIS MARCH MILITIA

The Bryceland DMM has a reputation for pragmatism, subsuming politics and individual ambition to the needs of defending their PDZ against the Draconis Combine and the Outworlds Alliance. Though the militia remains concentrated on its homeworld, it has established a corps of observers on every world under its protection, even those, like Pascagoula, that are more than a jump inside the Commonwealth. When combined with extensive knowledge of conventional and pirate jump points for each system, this intelligence network allows the DMM to respond quickly and efficiently to any incursion.

OFFICERS

Oci Begurnson has commanded the Bryceland DMM for almost two decades. In his opinion, the strangest—and most dangerous—development in that time is the emaciation of the defenses along the border with the Draconis Combine. Though he regards the troops under his command as highly skilled and motivated, he does not believe the AFFC is justified in leaving the defense of an entire front-line PDZ in their hands.

TACTICS

The Bryceland DMM specializes in small-unit operations, often deploying single companies or battalions to deal with threats to the region's security. They use a decentralized command structure that affords considerable flexibility, with task forces assembling around designated command companies.

Bryceland Draconis March Militia
Regiment/Regular/Reliable
CO: Leftenant General Oci Begurnson

Bryceland DMM Infantry Brigade
4 Regiments/Regular/Reliable
CO: Leftenant General Cherry Petty

Bryceland DMM Armor Brigade
2 Regiments/Regular/Reliable
CO: Kommandant Nial Dune

Bryceland DMM Wing
Wing/Regular/Reliable
CO: Leftenant General Alexandra Palais

CLOVIS DRACONIS MARCH MILITIA

The Kentares PDZ has historically been a hotbed of anti-Combine sentiments. Most families can recount incidents of terror and brutality visited upon them by members of the DCMS, while those few that can't have only to look at the masses of unmarked gravestones and monuments on each world. While the members of the Clovis DMM have had numerous opportunities to repay the Combine throughout the years, they are far from sated. Only the deliberate reassignment of their JumpShip squadron long ago and the heavy-handed direction of their commanders have kept them from reigniting war along the Draconis front.

OFFICERS

Just six months ago, Leftenant General Brian Bruning was found brutally murdered, the victim of an as-yet-unsolved crime. Leftenant General Jamil Ahmad was promoted in his place and immediately began to clean up his new command. Already, seventeen officers have been indicted for various violations of the AFFC Code of Conduct, while scores more have been discharged from service. Additionally, Ahmad brought four officers into the unit with him, each of whom has been tasked with enforcing discipline. To date, Ahmad has received eleven anonymous death threats and numerous other threats of violence.

TACTICS

Since the breakdown of discipline over the past year, the unit has lost most of its combat edge, though Ahmad's reformations are slowly making progress.

Clovis Draconis March Militia
Regiment/Green/Reliable
CO: Leftenant General Josephus Romil

Clovis DMM Infantry Brigade
3 Regiments/Regular/Reliable
CO: Leftenant General Colby Springsteen

Clovis DMM Armor Brigade
2 Regiments/Green/Reliable
CO: Leftenant General Beryl Richards

Clovis DMM Aerospace Wing
Wing/Green/Reliable
CO: Kommandant Lexi Roth

DAHAR DRACONIS MARCH MILITIA

Harsh and deadly, the world of Dahar IV creates similar characteristics in the native population, which has a reputation as excellent if undisciplined fighters much sought after as infantry and MechWarriors. Together with the Daharian people's desire to escape the rigors of their homeworld, this results in their composing almost 80 percent of the Dahar DMM. The Galtor Campaign destroyed the Dahar DMM as a fighting force, and they have struggled to recover.

OFFICERS

Leftenant General Brian De Waters began his career with the Dahar DMM, joining the unit as an aerospace pilot shortly before the unit's heroic but ultimately futile defense of New Derry on Galtor. He helped re-form the DMM in the late 3020s but eventually transferred out of the unit to fight with the Crucis Lancers in the Fourth Succession War. His return to the unit as head of the aerospace wing in 3040 was seen as a homecoming by many Daharis, and his popularity guaranteed his promotion to head the DMM in 3046.

TACTICS

Inspired by the sacrifice of their predecessors, the Dahar DMM spends considerable effort on perfecting the difficult art of city fighting. However, this has degraded the militia's abilities in other areas, leaving it particularly vulnerable in open terrain.

Dahar Draconis March Militia
Regiment/Regular/Reliable
CO: Leftenant General Brian De Waters

Dahar DMM Aerospace Wing
Wing/Veteran/Reliable
CO: Leftenant General Jacoba Witt

Dahar Infantry Brigade
8 Regiments/Regular/Reliable
CO: Leftenant General Jason Greening

AFFC MARCH MILITIAS

KILBOURNE DRACONIS MARCH MILITIA

Responsible for the Draconis March's border with the wild Periphery, much of the Kilbourne DMM's efforts are focused on countering the activities of pirates and smugglers. Their unceasing war against the pirates has earned them considerable respect among the local population but has also resulted in the highest casualty rate in any DMM unit. Several notable failures have resulted in claims that some Kilbourne DMM teams are working in conjunction with the raiders, prompting a succession of internal investigations. No solid evidence of wrongdoing has been uncovered, but the cloud of suspicion has caused the unit's morale to plummet.

OFFICERS

Mariva Kelly served with the Davion Assault Guards for more than a decade before her attempts to manipulate the political situation on Frazer led to her being "promoted" to head the Kilbourne DMM. The leftenant general regards her de facto exile with stoicism and has worked hard to turn the PDZ into her personal fiefdom, subverting several planetary dukes and PDZ Marshal Lisa Talrude. This influence has allowed Kelly to deflect the AFFC investigation, and she hopes to exploit the chaos that seems set to engulf the Federated Commonwealth.

TACTICS

The Kilbourne DMM's focus on anti-pirate operations means they excel in small-unit operations but have little experience fighting large-scale engagements. Indeed, exercise results suggest that deployments over battalion size may be counterproductive, with poor coordination in the DMM working in their opponents' favor.

Kilbourne March Militia
Regiment/Green/Reliable
CO: Leftenant General Mariva Kelly

Kilbourne DMM Infantry Brigade
5 Regiments/Green/Reliable
CO: Leftenant General Andrea James

Kilbourne DMM Armor Brigade
2 Regiments/Regular/Reliable
CO: Leftenant General Kolino Black

Kilbourne DMM Aerospace
2 Wings/Regular/Questionable
CO: Leftenant General Cerice May

MAYETTA DRACONIS MARCH MILITIA

Far from any hostile borders, the Mayetta DMM has become lazy and complacent. The unit rarely takes part in exercises and is considered borderline operational by the AFFC. Corruption is prevalentbut despite attempts by the AFFC to "clean up" the Mayetta DMM, support for the Yalos family is too strong. The AFFC has considered disbanding the Militia but believe doing so would result in major casualties in the Mayetta DMM and any unit sent to enforce the order.

OFFICERS

While Prince Victor was off fighting the Smoke Jaguars, Leftenant General Kirk Yalos set about establishing his own fief in the Draconis March. He engineered his son's rapid promotion to command the militia, allowing him to retain influence in the unit while concentrating on political and economic matters.

Jason Yalos is an ineffectual leader, promoted far beyond his abilities and heavily reliant on his father and close associates. Though he does not see it, many of his "friends" are merely using him to advance their own careers.

TACTICS

Only the most generous observer would claim the Mayetta DMM is proficient in any form of combat tactics. Indeed, the near lack of authority in the unit and the self-importance of many sub-unit commanders mean any attempt to deploy the unit en masse is unlikely to succeed and could trigger a miniature civil war in the DMM.

Mayetta March Militia
Regiment/Green/Reliable
CO: Leftenant General Jason Yalos

Mayetta DMM Infantry Brigade
6 Regiments/Green/Questionable
CO: Leftenant General Eithene Caleb

Mayetta DMM Armor Regiment
Regiment/Green/Questionable
CO: Leftenant General Anders Holbein

Mayetta DMM Aerospace Wing
Wing/Green/Questionable
CO: Leftenant General John Santos

MILLIGAN DRACONIS MARCH MILITIA

Leftenant General Juan Nishioka ensured the Milligan DMM was well-trained, building a formidable reputation in spite of the unit's green status. However, Nishioka was also a staunch supporter of Prince Victor, and following the ascension of Katherine Steiner-Davion, forces loyal to the Archon engineered his removal by promoting him to deputy PDZ commander. He was replaced by a political appointee, Jill Farlon, whose lackluster command abilities have set the DMM on a downward spiral.

OFFICERS

Leftenant General Jill Farlon is the daughter of the Duke of Jesup. Her promotion to head the DMM is one of many changes in the AFFC instigated by the Archon-Princess to weaken her brother and strengthen her own position. Though charismatic, Farlon has little command experience and has made a series of poor decisions that have wasted the few resources assigned to the unit. The shortage of parts and ammunition seriously hampers the RCT's abilities, and it is estimated that only 40 percent of the unit would be able to mobilize should a crisis strike the region.

TACTICS

The DMM favors mobile war, using the unit's speed and flexibility to counter enemy actions. However, the AFFC judges them too reliant on the intelligence-gathering and ground-attack abilities of their fighter wing. Indeed, in the unit's 3060 summer exercises, the DMM was performing admirably until the midpoint of the contest, when the umpires ruled the DMM air wing destroyed on the ground by sabotage. Without this support, the rest of the DMM position quickly collapsed.

Milligan March Militia
Regiment/Green/Reliable
CO: Leftenant General Jill Farlon

Milligan DMM Infantry Brigade
5 Regiments/Green/Reliable
CO: Leftenant General Jessica Corrino

Milligan DMM Armor Brigade
2 Regiments/Green/Reliable
CO: Leftenant General Judge Kynes

Milligan DMM Aerospace Wing
Wing/Regular/Reliable
CO: Leftenant General Paul Jereboa

RAMAN DRACONIS MARCH MILITIA

Having lost many of the worlds in their care to the Draconis Combine over the past several decades, there is no love lost between the Raman DMM and House Kurita. The militia unit has long opposed the peace between the Federated Commonwealth and their historical enemies. With the Combine military struggling with the threat to its coreward border, General Mazner has advocated a preemptive strike to recover worlds such as Galtor and Klathandu, lost to the Combine in the Fourth Succession War.

OFFICERS

In most circumstances, Renée Mazner is a calm and conscientious officer, a credit to the AFFC. On closer inspection, the leftenant general is a driven woman. Her one goal, which has shaped her existence since she was a young girl on Harrow's Sun, is to bring the Draconis Combine to its knees. Mazner took command of the DMM in 3049 and within hours of the Battle of Tukayyid advocated an assault on the Combine, the so-called nightmare scenario. Her proposal was rejected, but she keeps a close eye on the situation and believes there will never be a better time than now to strike at her old enemy.

TACTICS

Their many clashes with the DCMS have allowed the Raman DMM to establish a substantial cache of Kuritan equipment. Several companies have drilled in Combine tactics, and Mazner plans to use them—equipped with the captured gear—to infiltrate Combine positions and destroy them from within.

Raman Draconis March Militia
Regiment/Green/Reliable
CO: Leftenant General Renée Mazner

Raman DMM Armor Brigade
2 Regiments/Green/Reliable
CO: Leftenant General Jack Celliers

Raman DMM Infantry Brigade
5 Regiments/Green/Reliable
CO: Leftenant General Regan MacElhone

Raman DMM Aerospace Wing
Wing/Green/Reliable
CO: Leftenant General N'Bushe Sharpe

ROBINSON DRACONIS MARCH MILITIA

Few AFFC units have faced the internal conflicts the Robinson DMM is currently experiencing. In this age of resurgent regional loyalties and the changing face of the AFFC's highest command levels, the members of the Militia are finding themselves divided among the different camps. Though the majority of the senior officers in the Militia owe quite a bit of personal loyalty to Field Marshal Sandoval, most of the enlisted personnel and junior officers do not have the same respect for their duke as other units, like the Robinson Rangers. Likewise, many feel that a change in the leadership of the Federated Suns is a good thing, though neither Prince Victor nor Katherine appeals to the majority of the RCT's personnel.

OFFICERS

Passed over for promotion year after year, Leftenant General Jennifer Durrett has seemingly lost interest in her unit and her career in the AFFC. Instead, she has built quite a following as a critical political commentator on Le Blanc, writing a column for the local news service and frequently appearing on the weekly newsholo *The MacCullough Forum*. Though the AFFC does not condone her "hobbies," the fact that her latest book reached No. 15 on the New Avalon press prevents the pro-Katherine officers within the High Command from taking any official action against her.

TACTICS

The Robinson DMM is a capable defensive force, but it does not function well when forced to operate with most other AFFC units.

Robinson Draconis March Militia
Regiment/Green/Reliable
CO: Leftenant General Bryce LaGrange

Robinson DMM Armor Brigade
2 Regiments/Green/Reliable
CO: Leftenant General Saburo Hiroshi

Robinson DMM Infantry Brigade
3 Regiments/Regular/Reliable
CO: Leftenant General Odelia Farashue

Robinson DMM Aerospace Wing
Wing/Green/Reliable
CO: Leftenant General Timothy William Horus

ALBION MILITARY ACADEMY CADRE

The Albion Military Academy may have been supplanted as the AFFC's favored academy, but it continues to graduate class after class of outstanding cadets, just as it has for the past six centuries. Where the NAIS College of Military Sciences and its cadets rely on the high-tech simulators and brand-new BattleMechs assigned to them by the High Command, the members of the Albion Cadres are taught to rely on their own skills and on their teammates. While the constant pressure on the cadets to perform creates a highly stressful environment within the academy, the Cadres' staff is there solely to help its cadets reach their full potential, resulting in a far more relaxed air within the Cadres.

Where other academies rely heavily on veteran officers to command their training units, the Albion Cadres use their veteran combat officers more as advisors than as commanders, letting their cadets learn firsthand the trials of command. If the Cadres become involved in any actual combat situation, however, the advisors step in to ensure the safety of their charges.

COMMANDING OFFICER

As Albion's Commandant, Major General Frederick Hastings has total responsibility for the academy and its two Cadre units. A decorated veteran of the Fourth Succession War and the War of 3039, as well as other still-classified operations, Hastings is also a celebrated military historian who has authored quite a few texts in use by academies throughout the Inner Sphere. Since his posting as Commandant, he has hand-picked each new officer assigned to the academy or its Cadres. Though his critics say he is stacking the academy with the most pro-Davion officers he can lay his hands on, few can condemn the fact that his staff rates the highest of any AFFC academy.

FORCES

The Albion Cadres are not given the same preference as the CMS training units, but they nonetheless field relatively new combat equipment. Both Cadre units display an almost unnatural sense of teamwork, quite surprising for an academy training unit and nearly unheard of in the Regular Army. The academy focuses heavily on the 'Mech, cavalry and aerospace combat specialties, leaving little more than a perfunctory infantry detachment assigned to each Cadre.

COLORS AND INSIGNIA

The colors of the Albion Military Academy are white, blue and gold, and unless assigned to field operations, all Cadre equipment is painted in those colors. The academy's insignia is a green "A" set atop a stylized sunburst emblazoned upon a blue disc, while the insignia for the Cadres is a white and green diamond upon a black shield. Each Cadre adds its numeric designation below its crest.

ALBION MILITARY ACADEMY CADRE

1ST ALBION TRAINING CADRE

Competition for a position in the First Cadre is fierce. Though fielding primarily light- and medium-class vehicles, the First's advisor/trainers see their duty as preparing their charges for the unexpected. So in addition to a strenuous mental and physical regimen, the First's cadets are expected to complete vehicular and 'Mech obstacle courses that would make veteran officers sweat. By the time a cadet graduates from the First Cadre, he or she is a skilled 'Mech or vehicle pilot and a capable and confident officer.

OFFICERS

Because he has overall responsibility for the First Cadre, Leftenant General Derrick Gray believes he also has overall responsibility for its graduates. He developed the basis for the First's unique training schedule soon after accepting command of the Cadre and has been modifying the plan ever since to better train his cadets and to adjust to the ever-changing face of the Federated Suns.

TACTICS

The First Cadre often keeps an enemy off-balance by using its mobility to suddenly appear from an unexpected quarter.

1st Albion Cadet BattleMech Regiment
Regiment/Green/Fanatical
CO: Leftenant General Derrick Gray
Aide: Colonel Natalia Lukhoi

1st Albion Cadet Wing
Wing/Green/Fanatical
CO: Major Arimus Kass

1st Albion Cadet Combat Auxiliary
Regiment/Green/Fanatical
CO: Colonel Pablo Lee

2ND ALBION TRAINING CADRE

The Second Cadre is Albion's heavier cadet unit and is trained in a completely different subset of skills than the First. With graduates of this Cadre assigned to positions throughout the AFFC, the Second's advisors must thoroughly train their charges in a multitude of combat styles and tactics. Offensive actions, especially combat drops, are heavily emphasized, and cadets can expect to participate in at least four major simulated offensives each year, often lasting weeks at a time. At the same time, defensive operations are not ignored; the Cadre is commonly tasked to help build or repair dams and other major civilian architectural structures, giving the cadets a chance to practice constructing fortifications and field works.

OFFICERS

Leftenant General Giovanni Leonard took command of the Second Cadre nine months ago, succeeding Leftenant General Vincent Rasmussen, who finally retired after thirty years of service. Himself a relative upstart, Leonard is a bold strategist and yet an extremely patient officer, qualities that Major General Hastings could not afford to pass up. Leonard has not quite settled into his new position, however, and is taking far more interest in the day-to-day operations of the unit than did Rasmussen.

TACTICS

The Second Cadre may not have the same speed as the First, but its cadets push their machines to their absolute maximum, relying on mobility almost as much as sheer firepower.

2nd Albion Cadet BattleMech Regiment
Regiment/Green/Fanatical
CO: Leftenant General Giovanni Leonard
Aide: Colonel Lucy Rich

2nd Albion Cadet Wing
Wing/Green/Fanatical
CO: Major Soon-Min Oh

2nd Albion Cadet Combat Auxiliary
Regiment/Green/Fanatical
CO: Colonel Brendan Liu

COLLEGE OF MARTIAL SCIENCES CADET CADRE

The NAIS College of Martial Sciences is, without a doubt, the premier military academy in the Inner Sphere, and likely in known space. It also graduates the largest classes of any academy, with some 1,200 new members entering the AFFC every year. While the top cadets in each combat specialty are assigned directly to an AFFC Regular Army unit following graduation, the rest find themselves transferred to one of the CMS' three Cadres.

While CMS cadets make up some 90 percent of these Cadres, every command slot, from company commander on up, is staffed with a veteran AFFC officer. Where the CMS is tasked with providing its cadets with the best education available, the Cadres take those new graduates and give them the training that will keep them alive on the battlefield. Since each cadet serves a single two-year tour with the Cadres, each Cadre unit experiences a nearly complete turnover every two years.

COMMANDING OFFICER

As Commandant of the College of Military Sciences, General Ranier Wolfgram commands the academy's three Cadet Cadres as well. A former armored brigade commander in the Royal Brigade, Wolfgram is no stranger to combat, having served in every major action since the Fourth Succession War. He has assembled extremely experienced and capable instructors and staff, giving the cadets their best chance of surviving their first combat and learning something from it.

FORCES

Though only a training organization, the Cadres are assigned some of the best and newest equipment in the AFFC and are often treated as a Regular Army unit in the assignments they receive.

COLORS AND INSIGNIA

For ceremonial purposes, the Cadres paint their equipment in the standard Davion green. The insignia for the CMS Cadres is two diamonds, one white and one black, set upon a green shield with the words "College of Military Sciences" emblazoned above it. Each Cadre places its numeric designation below the insignia.

1ST NAIS CADET CADRE: WARRIORS OF VIRTUE

The First Cadre is the CMS' most specialized cadet unit. Consisting solely of light-class BattleMechs and vehicles, the First is far from the most prominent and publicized Cadre unit. Nevertheless, roughly half of the cadets in the unit specifically requested the assignment.

The First Cadre is one of the few light units in the Regular Army and thus is often called upon to assist in battlefield reconnaissance and scouting missions. Because of its light composition, the First rarely finds itself in anything but a support situation and is in fact under standing orders to withdraw in the face of any significant threat. The unit's record speaks for itself: to date, the First has lost more cadets to training accidents than it has to actual combat.

OFFICERS

A long-time academy staff member, Leftenant General Jonathan Sanchez has used his time at the CMS to advance his own education. Since 3044, he has earned two doctorates in history and published more than a dozen works on historical and tactical topics.

TACTICS

When pressed into combat, the Cadre will attempt to isolate enemy elements and then encircle and destroy them.

1st NAIS Cadet BattleMech Regiment
Regiment/Green/Fanatical
CO: Leftenant General Jonathan Sanchez
Aide: Colonel William "Wild Willie" Vermillion

1st NAIS Cadet Cavalry Regiment
Regiment/Green/Fanatical
CO: Colonel Tessa Lee-Davies

1st NAIS Cadet Armored Infantry Battalion
Battalion/Green/Fanatical
CO: Major Mel Goldsmith

1st NAIS Cadet Wing
Wing/Green/Fanatical
CO: Major Romil Sheth

ALBION MILITARY ACADEMY CADRE

2ND NAIS CADET CADRE: WARRIORS OF SKILL

Consisting primarily of fast medium-class 'Mechs and vehicles, the Second Cadre often receives assignments reserved for cavalry units. While its composition dictates the types of combat missions it can accomplish, the Second's staff nevertheless drills its cadets in the tactics and procedures used by every type of AFFC combat force, preparing them for service in any unit.

Assigned to Kathil, the Second has been given a number of opportunities to train in equipment on loan from GM. They have also been unwillingly embroiled in the political battle between Duke George Hasek and the Eighth F-C's General Weintraub.

OFFICERS

Though many consider her career over as a result, Leftenant General Helen Sanderson has willingly stayed with the Second Cadre for some fifteen years, training the best and brightest the Federated Commonwealth has to offer. Some of her students have since surpassed her in rank, but few have ever shown anything but the highest regard for her. Her pro-Victor stance, however, has earned her more than a few enemies in the AFFC.

TACTICS

Though the unit practices tactics emulating a number of different combat units, it specializes in those of cavalry units. By using its highly mobile elements, the Second can often isolate and surround an enemy force or simply escape to fight another day.

2nd NAIS Cadet BattleMech Regiment
Regiment/Green/Fanatical
CO: Leftenant General Helen Sanderson
Aide: Colonel Thurston von Thule

2nd NAIS Cadet Cavalry Regiment
Regiment/Green/Fanatical
CO: Colonel Richard Scherzberg

2nd NAIS Cadet Armored Infantry Battalion
Battalion/Green/Fanatical
CO: Major Jane Ariel Tohm

2nd NAIS Cadet Wing
Wing/Green/Fanatical
CO: Major Fritz D'Arby

3RD NAIS CADET CADRE: WARRIORS OF HONOR

Many CMS cadets dream of assignment to the Third Cadre when they enter the academy. Indeed, the Cadre is the most potent training unit in the AFFC, and cadets who successfully complete their assignment in the Cadre are often rewarded with a billet in one of the AFFC's premier units. As a result, the atmosphere within the Cadre is competitive and tense.

Further adding to the tension is the nature of its assignment. Stationed along the Capellan border, the Cadre is just a few short jumps away from the ongoing war in the St. Ives Compact, a conflict that could easily engulf the entire Capellan March. The cadet corps is also well aware of the polarization gripping the AFFC.

OFFICERS

Leftenant General Carlos Post twice turned down a posting that could have led to his eventual command of the Davion Assault Guards to remain with the CMS Cadre. He alone kept the Cadre together when a simple raid on a Steel Viper world in 3054 nearly turned into a massacre following a fumbled drop. Post is also an Olympic-class swordsman and holds the honorary title of Crown Champion of the Sabre, a position once granted to the Prince's most skilled and trusted bodyguard.

TACTICS

The members of the Third Cadre use its withering long-range firepower to cut down opponents.

3rd NAIS Cadet BattleMech Regiment
Regiment/Green/Fanatical
CO: Leftenant General Carlos Post
Aide: Leftenant General Rico Seville

3rd NAIS Cadet Cavalry Regiment
Regiment/Green/Fanatical
CO: Leftenant General Judith Prehn

3rd NAIS Cadet Armored Infantry Battalion
Battalion/Green/Fanatical
CO: Kommandant Laban Devers

3rd NAIS Cadet Wing
Wing/Green/Fanatical
CO: Leftenant General Hohiro Kira

TRAINING BATTALIONS

1ST BROCKTON TRAINING BATTALION

Drawing its recruits from the Broken Wheel and Anjin Muerto Combat Regions, the Brockton Training Battalion contains a diverse mix of cultures and personalities. Some, like the anarchists of Cogdell, do not fit in well with the military lifestyle and are expelled within the first few days. Others, like the aquafarmers of Lackland, are keen to learn, knowing that the skills they gain can be applied to the subaquatic agromechs on their homeworld. Weeding out the wasters from the potential gems is a difficult process, carried out by the unit's sergeant-instructors during the initial weeklong evaluation period. Of those who graduate to the battalion proper—roughly half of those who attend the evaluation—the pass rate is 65 percent, the highest in the regional cadres.

OFFICERS

Leslie Stokoi regards harsh discipline as the only way of ensuring those who join the battalion fulfill their potential. Though rarely involving violence, her methods have drawn criticism from several quarters as being brutal and inhumane. However, the kommandant refuses to compromise, stating that her results speak for themselves. So far the AFFC has agreed, though the Department of Military Education keeps the battalion under close watch.

1st Brockton Training Battalion
Battalion/Green/Reliable
CO: Kommandant Leslie Stokoi

All the 'Mechs used by the Brockton Battalion are rigged with command overrides, allowing the training officers to simulate a wide range of battle damage and system failures. The sergeants take a perverse pleasure in triggering such malfunctions at the worst possible time for the pilot and observing their attempts to deal with the problem.

1ST BELL TRAINING BATTALION

Though operating from permanent bases on Bell and Axton for the past several decades, the battalion spends two-thirds of each year away from its permanent facilities, training instead on the homeworlds of its members. The unit is forced to rely on commercial transport for its journeys, but that cost is more than offset by the additional experience afforded its trainees, both in operating on unfamiliar worlds and in dealing with the stresses of interstellar transit.

With its trainees hailing from the worlds of the Valexa PDZ, the atmosphere in the battalion has become quite tense over the past several years. Suddenly sitting on the border of a reinvigorated Capellan Confederation, the battalion's members are naturally concerned with the lack of response from their leaders. While that attitude will not get a graduate far in most of the AFFC, it is prevalent in the Capellan March, ensuring that successful graduates of the Bell Training Battalion will find positions in the Federated Suns military.

OFFICERS

Field Marshal George Hasek appointed Leftenant Colonel Milko Pochtarev to lead the battalion a year ago, following the death of its former commander, Paula Quarnry. Kommandant Quarnry had allowed discipline to lapse in the last years of her life, and many of the battalion's recent graduates have either been dismissed from service or transferred into a new specialty. Just months after taking command, Pochtarev dismissed a third of the battalion. After finally getting the unit under control, Pochtarev is expecting to graduate his first group of cadets soon.

1st Bell Training Battalion
Battalion/Green/Reliable
CO: Leftenant Colonel Milko Pochtarev

The battalion is officially operating at full strength, though a number of 'Mechs still exhibit the mechanical problems that plagued the unit prior to Pochtarev's appointment.

ALBION MILITARY ACADEMY CADRE

1ST CONROE TRAINING BATTALION

Created in 3024, the Conroe Training Battalion was the first of the regional training units created on the orders of Hanse Davion. Its first graduates vindicated the concept, proving that competent pilots could be created without extensive (and expensive) schooling, albeit without much of the ancillary education (such as tactics and technical matters) normally evident in academy graduates.

The recruits for the Conroe Battalion come from worlds in the Bryceland PDZ, so there is a strong anti-Combine bias. Most in the current class are disappointed that, rather than immediately jumping the border to "punish the Drac scum," they spend several weeks in a boot camp instilling basic discipline before being allowed near a 'Mech.

OFFICERS

Herself a product of the Conroe Training Battalion shortly before the Fourth Succession War, Leona Peterson has worked hard to overcome the prejudice that surrounds the regional training units. Having received the Robinson Medal of Valor for her courage during the doomed defense of Niles in the Fourth Succession War, and also because of her subsequent rise to command the Conroe Battalion, Peterson is often held up as an example of the system's effectiveness.

1st Conroe Training Battalion
Battalion/Green/Reliable
CO: Kommandant Leona Peterson

The equipment used by the Conroe Training Battalion is old and decrepit but is kept in working order by a team of dedicated techs. The appearance of the machines is not helped by the steady accumulation of graffiti inside their cockpits, added by successive generations of recruits.

FILTVELT ACADEMY TRAINING BATTALION

Most AFFC officers lump the graduates of the Filtvelt Academy in with those who find their way out of the Conroe Training Battalion. While both training programs recruit primarily from the same worlds, members of the Filtvelt Battalion are still graduates of one of the AFFC's academies and are professional military personnel.

Nevertheless, the academy's cadets and trainees feel like second-class citizens. Even the arrival of Major General Cluey Montserrat and his hand-picked staff at the academy has done little to boost Filtvelt's reputation. At least the academy has received some additional funding and support since Archon Katherine's rise to power. Most of the academy's personnel are looking to the Archon in hopes of finally eliminating the prejudices against them.

OFFICERS

Kommandant Asumi Tomisawa followed Major General Montserrat to the academy from their prior assignment with the Second Robinson Rangers. Here she holds much the same position as she did with the Second Rangers, though most of the trainees she gets are far less opinionated than those she encountered in the Rangers. Like Montserrat, she often speaks her mind plainly and publicly.

Filtvelt Academy Training Battalion
Reinforced Battalion/Green/Reliable
CO: Kommandant Asumi Tomisawa

Stationed as it is in the "Outback," the battalion is often called on to deal with pirate bands. The unit's five companies of BattleMechs are often required to operate independently for weeks or even months at a time. A mixed regiment of militia infantry and armor is also available to assist the battalion on these assignments, as are the academy's three DropShips, two *Unions* and an ancient *Lion.*

GOSHEN WAR COLLEGE TRAINING BATTALION

While the cadets who graduate the Goshen War College are among the most skilled in the AFFC, many feel they could benefit from additional training before moving on to a Regular Army unit. The Goshen War College Training Battalion exists to fill that need. Assignment to the battalion is not compulsory; cadets graduating from the academy must request the assignment. Even then, the transfer is made only after the cadet meets with his advisor and both agree that assignment to the battalion would be more beneficial than moving on to a Regular Army unit.

Because of its unique nature, the tone in the battalion is far more relaxed and friendly than in any other training unit, approaching that of a unit that has fought together for decades. Advisor/trainers take a personal interest where needed, but by and large they stay out of the cadets' way, letting them hone their own fine edge.

OFFICERS

Leftenant General Archibald Enoch is the college's Commandant of Cadets and the Director of Training, and by extension the operational commander of the training unit. Himself a graduate of Goshen, he returned to the academy almost two decades ago to assist with restructuring the College's training program under the newly formed AFFC. He remained on when offered the Commandant position and has raised his six children on Goshen, the two oldest of which just graduated from the College.

Goshen War College Training Battalion
Reinforced Battalion/Green/Reliable
CO: Leftenant General Archibald Enoch

Unlike any other training battalion in the AFFC, the structure of this unit changes with each class of cadets. The college's patrons have been generous enough that the unit can field nearly two battalions of 'Mechs, plus an additional two wings of fighters, if necessary. The unit currently consists of twenty-three MechWarrior and five aero pilot trainees, along with an additional eight advisor/trainers.

KILBOURNE ACADEMY TRAINING BATTALION

The Kilbourne Academy came under initial DMI investigation in late 3059, apparently after several graduates serving within the Draconis March were indicted on charges of sedition and conspiracy. The investigation expanded when several more graduates, both military and civilian, were found to be key members of anti-Davion rebel cells scattered throughout the Federated Suns Outback—the region of space bordering the largely uncolonized Periphery. Though a handful of academy personnel were later indicted on unrelated charges, the investigation into the academy continues today, creating increased hostility toward the Davion family and the AFFC High Command.

With the Kilbourne DMM headquartered on the same base, the battalion often trains against the militia, gaining much-needed experience in dealing with enemies that field a combined-arms force. Additionally, quite a few battalion cadets serve with the militia on its numerous anti-pirate missions every year. By the time cadets leave the battalion, they are as prepared as they could be, if somewhat jaded toward their High Command.

OFFICERS

During his years in the Lyran Command, Kommandant Sam Collin was a notorious critic of the AFFC military machine and eventually found himself promoted almost as far away from his homeworld of Tamar as possible. "Exiled" to the backwaters of Kilbourne, he almost gleefully transfers his distaste to his cadets, who mimic his attitudes but revile him for his near contempt.

Kilbourne Academy Training Battalion
Battalion/Green/Reliable
CO: Kommandant Sam Collin

With enrollment down some 30 percent since the beginning of the DMI investigation, the battalion began to take on locals who showed the potential to become MechWarriors. Nearly half of the unit's current personnel are raw trainees recruited from Kilbourne's farms and streets.

1ST KITTERY TRAINING BATTALION

Originally the most active training battalion in the AFFS, following the Fourth Succession War the Kittery unit was suddenly protected from the Capellan Confederation by the newly formed St. Ives Compact. With war raging just a few light-years away, however, the members of the Kittery Training Battalion are receiving the Confederation's lesson in interstellar politics loud and clear.

Because of its prime location in the Capellan March and the support of the Dukes of New Syrtis, this unit has always been the most prominent of the AFFC's training battalions. While the unit is highly visible, the extra attention paid to it earns graduates of the battalion no additional respect from academy graduates and even scorn from those who served in one of the Federated Suns' less prominent training units. The training is first-rate, however, with the staff of Kittery's Ft. Damien Hasek and the battalion itself receiving some of the DME's highest ratings.

OFFICERS

Major Raul Bethune is considered by many outside the battalion to be little more than a political appointee, but those who have served with him know that to be untrue. The husband of Duke Hasek's minister of finance, Bethune is not only a veteran combat officer, having served with the First FedSuns Armored Cavalry, but also a highly qualified educator. He runs an extremely tight ship but is respected by his students.

1st Kittery Training Battalion
Battalion/Green/Reliable
CO: Major Raul Bethune

The failure rate in the training program is the highest in the AFFC, but its dropout rate is the lowest, the result of an intensive but engaging cycle designed by Major Bethune. In a complete turnaround from the norm in the training battalions, almost all of the program's washouts go on to have highly successful careers in other specialties in the AFFC.

POINT BARROW ACADEMY TRAINING BATTALION

Once virtually blacklisted by the High Command because of ongoing terrorist activities linked to the academy and the battalion, with the assistance of Melissa Steiner-Davion, the Point Barrow Academy is once again a respected institution. With alumni filling command billets in units and offices throughout the AFFC, graduates of the academy and its training unit are often sought-after commodities.

Because of the support shown the academy by her mother, Archon Katherine has inherited the loyalty of the majority of the academy's staff and cadet corps. Though few are blind to the growing division in the AFFC, most are hopeful that the Archon can eventually calm the rising tempest the way her mother would have.

OFFICERS

Kommandant Laurie Empen succeeded Hauptmann General Severine Brint after she rose to command of the academy. With fifteen years of service under her belt, Empen is looking at command of the battalion as her final assignment, though retirement is still years away. She is hard on her cadets but is dedicated to making them the best they can be, even if it means "cracking some eggshells."

Point Barrow Academy Training Battalion
Battalion/Green/Reliable
CO: Kommandant Laurie Empen

As one of the first units to pledge her their loyalty, the Archon awarded the battalion a full company of *Enforcer IIIs* and *JagerMech IIIs*, giving the unit a much-needed equipment update. While most of the replaced 'Mechs have been sold as salvage, much of the unit's remaining complement is in the same condition.

ROBINSON BATTLE ACADEMY TRAINING BATTALION

With the academy located scant kilometers from the command center of the Draconis March and staffed by so many close associates and relatives of the Sandoval family, members of the Robinson Battle Academy's Training Battalion are under almost constant observation and evaluation. Historically, cadets who prove themselves worthy are well-rewarded by the Sandovals, while those who do poorly are shipped off for service in an Outback Militia or some other similar scut duty. At least, that is the myth that has been perpetuated over the years.

In reality, the battalion's cadets are evaluated only by their own training staff. Though Duke Sandoval often rewards the academy's best cadets with choice assignments, he relies on the evaluations of the academy's staff, not on a mythical network of spies. Nevertheless, the battalion's cadets are continually looking for opportunities to make an impression, leading to dozens of training accidents every year, including several major injuries and, unfortunately, the occasional fatality.

Further feeding the cadets' enthusiasm is their hereditary hatred of the Draconis Combine. With reenactments of the final battle for Robinson played out in a weeklong festival every year, the cadets are unlikely to forget the pain the Combine caused their ancestors.

OFFICERS

Kommandant Aki Jameson was an unlikely choice to head the battalion. A battalion commander in the First Robinson Rangers during that unit's deployment with Task Force Bulldog, Jameson was suspected of igniting several incidents with Combine personnel on Wolcott, and his overall performance was considered lackluster. Nevertheless he received a transfer into the academy's training battalion, where he has performed better than expected. He continues to feed the cadets' zeal, however, resulting in a more vigorous training cycle and an increase in training-related accidents.

Robinson Battle Academy Training Battalion
Battalion/Green/Reliable
CO: Kommandant Aki Jameson

The battalion commonly uses the academy's DropShips and JumpShip for transport to training facilities on nearby worlds.

SAKHARA ACADEMY TRAINING BATTALION

The Sakhara Academy is a privately run military academy and is widely considered the best private institution in the Federated Suns. Its training battalion retains a unique status in the AFFC. All of the unit's equipment is either privately owned or owned by the academy, and all of its members are either academy graduates who have attained the AFFC rank of subaltern or academy staff. All battalion staff members hold commissions in the AFFC, whether current or retired officers (on special reserve status), and as such allow the battalion to be retained on the AFFC's rolls.

With most of the cadets hailing from affluent families, the tone in the academy and its training unit has been slowly changing over the course of the past few years. Many families have profited greatly since the Archon's rise to power on New Avalon, leading to a natural feeling of loyalty to her in the academy's cadet corps. On the other hand, a majority of the staff has nothing but contempt for their new leader. Tensions in the battalion are reaching the boiling point as a result.

OFFICERS

Elkin Odds retired from active service as an AFFS leftenant colonel but was awarded the AFFC rank of Kommandant when he took charge of the battalion. Already in his mid-sixties, Odds vigorously maintains his health and fitness, setting an example for his trainees and staff, and is still a rated MechWarrior. He is also an exceedingly wealthy man, with a family fortune valued at several hundred million C-bills.

Sakhara Academy Training Battalion
Battalion/Green/Reliable
CO: Kommandant Elkin Odds

Nearly a third of the battalion's complement is made up of 'Mechs brought into the unit by cadets who want to train in their family BattleMech.

TIKONOV MARTIAL ACADEMY TRAINING GROUP

Once, the Tikonov Academy Training Group was considered one of the finest training units in the Federated Commonwealth. That was before the Marik-Liao Offensive of '57 and the subsequent dissolution of the Sarna March. Today, the training group suffers from a number of problems. With the rampant rise of terrorism in the Tikonov Reaches, the academy is finding itself understaffed. Because enrollment has dropped sharply, both the DME and the DQ have cut support to the academy and its training group to bare-bones levels.

Attitudes in the training group have also plummeted drastically. Though many of the group's remaining officers are doing their best to reinforce the concepts of duty, honor and loyalty to country, the cadets are simply no longer listening. The call for loyalty to Tikonov, supported by Chancellor Liao through his Zhanzeng de Guang organization, is apparently too much for most of the cadets to resist.

OFFICERS

Leftenant General Greg Murray is a native of New Hessen and understands the conflicting loyalties tormenting his cadets. He is dedicated to his job and to preparing his cadets for service to their nation—whatever that nation might be. With the continuing terrorism interfering with his own sense of duty, honor and loyalty, however, it is all he can do to suppress the unrest inside the academy and out.

Tikonov Martial Academy Training Group
Battalion/Green/Reliable
CO: Leftenant General Greg Murray

The battalion is currently operating at only 75 percent strength. While Murray and his staff are doing their best to keep the battalion together, they are expecting the next round of dropouts to bring the unit down past 50 percent, while the next academy class is not scheduled to provide any replenishments for almost ten months.

FEDERATED SUNS FORCES

(Deployment as of 12 November 3062)

BattleMech Strength: 86 Regiments, 2 Battalions

[Units in italics are currently stationed in the Lyran Alliance]

UNAFFILIATED AFFC UNITS

Unit Name	Exp	Loyalty	Faction	Homeworld
1st Aragon Borderers	Vet	Rel	Victor	New Aragon
1st Argyle Lancers	*Vet*	*Rel*	*Victor*	*Crimond*
1st Capellan Dragoons	Reg	Ques	Neutral	Kathil
1st Federated Suns Armored Cavalry	Elite	Fan	Victor	Ziliang
1st Kestrel Grenadiers	Elite	Fan	Victor	Zurich
1st Kittery Borderers	Reg	Rel	Neutral	Kittery

AVALON HUSSARS

CO: Marshal Roger William Waters

Unit Name	Exp	Loyalty	Faction	Homeworld
11th Avalon Hussars	Reg	Rel	Katherine	Brockway
17th Avalon Hussars	Reg	Rel	Neutral	Cassias
20th Avalon Hussars	Vet	Fan	Victor	Alcyone
22nd Avalon Hussars	Vet	Rel	Katherine	Bromhead
39th Avalon Hussars	*Reg*	*Rel*	*Victor*	*Newtown Square*
41st Avalon Hussars	Reg	Rel	Victor	Benet III
42nd Avalon Hussars	*Reg*	*Rel*	*Neutral*	*Poulsbo*

CHISHOLM'S RAIDERS

CO: Marshal Amelia Fitzwilliam

Unit Name	Exp	Loyalty	Faction	Homeworld
1st Chisholm's Raiders	Grn	Ques	Katherine	Breed
2nd Chisholm's Raiders	Vet	Rel	Katherine	Verlo

CETI HUSSARS

CO: Marshal Nebuchadnezzar Talles

Unit Name	Exp	Loyalty	Faction	Homeworld
1st Ceti Hussars	Reg	Fan	Victor	Pajarito
2nd Ceti Hussars	Vet	Rel	Neutral	Frazer
3rd Ceti Hussars	Reg	Rel	Neutral	Manapire

CRUCIS LANCERS

CO: Marshal Ivor Wasjinji

Unit Name	Exp	Loyalty	Faction	Homeworld
1st Crucis Lancers	Reg	Fan	Victor	Bastrop
2nd Crucis Lancers	*Reg*	*Rel*	*Victor*	*Timbiqui*
3rd Crucis Lancers	Ellite	Rel	Victor	Cassias
4th Crucis Lancers	*Vet*	*Rel*	*Neutral*	*Giausar*
5th Crucis Lancers	Vet	Fan	Victor	Demeter
6th Crucis Lancers	*Vet*	*Rel*	*Neutral*	*Langhorne*
7th Crucis Lancers	*Elite*	*Fan*	*Neutral*	*Winter*
8th Crucis Lancers	Reg	Rel	Neutral	Mara

DAVION BRIGADE OF GUARDS

CO: Marshal Bishop Sortek

Unit Name	Exp	Loyalty	Faction	Homeworld
Davion Assault Guards	Vet	Rel	Victor	Tigress
Davion Heavy Guards	Elite	Fan	Victor	Galax
Davion Light Guards	Vet	Rel	Neutral	Mendham
1st Davion Guards	Elite	Fan	Victor	New Avalon
2nd Davion Guards	Vet	Fan	Victor	Bromhead
3rd Davion Guards	Vet	Fan	Victor	Kesai IV
4th Davion Guards	*Elite*	*Fan*	*Neutral*	*Ft. Loudon*
5th Davion Guards	Reg	Fan	Victor	Alcyone

DENEB LIGHT CAVALRY

CO: Marshal Andrew Terlecki, Jr.

Unit Name	Exp	Loyalty	Faction	Homeworld
4th Deneb Light Cavalry	*Reg*	*Rel*	*Neutral*	*Dalkeith*
8th Deneb Light Cavalry	*Vet*	*Rel*	*Neutral*	*Kikuyu*
10th Deneb Light Cavalry	Vet	Rel	Katherine	New Avalon
12th Deneb Light Cavalry	Reg	Rel	Neutral	Tybalt
15th Deneb Light Cavalry	Reg	Rel	Katherine	Demeter

FEDERATED COMMONWEALTH CORPS

CO: Marshal Alistair Venger

Unit Name	Exp	Loyalty	Faction	Homeworld
1st FedCom RCT	Vet	Rel	Neutral	Nanking
3rd FedCom RCT	Reg	Rel	Katherine	New Aragon
5th FedCom RCT	Grn	Rel	Katherine	Chesterton
6th FedCom RCT	Reg	Rel	Katherine	Addicks
7th FedCom RCT	Reg	Rel	Neutral	Spica
8th FedCom RCT	Reg	Rel	Katherine	Kathil

NEW IVAARSEN CHASSEURS

CO: Hauptmann General Kristi Johnson

Unit Name	Exp	Loyalty	Faction	Homeworld
1st New Ivaarsen Chasseurs	Elite	Fan	Victor	New Ivaarsen
2nd New Ivaarsen Chasseurs	Reg	Fan	Victor	New Ivaarsen

ROBINSON RANGERS

CO: Field Marshal James Sandoval

Unit Name	Exp	Loyalty	Faction	Homeworld
1st Robinson Rangers	Vet	Fan	Neutral	Robinson
2nd Robinson Rangers	*Reg*	*Rel*	*Neutral*	*Phecda*
3rd Robinson Rangers	Reg	Rel	Katherine	New Avalon

SYRTIS FUSILIERS

CO: Field Marshal George Hasek

Unit Name	Exp	Loyalty	Faction	Homeworld
5th Syrtis Fusiliers	Grn	Rel	Neutral	Lee
6th Syrtis Fusiliers	Elite	Rel	Neutral	Hadnall
8th Syrtis Fusiliers	Reg	Rel	Neutral	Lothair

CAPELLAN MARCH MILITIA

CO: Marshal Christopher Tice

Unit Name	Exp	Loyalty	Faction	Homeworld
Alcyone Capellan March Militia	Reg	Rel	Katherine	Alcyone
Kathil Capellan March Militia	Grn	Rel	Neutral	Kathil
New Syrtis Capellan March Militia	Grn	Ques	Neutral	New Syrtis
Ridgebrook Capellan March Militia	Grn	Rel	Katherine	Ridgebrook
Sirdar Capellan March Militia	Reg	Rel	Katherine	Sirdar
Valexa Capellan March Militia	Reg	Rel	Victor	Valexa
Warren Capellan March Militia	Grn	Rel	Katherine	Warren
Achernar Sarna March Militia	Grn	Ques	Katherine	Achernar
Nanking Sarna March Militia	Grn	Ques	Katherine	Nanking

CRUCIS MARCH MILITIA

CO: Hauptmann General Maurice Konrad

Unit Name	Exp	Loyalty	Faction	Homeworld
Anjin Muerto Crucis March Militia	Reg	Rel	Katherine	Anjin Muerto
Islamabad Crucis March Militia	Grn	Rel	Katherine	Islamabad
Kestrel Crucis March Militia	Reg	Rel	Neutral	Kestrel
Malagrotta Crucis March Militia	Grn	Ques	Katherine	Malagrotta
Marlette Crucis March Militia	Reg	Rel	Katherine	Marlette
New Avalon Crucis March Militia	Reg	Rel	Katherine	New Avalon
Remagen Crucis March Militia	Reg	Rel	Katherine	Nunivak
Tsamma Crucis March Militia	Reg	Rel	Katherine	Tsamma

DRACONIS MARCH MILITIA

CO: Hauptmann General Katheryn Sandoval-Ito

Unit Name	Exp	Loyalty	Faction	Homeworld
Addicks Draconis March Militia	Reg	Rel	Neutral	Addicks
Bremond Draconis March Militia	Reg	Rel	Katherine	Bremond
Bryceland Draconis March Militia	Reg	Rel	Katherine	Bryceland
Clovis Draconis March Militia	Grn	Rel	Neutral	Kentares IV
Dahar Draconis March Militia	Reg	Rel	Katherine	Dahar IV
Kilbourne Draconis March Militia	Grn	Rel	Katherine	Kilbourne
Mayetta Draconis March Militia	Grn	Rel	Katherine	Mayetta
Milligan Draconis March Militia	Grn	Rel	Katherine	Milligan
Raman Draconis March Militia	Grn	Rel	Katherine	Raman
Robinson Draconis March Militia	Grn	Rel	Neutral	Le Blanc

ACADEMY AND TRAINING UNITS

Unit Name	Exp	Loyalty	Faction	Homeworld
1st Albion Training Cadre	Grn	Fan	Vic	Broken Wheel
2nd Albion Training Cadre	Grn	Fan	Vic	Enchi
1st NAIS Cadet Cadre	Grn	Fan	Vic	Schedar
2nd NAIS Cadet Cadre	Grn	Fan	Vic	Kathil
3rd NAIS Cadet Cadre	Grn	Fan	Vic	Lee
1st Brockton Training Battalion	Grn	Rel	Katrina	Brockton
1st Bell Training Battalion	Grn	Rel	Neu	Axton
1st Conroe Training Battalion	Grn	Rel	Kat	Tancredi IV
Filtvelt Academy Training Battalion	Grn	Rel	Kat	Filtvelt
Goshen War College Training Battalion	Grn	Rel	Neu	Goshen
Kilbourne Academy Training Battalion	Grn	Rel	Kat	Kilbourne
1st Kittery Training Battalion	Grn	Rel	Neu	Kittery
Point Barrow Academy Training Battalion	Grn	Rel	Kat	Point Barrow
Robinson Academy Training Battalion	Grn	Rel	Neu	Robinson
Sakhara Academy Training Battalion	Grn	Rel	Kat	Sakhara V
Tikonov Martial Academy Training Group	Grn	Ques	Kat	Tikonov

RULES

This section provides new rules for regular campaign play that reflect the unique strengths and weaknesses of the AFFC and associated forces. These rules are for use with *BattleTech, BattleForce 2 (BF2),* and *MechWarrior, Third Edition (MW3)* as noted in the text. References to the *BattleTech Master Rules* are abbreviated as *BMR.*

Unit Assignment Tables includes tables for quickly generating BattleTech forces for a pick-up game. *AFFC Regimental Abilities* provides optional rules for simulating the unique characteristics of the various forces described in this manual. *MechWarrior Rules* offers several new Life Paths and other supporting material intended for direct use with *MW3.* The last sections, including *New Equipment* and *New 'Mechs,* provide descriptions and game information for new weapons and vehicles.

UNIT ASSIGNMENT TABLES

The 'Mech and vehicle assignment tables in this section can be used alone, but were designed to be used as an expansion to the *Creating Scenarios* rules starting on page 97 of the *BMR.* Likewise, they may be used to supplement the Assignment Tables starting on page 90 in *MW3.* Though the Random 'Mech and Vehicle Assignment Tables in this book can be used to determine the composition of any Inner Sphere force, it is designed specifically for use with Federated Suns (and Lyran Alliance) regiments. When using non-Federated Suns regiments, use the Random 'Mech Assignment Table from the appropriate *BattleTech Field Manual,* the *BMR,* or *MW3* for more accurate force composition.

ASSIGNING 'MECHS AND VEHICLES

After determining the weight classes of the 'Mechs in each force (p. 104, *BMR*), use the *Random 'Mech Assignment Table,* p. 139, to determine the specific 'Mech designs to be used.

To use the table, the players must set the equipment ratings to be used in the scenario. If playing elements from specific Federated Suns regiments, consult the *AFFC Regiments,* pp. 140–149, to determine the equipment rating of each force. This section not only tells players what table to roll upon (use the Random 'Mech Assignment Table: Federated Suns, as found on p. 139 of this book, or the House Steiner column of the *Random 'Mech Assignment Table: Inner Sphere 2* (p. 106, *BMR*), or the *Random 'Mech Assignment Table: Lyran Alliance,* as found in the upcoming *Field Manual: Lyran Alliance*), but also what column to roll upon.

RANDOM VEHICLE ASSIGNMENT TABLE: FEDERATED SUNS

2D6	Light Vehicles (20 to 35 tons)	Medium Vehicles (40 to 55 tons)	Heavy Vehicles (60 to 75 tons)	Assault Vehicles (80 to 100 tons)
2	Warrior H8 Attack Helicopter [20] V	Regulator Hovertank [45] H	Von Luckner Heavy Tank [75] T	Schiltron B [80] W
3	Cavalry Attack Helicopter [25] V	Condor Heavy Hovertank [50] H	Pilum Heavy Tank [70] W	Behemoth Heavy Tank [100] T
4	Pegasus Scout Hovertank [35] H	Maxim Heavy Hover Transport [50] H	SRM Carrier [60] T	Schrek PPC Carrier [80] T
5	Hunter Light Support Tank [35] T	Goblin Infantry Support Vehicle [45] T	Brutus Assault Tank [75] T	Alacorn Mk VI Heavy Tank [95] T
6	Galleon Light Tank [30] T	Drillson Heavy Hovertank [50] H	Patton Tank [65] T	Partisan Air Defense Tank [80] T
7	Striker Light Tank [35] T	Vedette Medium Tank [50] T	Manticore Heavy Tank [60] T	Ontos Heavy Tank [95] T
8	Pegasus Scout Hovertank [35] H	Condor Heavy Hovertank [50] H	Rommel Tank [65] T	Demolisher Heavy Tank [80] T
9	Yellow Jacket Gunship [30] V	Goblin Medium Tank [45] T	LRM Carrier [60] T	SturmFeur Heavy Tank [85] T
10	Centipede Scout Car [20] H	Drillson Heavy Hovertank [50] H	Typhoon Urban Assault Vehicle [70] W	Challenger X MBT [90] T
11	Blizzard Hover Transport [25] H	Myrmidon Medium Tank [40] T	Manticore Heavy Tank [60] T	Partisan Air Defense Tank [80] T
12	Mantis Light Attack VTOL [15] V	Fulcrum Heavy Hovercraft [50] H	Tokugawa Heavy Tank [60] W	Schiltron Primary [80] W

H Hover T Tracked V VTOL W Wheeled

Alternatively, the players may simply set an equipment rating for their forces (Equipment Rating C is considered average) or may each roll 1D6 to determine their force's equipment rating; a result of 1 provides a rating of "A," a result of 2 a rating of "B," 3–4 a rating of "C," 5 a rating of "D" and 6 an "F."

Using similar equipment ratings for both sides will increase the likelihood of an evenly matched battle.

Because the Federated Suns and the Lyran Alliance have been so closely tied to each other, much of the equipment built within each state is in use within the other. When randomly assigning 'Mechs to either a Federated Suns or Lyran Alliance regiment, the controlling player may choose to roll one out of every four 'Mechs on the other affiliation's assignment table, at one equipment rating lower (unless otherwise stated). The choice to roll on the alternate table must be made before any 'Mech assignment rolls are made. Once the weight classes of these "alternate" 'Mechs are chosen, the player rolls on either the House Davion or House Steiner column of the *Random 'Mech Assignment Table: Inner Sphere 2* (p. 106, *BMR*), the Random *'Mech Assignment Table: Federated Suns* on p. 139 of this book or the *Random 'Mech Assignment Table: Lyran Alliance* in the *Field Manual: Lyran Alliance.*

Equipment Ratings

The name of each regiment is followed by a letter, which is followed by an "AFFC" or an "LAAF" in parentheses. The letter is the regiment's equipment rating, and represents its access to advanced 'Mechs and other equipment. The "AFFC" or "LAAF" represents which military command the regiment is primarily drawing supply from as of 3062. The rating is used when assigning 'Mechs to a unit while using the Random 'Mech Assignment Tables, but also represents the regiment's relationship with the Department of the Quartermaster and so determines how easily it can obtain equipment and supplies.

An equipment rating is often averaged out over the regiment's entire TO&E. Therefore, in a "B" rated company, there may actually be an "A" lance, a "B" lance and a "C" lance.

Assigning 'Mechs and Vehicles in MechWarrior

The *Random 'Mech Assignment Table: Federated Suns* can also be used to assign the starting 'Mech or Vehicle for a new MechWarrior character who comes from the Federated Suns. In this case, use this table instead of the standard Federated Suns column of the *Vehicle Assignment Table* (pp. 91–92, *MW3*). Base the character's equipment rating on the character's current regiment (or the regiment with which he last served).

If the character has not served in a regiment, assume he has a rating of C. Increase the rating by one level for every two of the following which the character possesses: Land Grant, Title, an appropriate Contact (such as AFFC), Well-Connected (Federated Suns), Social Standing of seven or higher, or a Tour of Duty: Inner Sphere (only one Tour of Duty: Inner Sphere path counts towards this total).

Assigning Pilots

Once the players have determined the 'Mechs and other battlefield units they will be using, they should then assign pilots to each. The players may make direct use of the *Random Experience Level Table* and the *Random Skills Table (Expanded)*, found on p. 108 of the *BMR*.

If playing elements from a specific Federated Suns regiment, consult the appropriate regiment writeup (beginning with AFFC Rosters, p. 43); each regiment (and sub-unit) is given a rating of Elite, Veteran, Regular or Green. If playing an Elite regiment, the controlling player should add a +4 modifier to the Random Experience Level roll and a +2 modifier to the Random

RANDOM 'MECH ASSIGNMENT TABLE: FEDERATED SUNS

Light 'Mechs (20 to 35 tons) — Equipment Rating

2D6	A	B	C	D	F
2	SDR-9K Venom [35]	NTK-2Q Night Hawk [35]	ZPH-1A Tarantula [25]*	DRT-3S Dart [25]	WLF-1 Wolfhound [35]
3	NTK-2Q Night Hawk [35]	DRT-3S1 Dart [25]	RVN-3L Raven [35]	COM-5S Commando [25]	FS9-H Firestarter [35
4	GRM-01C Garm [35]	SCB-9A Scarabus [30]	WLF-2 Wolfhound [35]	FS9-H Firestarter [35]	JVN-10F Javelin [30]
5	BZK-F3 Hollander [35]	ALM-7D Fireball [20]	FS9-S Firestarter [35]	JVN-10P Javelin [30]	HNT-171 Hornet [20]
6	SPR-5F Spector [35]	RTX1-0D Raptor [25]	JVN-10P Javelin [30]	FS9-S Firestarter [35]	FS9-H Firestarter [35]
7	RTX1-0A Raptor [25]	BH-K305 Battle Hawk [30]	ALM-7D Fireball [20]	JVN-10P Javelin [30]	JVN-10N Javelin [30]
8	ALM-7D Fireball [20]	ALM-7D Fireball [20]	JVN-10F Javelin [30]	FS9-S Firestarter [35]	JVN-10F Javelin [30]
9	OW-1C Owens [35]	BZK-F3 Hollander [35]	BH-K305 Battle Hawk [30]	WLF-2 Wolfhound [35]	FS9-H Firestarter [35]
10	GRM-01B Garm [35]	SPR-5F Spector [35]	FS9-S1 Firestarter [35]	JVN-10F Javelin [30]	COM-2D Commando [25]
11	JA-KL-1532 Jackal [30]	OW-1B Owens [35]	BZK-F3 Hollander [35]	WLF-1 Wolfhound [35]	JVN-10F Javelin [30]
12	RTX1-0B Raptor [25]	HM-1 Hitman [30]	SCB-9A Scarabus [30]	ALM-7D Fireball [20]	JR7-D Jenner [30]

Medium 'Mechs (40 to 55 tons) — Equipment Rating

2D6	A	B	C	D	F
2	SR1-O Strider [40]	SR1-O Strider [40]	FS9-O Firestarter [45]	CN9-D Centurion [50]	HCT-3F Hatchetman [45]
3	CN10-B Centurion [55]	LNX-9Q Lynx [55]	SNT-04 Sentry [40]	BJ-2 Blackjack [45]	ENF-5D Enforcer [50]
4	STH-1D Stealth [45]	ENF-6M Enforcer III [50]	BJ-2 Blackjack [45]	WTH-2 Whitworth [40]	BJ-2 Blackjack [45]
5	FS9-O Firestarter [45]	SNT-04 Sentry [40]	WCT-4M Watchman [40]	ENF-4R Enforcer [50]	WTH-2 Whitworth [40]
6	BJ2-OD Blackjack [50]	WCT-4M Watchman [40]	BSW-X1 Bushwacker [55]	DV-7D Dervish [55]	DV-6M Dervish [55]
7	ENF-6M Enforcer III [50]	STH-1D Stealth [45]	ENF-5D Enforcer [50]	ENF-5D Enforcer [50]	ENF-4R Enforcer [50]
8	FS9-OB Firestarter [45]	ENF-5D Enforcer [50]	TBT-7M Trebuchet [50]	TBT-7M Trebuchet [50]	VL-2T Vulcan [40]
9	BJ2-OF Blackjack [50]	ENF-6M Enforcer III [50]	STH-1D Stealth [45]	DV-7D Dervish [55]	TBT-5N Trebuchet [50]
10	ENF-6T Enforcer III [50]	BJ2-OC Blackjack [50]	ENF-5D Enforcer [50]	HCT-5S Hatchetman [45]	HBK-4G Hunchback [50]
11	BSW-S2 Bushwacker [55]	NGS-4S Nightsky [50]	BJ-3 Blackjack [45]	BSW-X1 Bushwacker [55]	VL-5T Vulcan [40]
12	NGS-5T Nightsky [50]	STY-3C Starslayer [50]	NGS-4S Nightsky [50]	VT-5M Vulcan [45]	TBT-5N Trebuchet [50]

Heavy 'Mechs (60 to 75 tons) — Equipment Rating

2D6	A	B	C	D	F
2	CTF-3D Cataphract [70]	BNDR-01A Bandersnatch [75]	WR-DG-02FC War Dog [75]	AXM-2N Axman [65]	JM6-DD JagerMech [65]
3	CES-3D Caesar [70]	MDG-1A Rakshasa [75]	FLC-8R Falconer [75]	ON1-K Orion [75]	CPLT-C1 Catapult [65]
4	JM7-F JagerMech [70]	CTS-6Y Cestus [65]	QKD-5K Quickdraw [60]	CPLT-C4 Catapult [65]	JM6-A JagerMech [65]
5	JM7-D JagerMech [70]	WR-DG-02FC War Dog [75]	PTR-4D Penetrator [75]	CES-3R Caesar [70]	GHR-5H Grasshopper [70]
6	CTS-6Y Cestus [65]	FLC-8R Falconer [75]	CES-3R Caesar [70]	CTF-3D Cataphract [70]	CTF-1X Cataphract [70]
7	FLC-8R Falconer [75]	CES-3R Caesar [70]	JM6-DD JagerMech [65]	JM6-DD JagerMech [65]	JM6-S JagerMech [65]
8	PTR-4D Penetrator [75]	PTR-4D Penetrator [75]	AXM-1N Axman [65]	QKD-5K Quickdraw [60]	GHR-5H Grasshopper [70]
9	MDG-1A Rakshasa [75]	CTF-3D Cataphract [70]	CTF-3D Cataphract [70]	GHR-5H Grasshopper [70]	QKD-5K Quickdraw [60]
10	GAL-1GLS Gallowglas [70]	JM6-D3 JagerMech III [65]	PTR-4D Penetrator [75]	CTF-1X Cataphract [70]	JM6-A JagerMech [65]
11	PTR-6S Penetrator [75]	GAL-1GLS Gallowglas [70]	JM6-D3 JagerMech III [65]	AXM-1N Axman [65]	GHR-5H Grasshopper [70]
12	HEL-3D Helios [60]	CAG-2GLS Gallowglas [70]	CTS-6Y Cestus [65]	FLC-8R Falconer [75]	ON1-K Orion [75]

Assault 'Mechs (80 to 100 tons) — Equipment Rating

2D6	A	B	C	D	F
2	TLR1-0B Templar [85]	AWS-9Q Awesome [80]	BNC-3S Banshee [95]	DVS-2 Devastator [100]	BNC-3S Banshee [95]
3	TLR1-0A Templar [85]	MR-V2 Cerberus [95]	GUN-1ERD Gunslinger [85]	VTR-9B Victor [80]	CP-10-Z Cyclops [90]
4	PPR-5S Salamander [80]	NSR-9J Nightstar [95]	AWS-9M Awesome [80]	STK-5S Stalker [85]	CGR-1A1 Charger [80]
5	NSR-9FC Nightstar [95]	LGB-7Q Longbow [85]	ZEU-9S Zeus [80]	AWS-9M Awesome [80]	CGR-1AS Charger [80]
6	DVS-2 Devastator [100]	VTR-9K Victor [80]	AWS-8Q Awesome [85]	AS7-S Atlas [100]	AS7-D Atlas [100]
7	NSR-9J Nightstar [95]	AWS-8Q Awesome [85]	AS7-S Atlas [100]	VTR-9K Victor [80]	VCT-9B Victor [80]
8	GUN-1ERD Gunslinger [85]	AS7-S Atlas [100]	VTR-9K Victor [80]	AWS-8Q Awesome [85]	AWS-8Q Awesome [80]
9	EMP-6A Emperor [90]	DVS-2 Devastator [100]	BNC-5S Banshee [95]	ZEU-9S Zeus [80]	STK-5S Stalker [85]
10	LGB-7Q Longbow [85]	GUN-1ERD Gunslinger [85]	LGB-7Q Longbow [85]	LGB-7Q Longbow [85]	AWS-8Q Awesome [85]
11	TLR1-0 Templar [85]	EMP-6A Emperor [90]	DVS-2 Devastator [100]	BNC-5S Banshee [95]	ZEU-6S Zeus [80]
12	BRZ-A3 Berserker [100]	PPR-5S Salamander [80]	NSR-9J Nightstar [95]	AS7-K Atlas [100]	BNC-3S Banshee [95]

* Indicates a four-legged (quad) BattleMech.

Skill roll. If playing a Veteran regiment, add +2 to the Experience roll and +1 to the Skill roll. If playing a Regular regiment, apply no modifiers, and if playing a Green regiment, apply a –2 to the Experience roll and –1 to the Skill roll.

AFFC REGIMENTS

Each AFFC regiment possesses certain specialties and skills based on its experience and the abilities of its commanders. The regiment descriptions earlier in this book discuss these characteristics in detail. The following game rules simulate these characteristics and are intended to supplement existing *BattleTech*, *BF2* and *MW3* rules. These rules are intended to enhance game play, but not to give any unfair advantages. Gamemasters and players should all agree on the use of any supplemental rules before game play. Similarly, players should feel free to modify any rule that seems inappropriate for their campaign. All of the following rules are considered Level 3 rules, and are ineligible for tournament play.

Many of the following rules are specifically designed for use with the *Creating Scenarios* rules (p. 97, *BMR*). However, players should feel free to use them in other game situations. Unless otherwise stated, the following rules apply to *BattleTech* games only.

EQUIPMENT RATINGS

Rating	Description	Example
A	Top Priority	Davion Heavy Guards
B	Very Good	1st FedCom RCT
C	Good	1st Ceti Hussars
D	Adequate	8th Deneb Light Cavalry
F	Deficient	Alcyone Capellan March Militia

USING HIDDEN UNITS RULES

Certain regiments can use the *Hidden Units* rules (p. 77, *BMR*) to set up part or all of their forces when scenario circumstances allow. If the opportunity does not exist, the regiment cannot exercise this option. For example, in a chase scenario both sides move at the start of the game, so the Hidden Units rules are void. In a stand-up fight scenario, however, one side can use the *Hidden Units* rules. Players must decide if a scenario accommodates the use of hidden units; in all cases, both players must agree before using the rules in a scenario.

If the rules for a scenario type already call for hidden units, ignore the regiment's special ability to set up hidden units.

OVERLAPPING ABILITIES AND ADVANTAGES

If opposing forces possess identical abilities or advantages, the advantages cancel each other out. For example, if two opposing regiments both have the advantage of choosing the side of the map from which they enter, ignore the advantage and instead determine the edge from which a side will enter using standard rules.

FORCING THE INITIATIVE

Several Federated Suns regiments have the ability to "force the Initiative" and gain bonuses to their Initiative rolls in *BattleTech* games. The bonus is equal to the number of enemy units the side destroyed in the previous turn, reduced by the number of its own destroyed units.

For example, a side with this capability receives a +2 Initiative bonus if it destroyed 4 enemy units and lost 2 of its own during the previous turn.

OFF-MAP MOVEMENT

Some regiments are proficient in tactical movement behind the lines of combat. Regiments with the off-map movement ability may move units off the map during play. These units re-enter play at a predetermined map edge. This ability may only be used if the scenario type allows for it. For example, an offensive force may not use this ability in breakthrough scenarios; in a hold-the-line scenario, the defending force cannot use off-map movement.

To use this ability, the controlling player writes down the exit hex and mapsheet of each unit moving off-map. The controlling player then writes down the hex, mapsheet and turn at which each unit will re-enter. Before re-entering the map, an exiting unit must remain off the map for the number of turns it would take to travel from its exit hex to its entrance hex. To calculate this travel time, count the number of blank half hexes along the map edges between the entrance and exit hexes. Then divide this number by the unit's Running/Flank MP rating (rounding fractions up). The result is the *minimum* number of turns the unit must remain off the map. The controlling player can delay the unit's re-entry beyond this period if he desires, but must select the re-entry turn when the unit leaves the map.

Each returning unit re-enters the map in the End Phase of the turn noted on its "written orders." At that time, the opposing player can check the unit's orders to make sure the move was legal. If the move was illegal, the unit is considered destroyed and removed from play. (A move is illegal if the unit does not remain off the map for the required travel time or does not enter at the location and turn noted on its written orders.) In a campaign setting, these "destroyed" units may be brought back into play in a later battle, because they are actually lost rather than destroyed.

If a returning unit is supposed to enter the map in a hex containing an enemy unit, or containing a friendly unit that would violate the stacking limits of the hex, the returning unit is placed in a hex of the opposing player's choice adjacent to the intended entry hex.

Once the returning unit has re-entered the map, any enemy unit adjacent to it may immediately attack it using the rules for

Pointblank Shots from Hidden Units (p. 77, *BMR*). This attack does not restrict the attacking unit's ability to make attacks in the following turn.

Units off the map cannot engage in combat. If the opposing force destroys all the units remaining on the map before off-map units return, the battle is lost.

OVERRUN COMBAT

Some regiments are capable of "overrun" combat during *BattleTech* games. Under certain circumstances, these forces may fire before enemy forces can move or fire.

When rolling Initiative for overrun-capable regiments, note the die roll result's Margin of Success. Divide the Margin of Success by 2 and round fractions down. The result equals the number of individual units that may move, designate and resolve their firing before any other units move and fire. "Extra" successes do not carry over into later turns (i.e., if the player has more successes than 'Mechs available to move). For example, an overrun-capable force makes an Initiative Roll of 11, while its opponent rolls only a 6. That gives the overrun-capable force a Margin of Success of 5, which means that two of its units may move and fire before any other units move (5 divided by 2 = 2.5, rounded down to 2).

Units employing the overrun capability do not take part in the normal move/fire sequence. In other words, overrunning units do not count for determining how many units each side must move when alternating turns during the Movement, Reaction and Weapon Attack phases. Overrunning units make physical attacks as normal during the Physical Attack Phase.

BANKING INITIATIVE

Some regiments command an innate grasp of the ever-changing battlefield. They can take advantage of this by electing to give up the initiative only to seize it later in combat.

In game play, the player controlling these regiments may, *before* making the Initiative roll, choose to give the Initiative to his opponent. For each two turns the controlling player elects to lose initiative he may, at a later Turn and before making the Initiative roll, choose to take the Initiative. The controlling player may only "bank" up to two Initiatives (giving up four Initiatives) at any one time in this way. For purposes of any other special abilities, consider the opponent to have won or lost the initiative (as appropriate) by a margin of 1.

EQUIPMENT RATINGS

The name of each regiment is followed by a letter in parentheses. This letter is the regiment's equipment rating, and represents its access to advanced 'Mechs and other equipment. The rating is used when assigning 'Mechs to a regiment using the random 'Mech assignment tables, but also indicates the regiment's relationship with the procurement division of its military and so determines how easily it can obtain equipment and supplies.

UNAFFILIATED AFFC UNITS

1st Aragon Borderers (C , AFFC)

The Borderers are adept at maneuver. Up to half of the Borderers' units may make use of the off-map movement special ability if the total number of enemy units on the map is greater than the off-map force. When randomly assigning 'Mech forces, one out of every eight may be Clan. Once the weight classes of these 'Mechs are randomly determined, the Borderers player chooses which 'Mech will be Clan and rolls on the Smoke Jaguar column of the *Random 'Mech Assignment Table: Clans*, found on p. 107 of the *BMR*.

1st Argyle Lancers (D, LAAF)

When randomly assigning 'Mech forces, apply a –3 modifier when rolling to determine Random Lance Type (p. 103, *BMR*). Additionally, in scenarios where the Lancers are prevented from using their aerospace fighters because of weather (or because of some other outside force, not if the scenario simply does not allow for aerospace fighter integration), the Lancers receive a –1 Initiative penalty.

1st Capellan Dragoons (D, AFFC)

The Dragoons receive a +4 Initiative bonus on the first turn of any battle, but thereafter subtract one from that bonus for every two turns its units remain on the map and continue to do so until all of the Dragoons' elements are either destroyed or leave the map (this "bonus" may continue on into negative numbers).

1st Federated Suns Armored Cavalry (A, AFFC)

When randomly assigning 'Mech forces, apply a –2 modifier when rolling to determine Random Lance Type (p. 103, *BMR*). The Armored Cavalry may make use of the off-map movement and banking initiative special abilities.

1st Kestrel Grenadiers (B, AFFC)

The Grenadiers field a far heavier force than their status as a "household guard" would otherwise indicate. Before the scenario begins, the controlling player of the Grenadiers rolls a D6. On a 5 or 6, he has never faced the enemy regiment before and the Grenadiers receive a –3 Initiative penalty for the entire game. On any other roll, there is no effect.

1st Kittery Borderers (D, AFFC)

The Kittery Borderers receive a +1 Initiative bonus when fighting against Capellan regiments. Furthermore, they receive an additional +1 Initiative bonus while fighting within the Capellan March, but suffer a –1 Initiative penalty when operating outside that region.

AVALON HUSSARS

11th Avalon Hussars (B, AFFC)

At the beginning of the game, the Eleventh Avalon Hussars player must designate a command unit. As long as the command unit remains in play, the player controlling the unit in a *BF2* game will not use the standard Command "cup." Instead, the player may arrange the Commands in any order he or she wishes before the game starts, drawing the commands in that exact order. Once all of the commands have been drawn, the player will rearrange them again in any order before the next turn begins. The process will continue in this way until the game ends.

17th Avalon Hussars (B, AFFC)

The Seventeenth may make use of the overrun combat and banking initiative special abilities.

20th Avalon Hussars (A, AFFC)

The Twentieth is well-known for rapidly building field works and defensive fortifications. In any scenario where the Twentieth is the defender, the player controlling the Twentieth may place up to one half as many one-level Medium buildings as it will deploy units, so long as the scenario's circumstances allow it. In addition, the Twentieth may make use of the off-map movement special ability with up to a quarter of its deployed forces.

22nd Avalon Hussars (B, AFFC)

The Twenty-second is a unit that has not only experienced extremely good luck on the battlefield, but also incredibly bad luck. Prior to the start of any *BattleTech* game, roll 2D6. On a roll of 12, all the Twenty-second units receive –1 to-hit modifier. On a roll of 2, the unit is subject to a –4 Initiative penalty. On any other roll, there is no effect.

39th Avalon Hussars (C, LAAF)

The Thirty-ninth Hussars use reconnaissance elements extensively to locate and surprise any enemy forces. In scenarios where the Thirty-ninth is the attacker, the player controlling the Thirty-ninth will receive a +3 Initiative bonus so long as he or she keeps more than half of the force off-board. The remainder of the regiment may enter the Thirty-ninth's home edge at any time during the game, but the controlling player forfeits any Initiative bonus at the beginning of the turn in which any of the remaining forces enter the battle.

Additionally, when randomly assigning 'Mech forces, apply a –2 modifier when rolling to determine Random Lance Type (p. 103, *BMR*).

41st Avalon Hussars (A, AFFC)

The members of the Forty-first Hussars are at the same time some of the most reviled and revered in the AFFC. *MW3* characters assigned to the Forty-first are considered to be *Well Connected* within the pro-Victor Davion camp and to have *Stigma/Pro-Victor* within the pro-Katherine camp.

42nd Avalon Hussars (A, LAAF)

Every turn the Forty-second Hussars make use of pre-planned movement (the controlling player(s) write down each unit's movement before the turn begins, with no deviations once the turn begins), each unit of the Forty-second receives a –1 to-hit modifier for any single weapon attack during that Weapon Attack Phase only. If a Hussar unit cannot move through a hex because of an enemy unit's move or would violate the *Stacking* rule (p. 20, *BMR*), the controlling player of the Hussars can change that unit's movement but must adhere to the original plan as much as possible.

CHISHOLM'S RAIDERS

1st Chisholm's Raiders (F, AFFC)

At the beginning of the game, the First player must designate a command unit. As long as the command unit remains in play, the Raiders receive a –1 to-hit modifier to all physical attacks.

Additionally, the First Raiders lack any significant tactical cohesiveness. For every additional lance beyond the first, they receive a cumulative –1 Initiative penalty.

2nd Chisholm's Raiders (C, AFFC)

When randomly assigning 'Mech forces, apply a –2 modifier when rolling to determine Random Lance Type (p. 103, *BMR*). The Second Raiders may make use of the overrun combat special ability and may also make use of the *Sprinting* movement ability (see p. 13, *Maximum Tech, Revised*).

CETI HUSSARS

1st Ceti Hussars (C, AFFC)

The First Hussars may use the forcing the initiative and overrun combat special abilities. Any regiment facing the First Hussars may use off-map movement with up to a quarter of its forces, unless its abilities allow for more.

2nd Ceti Hussars (B, AFFC)

Opponents facing the Second Hussars may not use the overrun combat special ability and may not place any hidden units.

3rd Ceti Hussars (C, AFFC)

At the beginning of the game, the First player must designate a command unit. As long as the command unit remains in play, the Third receives a –3 Initiative penalty. Finally, all of the Third's artillery units gain a –1 to-hit modifier.

CRUCIS LANCERS

1st Crucis Lancers (A, AFFC)

When randomly assigning 'Mech forces, apply a +2 modifier when rolling to determine Random Lance Type (p. 103, *BMR*). Additionally, for every two units the First loses, the Lancers receive a cumulative +1 Initiative bonus (to a maximum of +4).

2nd Crucis Lancers (C, LAAF)

The Second Lancers can deploy up to half of its units in hidden positions at the beginning of any battle. In addition, when using the dropping troops rules (p. 72, *BMR*), all units receive a –1 modifier to the Piloting Skill roll to determine if they scatter. When randomly assigning 'Mech forces, one out of every eight may be Clan. Once the weight classes of these 'Mechs are randomly determined, the Lancers player chooses which 'Mech will be Clan and rolls on the Clan Wolf column of the *Random 'Mech Assignment Table: Clans* (p. 107, *BMR*).

3rd Crucis Lancers (A, AFFC)

Consider all of the Third Lancers' infantry forces as anti-'Mech trained. Additionally, those infantry units may deploy towed weapons (see *Field Guns*, p. 38, *Maximum Tech, Revised)*. If the scenario allows, any or all of the Third's infantry may deploy hidden. When randomly assigning 'Mech forces, one out of every eight may be Lyran Alliance. Once the weight classes of these 'Mechs are randomly determined, the Lancers player chooses which 'Mech will be Lyran and rolls on the House Steiner column of the *Random 'Mech Assignment Table: Inner Sphere 2* (p. 106, *BMR*) or the "A" column of the *Random 'Mech Assignment Table: Lyran Alliance*, found in the *Field Manual: Lyran Alliance*.

4th Crucis Lancers (C, LAAF)

When randomly determining maps, the Fourth Lancers gain a +2 Initiative bonus in any scenario, but the opposing player may choose the maps instead of rolling and may choose the Fourth Lancers' home edge.

5th Crucis Lancers (B, AFFC)

When in combat against an opponent connected to the Zhanzeng de Guang, the Thuggee or any other Capellan terrorist group, the Fifth Lancers gain a +2 Initiative bonus and each unit in the Fifth receives a –1 to-hit modifier to all weapon attacks. After experiencing firsthand some of the horrors apparently wrought by the terrorists, many of the Fifth's personnel are beginning to experience psychological problems. Roll 2D10 each month any *MW3* character is assigned to the Fifth; on a result of 4 or less, the character receives a one-point negative Trait, such as *Addiction*, *Introvert* or *Quirk/Psychosis*. On the other hand, all Fifth characters receive the Trait *Combat Sense* after a year of continuous service.

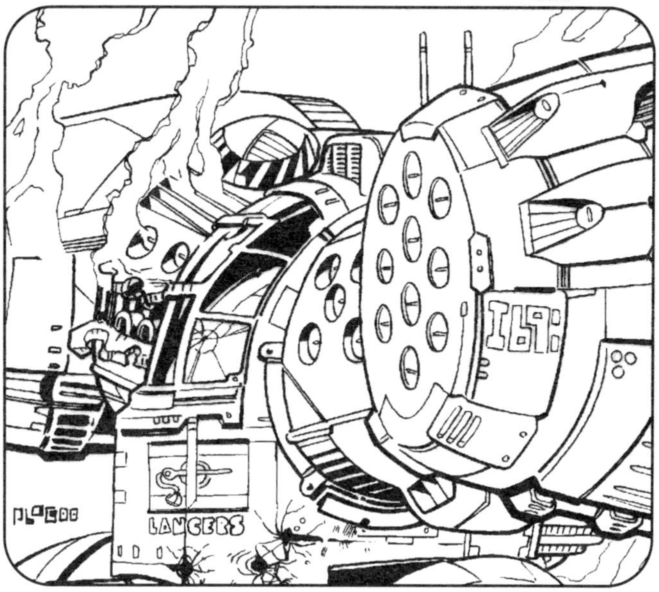

6th Crucis Lancers (B, LAAF)

If the Sixth Lancers are fighting in urban terrain and are the attackers in the scenario, the opponent cannot set up any units using the Hidden Units rule (p. 77, *BMR*).

7th Crucis Lancers (A, LAAF)

When randomly determining maps, if the player controlling the Seventh Lancers willingly allows his or her opponent to choose all of the maps involved in the scenario and choose the home edge for the Seventh, each of the Seventh's units gains one "Edge" point. Edge points may be used to force a re-roll of one single roll directly affecting the unit which spends the Edge point (a to-hit roll, either against itself or in an attack made by the unit in question, a PSR, etc.).

8th Crucis Lancers (B, AFFC)

Whenever facing a Draconis Combine regiment in battle, the Eighth receives gains a –3 Initiative penalty but also a –1 to-hit modifier for all weapons attacks.

DAVION BRIGADE OF GUARDS

Davion Assault Guards (A, AFFC)

When randomly assigning 'Mech forces, apply a +4 modifier when rolling to determine Random Lance Type (p. 103, *BMR*). In addition, the Guards may also use the forcing the initiative and overrun combat special abilities and may use the *Sprinting* movement ability (see p. 13, *Maximum Tech, Revised*).

Davion Heavy Guards (A, AFFC)

When randomly assigning 'Mech forces, apply a +3 modifier when rolling to determine Random Lance Type (p. 103, *BMR*). In addition, the Guards may also use the forcing the initiative, the banking initiative and overrun combat special abilities. All *MW3* characters assigned to the Heavy Guards gain the 2-point Traits of *Good Reputation* and *Well-Connected/New Avalon*, as long as they remain in the Guards.

Any *MW3* character assigned to the Armor Brigade gains 5 XP per month in either Tactics/Armor or Strategy.

Davion Light Guards (A, AFFC)

When randomly assigning 'Mech forces, apply a –3 modifier when rolling to determine Random Lance Type (p. 103, *BMR*). The Guards gain a +1 Initiative bonus if they outnumber their enemy, increasing to +3 if they outnumber their enemy by a ratio of 2:1 or greater.

1st Davion Guards (A, AFFC)

All *MW3* characters assigned to the First Guards RCT gain the 2-point Traits of *Good Reputation* and *Well-Connected/Federated Suns,* as long as they remain in the Guards. In addition, during any *BF2* games, the player controlling the First Guards may substitute any one command with Luck of the Fox prior to the start of the game; no matter what command was substituted, the new command is considered a "special command."

2nd Davion Guards (B, AFFC)

Before the start of any game, the Second Guards may preplot up to fifteen separate hexes on the map board for artillery strikes (see p. 70, *BMR*), unless the scenario would prevent it. In any event, the members of the Second Guards work extensively with their artillery elements and gain a –2 "spotter" modifier, instead of the standard –1 modifier (see p. 70, *BMR*). If the Second Guards are deprived of either their artillery support or their aerospace support (due to weather, terrain conditions, etc.; the fighters must be grounded or for some reason unable to survey the battle in any way), they gain a –2 Initiative penalty.

When randomly assigning 'Mech forces, one out of every eight may be Clan. Once the weight classes of these 'Mechs are randomly determined, the Guards player chooses which 'Mech will be Clan and rolls on the Clan Jade Falcon column of the *Random 'Mech Assignment Table: Clans* (p. 107, *BMR*).

3rd Davion Guards (C, AFFC)

All *MW3* characters assigned to the Third Guards automatically gain the negative Traits *Bad Reputation* and *Stigma/Hesperus Secessionist* (obviously a Trait significant to any Lyran characters and NPCs), but also gain the four-point Trait *Well-Connected/Federated Suns* (due to the unit's connections with ComStar); *Well-Connected* only applies while the character is in the Third Guards. In any *BF2* game, the Third may replace up to four commands with two each of Alpha Strike! and Stand and Shoot.

4th Davion Guards (B, LAAF)

When randomly determining maps, the opposing player may choose all of the maps for any scenario, but the Fourth Guards player will choose his or her home edge first. When randomly assigning 'Mech forces, one out of every eight may be Clan. Once the weight classes of these 'Mechs are randomly determined, the Guards player chooses which 'Mech will be Clan and rolls on the Clan Wolf column of the *Random 'Mech Assignment Table: Clans* (p. 107, *BMR*).

In a *MW3* game, if any Tactics or Strategy rolls (opposed or not) are required of the Fourth Guards prior to the start of a scenario, the controlling player receives a +3 bonus to the roll.

5th Davion Guards (A, AFFC)

Since their action on La Grave, the Fifth Guards have practiced extensively in the forests of the worlds on which they have been assigned. Any Fifth unit beginning its turn on a Wooded hex, and continuing to travel in Wooded hexes for the entire Movement Phase, gains 1 additional MP.

When randomly assigning 'Mech forces, one out of every eight may be Clan. Once the weight classes of these 'Mechs are randomly determined, the Guards player chooses which 'Mech will be Clan and rolls on the Clan Jade Falcon column of the *Random 'Mech Assignment Table: Clans* (p. 107, *BMR*).

DENEB LIGHT CAVALRY

4th Deneb Light Cavalry (C, LAAF)

At the beginning of the game, the Fourth Deneb Light Cavalry player must designate a command unit. As long as the command unit remains in play, the Fourth gains a unique attribute when playing in a *BF2* scenario. Whenever the Hello, HQ? command is drawn, the Fourth's controlling player should roll 1D6. On a roll of one, apply the effects of the command normally; if any other result is rolled, the controlling player may disregard the command's effects and instead apply the effects of any other command available in the game. No matter what the effect is applied, this command is still considered a "special command" and is thus set aside for one turn before being returned to the cup.

8th Deneb Light Cavalry (D, LAAF)

The Eighth has a deep connection with Duke Robert Steiner-Kelswa. All characters assigned to the RCT gain the four-point Trait *Well-Connected/Free Skye Movement*, while in the Eighth. When operating in forces larger than two companies, the Eighth receives a –1 Initiative penalty.

Additionally, the Eighth is constantly experiencing equipment breakdowns. For each lance the Cavalry field, roll 1D6 –3 prior to the start of the game. If the result is negative, randomly determine one 'Mech in the lance and roll 1D6 to determine which of the following critical hits to apply to that 'Mech: 1–Engine, 2–Sensor, 3–Foot Actuator, 4–Jump Jet, 5–Shoulder

Actuator, 6–Gyro. Once the specific critical hit has been determined, randomly assign it to an appropriate slot on the Critical Hit Table of the 'Mech's record sheet.

10th Deneb Light Cavalry (C, AFFC)

The Tenth may use the off-map movement special ability. Additionally, the regiment may place up to half of its units hidden, prior to the start of any scenario, so long as the circumstances of the scenario permit.

12th Deneb Light Cavalry (D, AFFC)

When using the dropping troops rules (p. 72, *BMR*), all units in the Twelfth receive a –1 modifier to the Piloting Skill Roll to determine if they scatter. In addition, the BattleMechs in the Twelfth may make a drop even when its DropShips are hovering at low altitude.

15th Deneb Light Cavalry (C, AFFC)

Seen by many within the AFFC as sympathizers, any characters assigned to the Fifteenth gain *Stigma/Katherine Supporter*, a negative Trait that could actually prove to be advantageous within certain circles. Additionally, the Fifteenth may make use of the overrun combat special ability.

FEDERATED COMMONWEALTH CORPS

1st FedCom RCT (B, AFFC)

The members of the First RCT are adept at small-unit tactics, but do not fight well in large groups. In any *BF2* game where the First's battalions fight independently (lances/platoons belonging to one battalion do not close within three hexes of one belonging to another), roll 1D6 for each command drawn. On a roll of one, the player controlling the First may chose *any BF2* command and apply its effects instead of the drawn command (no matter what command was drawn or chosen, treat it as a "special command"). On the other hand, if the First's battalions operate together, roll 1D6 for every command drawn; on a roll of one, apply the effects of the Hello, HQ? command.

3rd FedCom RCT (B, AFFC)

Prior to any battle in which the Third is the defender in the scenario, the Third's controlling player may place as many standard ten point minefields on the map as it has units deployed. Additionally, the Third may deploy up to one quarter of its units hidden, prior to the start of the game.

When randomly assigning 'Mech forces, the player controlling the Third FedCom may roll once per lance on the "A" column. The choice to roll on the "A" column must be made prior to any rolls.

5th FedCom RCT (C, AFFC)

In any Turn where the Fifth is outnumbered, the regiment receives a –3 Initiative penalty, but each unit gains a –1 to-hit modifier to all weapon attacks.

6th FedCom RCT (B, AFFC)

When randomly assigning 'Mech forces, apply a –2 modifier when rolling to determine Random Lance Type (p. 103, *BMR*). When randomly determining maps, the player controlling the Sixth may choose the maps for the scenario, and may choose its own home edge. Additionally, if using the Ice Special Case Rules (p. 78, *BMR*) or the Heavy Snowfall Weather Conditions (p. 11, *Maximum Tech, Revised*), the members of the Sixth gain a –2 modifier to all Piloting Skill rolls.

7th FedCom RCT (C, AFFC)

The members of the 7th RCT gain a –1 to-hit modifier on all weapon attacks when fighting against a regiment claiming loyalty to Katherine Steiner-Davion, but also suffer a –2 Initiative penalty.

8th FedCom RCT (D, AFFC)

Characters assigned to the Eighth RCT automatically gain the negative Trait of *Stigma/Katherine Supporter*. Additionally, in any turn in which the Eighth outnumbers an opponent, they gain a +1 Initiative bonus.

NEW IVAARSEN CHASSEURS

1st New Ivaarsen Chasseurs (C, AFFC)

The First Chasseurs gain a –1 to-hit modifier on all physical attacks when facing any Draconis Combine opponent. Additionally, the Chasseurs air units gain a –1 to-hit modifier on air-to-ground strafing runs. On the other hand, if the Chasseurs are deprived of their air cover (assuming the scenario conditions would normally allow for their inclusion), the entire regiment suffers a –1 Initiative penalty.

2nd New Ivaarsen Chasseurs (A, AFFC)

All *MW3* characters assigned to the Second Chasseurs gain 5 XP toward their *Strategy* skill every two months, so long as they remain with the unit.

ROBINSON RANGERS

1st Robinson Rangers (C, AFFC)

Prior to the start of any battle, the player controlling the First Rangers should divide his or her entire force into two separate groups, each fielding the same number of units. The player may devote one of those two groups to the battle; so long as the second group is kept off-map, the First Rangers gain a +2 Initiative bonus. Once the second group enters the board, the entire force may make use of the *Sprinting* movement ability (see p. 13, *Maximum Tech, Revised*), but they lose the +2 Initiative bonus.

2nd Robinson Rangers (D, LAAF)

The members of the Second Rangers suffer a +1 to-hit modifier on all long-range weapon attacks, but also gain a –1 to-hit modifier on all physical attacks. At the same time, the regiment gains a –2 modifier to all Piloting Skill rolls required when moving on pavement or entering or leaving buildings.

When randomly assigning 'Mech forces, once per Lance, the controlling player has the option of rolling on either the "A" column of the *Random 'Mech Assignment Table: Lyran Alliance* or the *Random 'Mech Assignment Table: Federated Suns.* The decision to roll on those alternate tables must be made before any random 'Mech assignment rolls.

3rd Robinson Rangers (A, AFFC)

If the controlling player of the Third Rangers moves all of his units in groups of four during the Movement Phase, the regiment gains a –2 modifier to all Piloting Skill rolls for that Turn only.

SYRTIS FUSILIERS

5th Syrtis Fusiliers (C, AFFC)

The Fifth Fusiliers are subject to a –1 Initiative penalty, but all their long-range weapon attacks receive a –1 to-hit modifier. The Fifth's infantry forces gain a –1 to-hit modifier to both weapon and Anti-BattleMech attacks.

6th Syrtis Fusiliers (C, AFFC)

When the Sixth is the defender in a scenario and maps are randomly determined, the player controlling the Sixth Fusiliers may choose the specific maps for the battle and may place up to a quarter of his or her units hidden.

8th Syrtis Fusiliers (C, AFFC)

When randomly assigning 'Mech forces, apply a –2 modifier when rolling to determine Random Lance Type (p. 103, *BMR*). The Eighth Fusiliers may use off-map movement and force the initiative.

When randomly assigning 'Mech forces, one out of every four may be Capellan Confederation. Once the weight classes of these 'Mechs are randomly determined, the Fusiliers player chooses which 'Mech will be Capellan and rolls on the House Liao column of the *Random 'Mech Assignment Table: Inner Sphere 1* (p. 105, *BMR*) or the "C" column of the *Random 'Mech Assignment Table: Capellan Confederation*, (p. 123, *Field Manual: Capellan Confederation*).

CAPELLAN MARCH MILITIA

Alcyone Capellan March Militia (F, AFFC)

The Alcyone CMM uses its ample artillery assets to both place minefields and to bombard an enemy force. Prior to the start of any scenario but a "Chase," the player controlling the CMM may place up to half as many standard 10-point minefields as he or she is fielding units. Additionally, the player may pre-designate up to that many hexes for artillery strikes (see p. 70, *BMR*).

Kathil Capellan March Militia (B, AFFC)

The members of the Kathil CMM are subject to a –1 Initiative penalty in any scenario, though all characters gain the *MW3* two-point Trait of *Well Connected/Kathil.*

When randomly assigning 'Mech forces, the player controlling the Kathil CMM may roll once per lance on the "A" column. The choice to roll on the "A" column must be made before any rolls are made.

New Syrtis Capellan March Militia (C, AFFC)

Prior to any scenario in which the Militia is the defender, the controlling player may position any or all of its units hidden.

When randomly assigning 'Mech forces, the player controlling the New Syrtis CMM may roll once per lance on the "B" column. The choice to roll on the "B" column must be made before any rolls are made.

Ridgebrook Capellan March Militia (F, AFFC)

Prior to the start of any *MW3* scenario in which the Militia is the defender, the players controlling each force involved should make an opposed *Tactics* roll, with the Ridgebrook CMM player receiving a +4 modifier. If the Militia is successful, it gains an Initiative bonus equal to its Margin of Success (to a maximum of +4).

Sirdar Capellan March Militia (F, AFFC)

The members of the Sirdar CMM are itching for any chance to fight the Capellan Confederation. If the unit is involved in a battle in which a Confederation regiment is participating and is not the enemy, the player controlling the Militia should roll 2D6. On a roll of 11 or 12, the Militia will fight normally. On any other result, the members of the Militia will immediately attack the Confederation troops, regardless of their mission or the circumstances of the battle.

Valexa Capellan March Militia (F, AFFC)

In any scenario where the Valexa CMM is the defender, the player controlling the Militia may position any or all of the regiment's infantry hidden. Additionally, the player may designate up to half of his or her infantry units as anti-'Mech trained, free of charge.

Warren Capellan March Militia (D, AFFC)

The Warren CMM may make use of the overrun combat special ability.

Achernar Sarna March Militia (D, AFFC)

The Achernar SMM gains a –1 to-hit modifier for both weapon and physical attacks made against all infantry units. Its 'Mech force is not in the best condition however. For each lance the

Achernar SMM fields, roll 1D6 – 3 prior to the start of the game. If the result is negative, randomly determine one 'Mech in the lance and roll 1D6 to determine which of the following critical hits to applies to that 'Mech: 1–Engine, 2–Sensor, 3–Foot Actuator, 4–Jump Jet, 5–Shoulder Actuator, 6–Gyro. Once the specific critical hit has been determined, randomly assign it to an appropriate slot on the Critical Hit Table of the 'Mech's record sheet.

Nanking Sarna March Militia (F, AFFC)

The Militia receives a –1 Initiative penalty. If the Nanking SMM is fighting on Nanking, its members gain a –1 to-hit modifier for all physical attacks.

CRUCIS MARCH MILITIA

Anjin Muerto Crucis March Militia (D, AFFC)

In any battle in which the CMM does not deploy any hovercraft or jumping BattleMechs, the entire regiment suffers a –1 Initiative penalty.

Islamabad Crucis March Militia (F, AFFC)

The CMM suffers a –1 Initiative penalty in any scenario in which the regiment deploys more than one company on the battlefield.

All *MW3* characters who profess an interest in the Messengers of Shiva must spend half of their earned XPs each month on the skills *Academic/Teachings of Shiva*, but increasing the WIL Attribute costs only 15 XP times the value you are raising it to, rather than 25 XP.

Kestrel Crucis March Militia (F, AFFC)

MW3 characters gain 5 XP per month they are assigned to the Kestrel CMM, applied directly to their *Academic/Military History* skill. Prior to each battle, all characters should make a *Tactics* and an *Academic/Military History* skill check with a +4 modifier. If both are successful, the character receives a +2 bonus to all Initiative rolls.

Malagrotta Crucis March Militia (D, AFFC)

Members of the Malagrotta CMM gain a –2 modifier to all Piloting Skill roll target numbers when fighting in lower than standard gravities (see pp. 78–79, *BMR*).

Marlette Crucis March Militia (F, AFFC)

Adept at pushing the offensive, the Marlette CMM may use the forcing the initiative and overrun combat special abilities. When randomly assigning 'Mech forces, the player controlling the New Syrtis CMM may roll once per lance on the "A" column. The choice to roll on the "A" column must be made before any rolls.

New Avalon Crucis March Militia (B, AFFC)

The members of the New Avalon CMM gain a –1 modifier to all Piloting Skill rolls required for movement on pavement.

Remagen Crucis March Militia (F, AFFC)

While fighting at night, the members of the Remagen CMM suffer only a +1 modifier in place of the standard +2 (see p. 81, *BMR*). When fighting at dusk or dawn, they suffer no penalties (see p. 11, *Maximum Tech, Revised*).

Tsamma Crucis March Militia (F, AFFC)

In any Turn in which the Tsamma CMM does not make weapon attacks, the entire unit gains a +4 Initiative bonus for the following Turn only and a –2 to-hit modifier on all physical attacks for the Physical Attack Phase of that Turn only.

DRACONIS MARCH MILITIA

Addicks Draconis March Militia (C, AFFC)

Prior to the start of any scenario, the Militia may place any or all of its force hidden.

Bremond Draconis March Militia (F, AFFC)

The Bremond DMM may use the banking initiative special ability.

Bryceland Draconis March Militia (F, AFFC)

Because of its decentralized command structure, the Bryceland DMM receives a –1 Initiative penalty. Opponents may not place hidden units on the board while fighting the Bryceland DMM on a world within the Bryceland PDZ.

During *BF2* games, the regiment is not susceptible to the effects of the Jam Transmission and the Hello, HQ? commands.

Clovis Draconis March Militia (F, AFFC)

MW3 characters assigned to the Clovis DMM automatically gain the two-point negative Trait *Quirk/Hates the Draconis Combine*.

Dahar Draconis March Militia (D, AFFC)

If the Dahar DMM is involved in any scenario in non-urban terrain, it suffers a –1 Initiative penalty. The members of the Militia's aerospace wing gain a –2 modifier to all control roll target numbers.

Kilbourne Draconis March Militia (F, AFFC)

If the Kilbourne DMM fields two companies or less, the Militia may make use of the overrun combat special ability. If the Kilbourne DMM fields more then two companies, however, the regiment loses this ability and further suffers a –1 Initiative penalty.

Mayetta Draconis March Militia (F, AFFC)

If the Mayetta DMM deploys in a force greater than a battalion, the player controlling the Militia should roll 1D6 for each company, or fraction thereof, he or she is committing to the battle. The player should randomly select that number of units;

RANK TABLE: NATIONS OF THE FEDERATED COMMONWEALTH

ENLISTED RANKS

Trait Points	AFFS	AFFC	LAAF
1	Recruit	Recruit	Recruit
2	Private	Private	Private
3	Private, First Class	[Private, First Class]	Private, First Class
4	Corporal	Corporal	Corporal
5	—	—	Senior Corporal
6	Sergeant	Sergeant	Sergeant
7	—	—	Staff Sergeant
8	Sergeant Major	Sergeant Major	Sergeant Major
9	—	—	Staff Sgt. Maj.
10	Command Sgt. Maj.	[Command Sgt. Maj.]	Senior Sgt. Maj.

OFFICER (COMMISSIONED) RANKS

Trait Points	AFFS	AFFC	LAAF
1	Subaltern	—	Leutnant
2	Leftenant	Leftenant	First Leutnant
3	Captain	Hauptmann	Hauptmann
4	Major	Kommandant	Kommandant
5	—	—	Hauptmann-Kommandant
6	Leftenant Colonel	[Leftenant Colonel]	Leutnant-Colonel
7	Colonel	Leftenant General	Colonel
8	Leftenant General	Hauptmann General	Leutnant-General
9	Major General	Marshal	Hauptmann-General
10	General/Marshal	—	Kommandant-General
11*	Field Marshal	Field Marshal	General
12*	Marshal of the Armies	Marshal of the Armies	General of the Armies

*These ranks are not normally attainable by player characters but are included here for comparison.

those units will be unavailable for battle.

Milligan Draconis March Militia (F, AFFC)

All air-to-ground attacks made by the Militia's aircraft gain a –1 to-hit modifier, so long as at least one Militia ground unit ('Mech, vehicle or infantry squad/platoon) has LOS to the unit(s) being attacked. If the Militia is deprived of its aerospace fighter support for any reason (and the scenario would otherwise have allowed it), the unit suffers a –2 Initiative penalty.

Raman Draconis March Militia (F, AFFC)

When randomly assigning 'Mech forces, one out of every four may be Draconis Combine. Once the weight classes of these 'Mechs are randomly determined, the Raman DMM player chooses which 'Mech will be Combine and rolls on the House Kurita column of the *Random 'Mech Assignment Table: Inner Sphere 1* (p. 105, *BMR*), or the "F" column of the *Random 'Mech Assignment Table: Draconis Combine*, (p. 142, *Field Manual: Draconis Combine*).

Robinson Draconis March Militia (F, AFFC)

If elements of the Robinson DMM share the battlefield with elements of another AFFC regiment, the entire force suffers a –1 Initiative penalty.

ACADEMY AND TRAINING UNITS

Albion Training Cadres (B, AFFC)

MW3 characters assigned to the Cadres as students receive 5 XP per month to be split up as the controlling player wishes between the Leadership skill *and* one or more Tactics skills.

Members of the First Cadre must possess at least 6 WIL before being accepted into the unit, but also gain an extra 4 XP per month to be used towards the increase of their Piloting skill.

So long as a member is assigned to a single class of 'Mech, vehicle or fighter for the entire two-year tour with the Cadre, he or she receives a +1 modifier to all piloting skill rolls when in control of that type of equipment.

NAIS Cadet Cadres (B, AFFC)

The Second Cadre may use the off-map movement special ability.

Members of the First Cadre receive 6 XP per month to be assigned to any Academic/History skill.

The atmosphere within the Third Cadre is so tense that members actually lose 1 XP per month assigned to the Cadre, but consequently gain the Good Reputation Trait after "graduation" from the cadre (so long as their actions within the Cadre would not garner them a Bad Reputation, in which case the two Traits would cancel each other out).

1st Brockton Training Battalion (F, AFFC)

Because of the varied backgrounds of its members and the lack of overall battalion cohesiveness, the battalion suffers a –1 Initiative penalty.

1st Bell Training Battalion (F, AFFC)

The Battalion gains a +1 Initiative bonus when operating on a world within the Valexa PDZ.

1st Conroe Training Battalion (F, AFFC)

For each lance the Conroe Training Battalion fields, roll 1D6 –4 prior to the start of the game. If the result is negative, randomly determine one 'Mech in the lance and roll 1D6 to determine which of the following critical hits to apply to that 'Mech: 1–Engine, 2–Sensor, 3–Foot Actuator, 4–Jump Jet, 5–Shoulder Actuator, 6–Gyro. Once the specific critical hit has been determined, randomly assign it to an appropriate slot on the Critical Hit Table of the 'Mech's record sheet.

Filtvelt Academy Training Battalion (F, AFFC)

Often called upon to deal with pirates operating within their area, members of the battalion gain a +2 to their Tactics roll in regards to pirate activity.

Goshen War College Training Battalion (Special, AFFC)

Because many of the battalion's 'Mechs are owned by the families of the cadets using them, the battalion fields an eclectic force composition. When randomly generating forces, the player(s) controlling the unit should roll 1D6 for each 'Mech. On a roll of "1" he or she should randomly roll on the appropriate "A" column, on a "2" roll on the "B" column, and so on. On a roll of "6" the player may choose what column he or she will roll on.

Kilbourne Academy Training Battalion (F, AFFC)

Apply a –5 to the Experience roll and –2 to the Skill roll when randomly determining Experience Levels and Skills (p. 108, *BMR*) for the battalion.

1st Kittery Training Battalion (F, AFFC)

Characters assigned to the battalion gain Contact (2)/Major Raul Bethune).

Point Barrow Academy Training Battalion (F, AFFC)

When randomly assigning 'Mech forces, the player controlling the battalion may roll once per lance on the "A" column. The choice to roll on the "A" column must be made before any rolls.

Robinson Battle Academy Training Battalion (F, AFFC)

When fighting a Draconis Combine unit, members of the Robinson Academy Training Battalion suffer a –3 Initiative penalty, but gain a –1 to-hit modifier on all physical attacks.

Sakhara Academy Training Battalion (B, AFFC)

The battalion will not prosecute any battles in a city. When fighting in urban terrain, the battalion suffers a –2 Initiative penalty.

Tikonov Martial Academy Training Group (F, AFFC)

For each lance the Conroe Training Battalion fields, roll 1D6 – 5 prior to the start of the game. If the result is negative, randomly determine one 'Mech in the lance and roll 1D6 to determine which of the following critical hits to apply to that 'Mech: 1–Engine, 2–Sensor, 3–Foot Actuator, 4–Jump Jet, 5–Shoulder Actuator, 6–Gyro. Once the specific critical hit has been determined, randomly assign it to an appropriate slot on the Critical Hit Table of the 'Mech's record sheet.

MECHWARRIOR RULES

RANK TRAIT

The table on p. 148 shows the various levels of the Rank Trait used throughout the former Federated Commonwealth. This replaces the Federated Suns and Lyran Alliance columns in the Rank Table (p. 87, *MW3*). The ranks shown in brackets were never officially awarded by the AFFC, yet were widely recognized.

ADDITIONAL LIFE PATHS

The following rules may be used with *MW3*. These will expand on the training opportunities open to citizens of the Federated Suns and cover opportunities from recent military events.

Players and gamemasters should follow all of the standard rules for determining character events (see p. 24, *MW3*), with one exception: instead of rolling 2D6, players should roll 2D10 to randomly select an event.

STAGE 3: MAJOR FEDERATED SUNS ACADEMY (GENERIC)

Affiliation Federated Suns only (Lyran Alliance may attend prior to the secession of the Alliance in 3057). Cannot have Combat Paralysis or any of the following Traits above the lowest possible level: Disabled, Lost Limb, Poor Vision, Poor Hearing. The character must not have any seriously "questionable" Events in his or her history (per the GM's judgment) unless the character has or purchases an appropriate Contact who can hide the Event.

This is not a full Life Path and should not be taken as one. Instead, the information contained below offers the generic information common to most Federated Suns military academies. The entries that follow round out a full and unique Life Path as offered by one of the Major Federated Suns Military Academies detailed in *Sharpening the Claws*, p. 35. Each Major Academy has its own entry, along with its own unique requirements, skill sets, etc. Simply replace or add to the requirements, traits and skills listed below with those unique ones listed in the specific academy entry.

Time: 1 Year
Attribute Minimums: INT 3, WIL 4, SOC 3
Attribute Thresholds: WIL +1, SOC +1
Traits: Promotion, Well Equipped
Skills: Academic/Federated Suns History +2, Protocol/Federated Suns +2, Perception +1
Fields: Basic Training
Previous Paths: Any Path or Event Roll which would normally allow admission to a Military Academy Path.
Next Path: Advanced Individual Training Subpath (mandatory, part of this path)

Federated Suns Academy Events

2 A simple prank went horribly wrong, and at least one cadet ended up in the morgue. [Enemy (2), Bad Reputation (2), lose all AIT skills, no more legitimate military Paths]
3 A camping trip turns into a life and death struggle. [BOD –2, RFL –2, Disabled, Lost Limb (2)]
4 "We were defending our teammate from a group of armed civilians, SIR!" [Bad Reputation, Bureaucracy/Federated Suns +2, Unarmed Combat +2, add 2 years to the time this Path takes]
5 It was supposed to be a simple training run. [+1 to all AIT skills; choose either Disabled, Introvert and Combat Sense *or* Brave and Quirk/Psychotic
6 The Commandant must not have liked the cut of your jib! [Add 1 year to the time this Path takes, Enemy]
7 Though no one can prove it, you were responsible. [Bad Reputation, Enemy]
8 That lower classman you liked to bully is a general's son! [Enemy (2)]
9 You drifted through the academy without standing out. To *anyone*. [CHA –1, lose one Contact]
10 Selling contraband on the side may not have been the most responsible extracurricular activity. [–2 to any two Military Fields or AIT skills, Appraisal +2, Scrounge +2, Stigma/"Go-to Man"]
11 Lucky you were a morning person. [Running +3, Swimming +3]
12 You lucky dog! You got the girl of your dreams and a career! [EDG +1, Good Reputation, Dependent]
13 You wiled away your free time with extracurriculars. [+2 in any three Interest/arts or sports skills]
14 Traded weekend passes for extra sim time. [+3 to any two AIT skills]
15 The Commandant served with your aunt "in the big one." [Contact (2), +1 to any three Military Field Skills]
16 Your scores were high enough to get you into an accelerated program. [Pick one Field from AIT or one Civilian Field, add one year to the time this Path takes]
17 The sarge barely survived the Clan front, but made it his business to make sure you would. [WIL +2, Toughness, Pain Resistance]
18 Congratulations, you graduated top in your class! [EDG +1, Good Reputation, Contact (2), Wealth (2), Well-Equipped *or* Vehicle (3), may take OCS next, then return and continue this Path]
19 [Choose one event or roll twice and apply both events]
20 [Choose two events or roll three times and apply each event]

Mandatory Subpath: Advanced Individual Training (AIT)

Basic Training Graduates only
Time: 2 years
Traits: MechWarriors and Aerospace Pilots receive Promotion to Sergeant (Rank 6), all others receive Promotion
Skills: +2 to any two Field Skills (combat fields marked with asterisk also gain Tactics/Any +1)
Fields: Choose one of the following primary fields, subject to the listed restrictions:

- Aerospace Pilot* (*minimum DEX 4, RFL 4*)
- Armored Infantry* (*minimum STR 6, BOD 5*)
- Basic Technician
- Cavalry* (*minimum RFL 3*)
- DropShip Pilot (*minimum DEX 4*)
- Infantry* (*minimum BOD 3, STR 3*)
- Marine* (*minimum BOD 4, may not have TDS Trait*)
- MechWarrior* (*minimum DEX 3, RFL 4*)

Events: Do not roll Events for Advanced Individual Training
Next Path: Tour of Duty: Federated Suns (4), Tour of Duty: Inner Sphere (4), Special Training (3, part of this pass; an option if character took Infantry, Armored Infantry, MechWarrior, DropShip Pilot or Basic Training in AIT) or Tour of Duty: Training Cadre (4, an option if the character took a Military Field and the school fields a training cadre or battalion)

Subpath: Special Training

AIT Graduates only
Time: 2 years
Attribute Minimums: WIL 5
Attribute Thresholds: Add +1 to any one threshold of the player's choice
Traits: Promotion, Well Equipped (2)
Skills: +1 to any three Basic Training Skills, +1 to half of the character's AIT skills (rounding up)
Fields: Choose one of the following Fields, subject to the listed restrictions:

- Aerospace Technician (*Basic Technician Field required*)
- JumpShip Pilot (*minimum INT 4, DropShip Pilot Field required, may not have TDS Trait*)
- 'Mech Technician (*Basic Technician Field required*)
- Military Scientist (*minimum INT 5*)
- Special Forces (*minimum BOD 4, RFL 4; Armored Infantry, Infantry, Marine or MechWarrior Fields required; may not have TDS Trait*)
- Vehicle Technician (*Basic Technician Field required*)

Next Path: Tour of Duty: Federated Suns (4), Tour of Duty: Inner Sphere (4) or Covert Ops (4, Special Forces or Military Scientist Fields only)

Subpath: Officer Candidate School (OCS)
Attainable by Event roll only. The character must not have any "questionable" Events in his or her history (per the GM's judgment) unless the character has or purchases an appropriate Contact who can hide the Event.
Time: 1 year
Attribute Thresholds: WIL +1
Traits: Commission (Rank 1)
Skills: Academic/Military History +2, Bureaucracy/Federated Suns +2, Protocol/Federated Suns +1, Strategy +1
Fields: Officer Training
Next Path: Return to last path at the same point as left for OCS

ALBION MILITARY ACADEMY

No Disabled, Glass Jaw, Poor Vision or Poor Hearing, in addition to the requirements above.
Attribute Minimums: INT 3, WIL 5, SOC 5
Attribute Thresholds: DEX +1, RFL +1, BOD +1, WIL +1
Traits: Promotion, Well-Connected/New Avalon, Well Equipped
Skills: Running +4, Career/Soldier +4, Protocol/Federated Suns +2, Academic/Federated Suns History +1, Leadership +1, Perception +1
Mandatory Subpath: Advanced Individual Training (AIT)
Skills: +2 to any two Military Field Skills, Leadership +1, Survival +1, Interest/Sport (any) +1
Subpath: Special Training
Attribute Minimums: WIL 5, EDG 5
Attribute Thresholds: +1 to BOD, DEX or RFL (player's choice)
Traits: Promotion, Well-Equipped, Contact (2)
Skills: +1 to any three Basic Training Skills, +1 to half of the character's AIT skills (rounding up), Career/Soldier +2, Survival +1
Fields: Only the following Fields are available: JumpShip Pilot, Military Scientist or Special Forces

NAIS COLLEGE OF MILITARY SCIENCES

No Slow Learners, in addition to the requirements above.
Attribute Minimums: INT 4, WIL 3, SOC 3
Attribute Thresholds: RFL +1, INT +1, WIL +1, SOC +1
Traits: Promotion, Well Equipped
Skills: Academic/Any +2, Bureaucracy/Federated Suns +2, Computers +1, Swimming +2

Mandatory Subpath: Advanced Individual Training (AIT)
Skills: Leadership +1, Strategy +2, +1 to any two relevant skills, +1 to any one field skill
Fields: Only the following Fields are available: Aerospace Pilot, Armored Infantry, Basic Technician, Cavalry, DropShip Pilot, Infantry, MechWarrior.

Subpath: Special Training
Time: 3 years
Attribute Thresholds: Add +1 to any one threshold of player's choice
Traits: Promotion, Wealth, Well-Equipped, Well-Connected/New Avalon
Skills: Strategy +2, Tactics/Any +2, +1 to three Basic Training Skills, +1 to half the character's AIT skills (round up), +2 to any one other skill
Fields: Only the following Fields are available: JumpShip Pilot, Military Scientist or Special Forces

WARRIOR'S HALL

Questionable events in a character's history will not necessarily bar a character from attending the Warrior's Hall, so long as the Event did not harm the Capellan March in any way (or, conversely, harmed the Capellan Confederation in some way). The GM should make the final determination on this.
Attribute Minimums: INT 3, WIL 4, SOC 3
Attribute Thresholds: WIL +1, SOC +1 (add +2 to the final SOC Attribute when operating within the Capellan March)
Traits: Promotion, Well Connected, Quirk/Hates Capellan Confederation
Skills: Academic/Capellan March History +6, Bureaucracy/Federated Suns +3, Academic/Federated Suns History (general) +2, Leadership +2, Perception +1
Subpath: Special Training
Attribute Minimums: WIL 4

Attribute Thresholds: SOC +1
Traits: Promotion, Well-Connected/Capellan March (2), Contact (2), Quirk/Loyalty to Hasek Family (GMs may treat this as a lesser form of In For Life)

ROBINSON BATTLE ACADEMY

Questionable events in a character's history will not necessarily bar a character from attending the Robinson Battle Academy, so long as the Event did not harm the Draconis March in any way (or, conversely, harmed the Draconis Confederation in some way). The GM should make the final determination on this.
Attribute Minimums: INT 3, WIL 4, SOC 3
Attribute Thresholds: WIL +1, SOC +1
Traits: Promotion, Well Connected, Quirk/Hates Draconis Combine
Skills: Bureaucracy/Federated Suns +1, Academic/Federated Suns History +1, Perception +1

Mandatory Subpath: Advanced Individual Training (AIT)

Skills: Strategy +2, Tactics/Any +2, +2 to any two Military Field Skills

Subpath: Special Training

Attribute Minimums: WIL 4
Attribute Thresholds: SOC +1
Traits: Promotion, Well-Connected/Draconis March (2), Contact (2), Quirk/Loyalty to Sandoval Family (GMs may treat this as a lesser form of In For Life)

SAKHARA ACADEMY

No Disabled, Poor Vision or Poor Hearing, in addition to the requirements above. Character must have at least one 2-point Contact (a graduate of Sakhara who can recommend the character).
Attribute Minimums: INT 6, WIL 5, SOC 5, EDG 3
Attribute Thresholds: WIL +1, SOC +1
Traits: Good Reputation (2), Well-Connected/Federated Suns (2), Natural Aptitude/Strategy, Poverty (4; character cannot have lower than Poverty (1) and must spend Character Points or have sufficient Wealth prior to play to offset any additional Poverty), Quirk/Honorable
Skills: Academic/Military History +6, Strategy +4, Perception +3, Tactics/Any two +3, Language/Any two the character does not already possess +2
Fields: Basic Training *and* Officer Candidate School (OCS)

Mandatory Subpath: Advanced Individual Training (AIT)

Skills: Career/Soldier +3, +3 to any three Military Field Skills, +1 to any two others
Fields: Only the following Fields are available: Aerospace Pilot, Cavalry, Infantry, MechWarrior.

Mandatory Subpath: Officer Candidate School (OCS)

Traits: Commission (Rank 1)
Skills: Bureaucracy/Federated Suns +4, Protocol/Federated Suns +4, Leadership +4, Training +3, Protocol/Any two the character does not already possess +2
Fields: Officer Training

Subpath: Special Training

Time: 3 years
Attribute Minimums: INT 6, WIL 5
Attribute Thresholds: EDG +1
Traits: Promotion, Well-Equipped, Contact (2), Good Reputation, Natural Aptitude/Tactics (Any), Poverty (2)
Skills: Leadership +4, +2 to half of the character's AIT skills (rounding up), +3 to any three other skills
Fields: Choose one of the following Fields, subject to the above listed restrictions: Military Scientist or Special Forces

ARMSTRONG FLIGHT ACADEMY

No Disabled, Glass Jaw, Night Blindness, Poor Vision, Poor Hearing or TDS, in addition to the requirements above.
Time: 2 years
Attribute Minimums: WIL 3, BOD 5, DEX 6, RFL 6, EDG 4
Attribute Thresholds: BOD +1, WIL +1
Traits: G-Tolerance, Promotion, Well Equipped, Vehicle (2), Custom Vehicle (2)
Skills: Perception +4, +1 to any five related Technician skills
Fields: Basic Training *and* either Aircraft Pilot or Basic Technician

Mandatory Subpath: Advanced Individual Training (AIT)

Fields: Only the following Fields are available: Aerospace Pilot or DropShip Pilot

Subpath: Special Training

Attribute Minimums: WIL 4
Attribute Thresholds: +1 to BOD, DEX or RFL (player's choice)
Traits: Promotion, Well-Equipped (2), Contact (2)
Skills: +1 to any three skills, +1 to any two Technician skills, +1 to any two Technician skills not already possessed.
Fields: Only the following fields are available: Aerospace Pilot, DropShip Pilot or Aerospace Technician

POINT BARROW MILITARY ACADEMY

Attribute Minimums: INT 4, WIL 4
Attribute Thresholds: BOD +1, DEX +1
Traits: Promotion, Well Equipped
Skills: Academic/Federated Suns History +2, Protocol/Federated Suns +2, Perception +2

Mandatory Subpath: Advanced Individual Training (AIT)

Skills: Scrounge +4, Computers +3, Engineering +3, +3 to any two Technician Skills, +2 to any two other Military Field Skills
Fields: Only the following Fields are available: Aerospace Pilot,

Basic Technician, Cavalry, Infantry or MechWarrior

Subpath: Special Training

Attribute Minimums: WIL 4

Attribute Thresholds: Add +1 to any one threshold of the player's choice

Traits: Promotion, Well-Equipped

Skills: +2 to half of the character's AIT skills (rounding up), +3 to any three other Military Field Skills

Fields: Only the following Fields are available: Aerospace Technician, JumpShip Pilot, 'Mech Technician, Vehicle Technician

WAR COLLEGE OF GOSHEN

No Disabled, Poor Vision or Poor Hearing, in addition to the requirements above. (Character may ignore these requirements if he has either Combat Sense or Sixth Sense, or two 2-point Contacts (alumni who can provide recommendations).

Attribute Minimums: INT 6, WIL 6, SOC 3, EDG 3

Attribute Thresholds: BOD +1

Traits: Natural Aptitude/Strategy, Quirk/Honorable

Skills: Academic/Military History +8, Academic/Philosophies of War +8, Academic/Philosophies of Peace +8, Strategy +6, Academic/Federated Suns History +4, Protocol/Federated Suns +4, Tactics/Any four +1

Mandatory Subpath: Advanced Individual Training (AIT)

Skills: Career/Soldier +3, +3 to any three Military Field Skills, +1 to any two others

Fields: Only the following Fields are available: Aerospace Pilot, Cavalry, Infantry, MechWarrior.

Subpath: Special Training

Time: 3 years

Attribute Minimums: INT 6, WIL 5

Attribute Thresholds: EDG +1

Traits: Promotion, Well-Equipped, Contact (2), Good Reputation

Skills: Academic/Military History +4, Academic/Philosophies of War +4, Academic/Philosophies of Peace +4, +2 to half of the character's AIT skills (rounding up), +3 to any three other skills

Fields: Only the following Fields are available: Military Scientist or Special Forces

KILBOURNE ACADEMY

Characters may possess "questionable" Events in their past so long as the Event in question is not a conviction for a serious crime (such as espionage, premeditated murder, etc.; the GM should use his or her best judgment).

Traits: Promotion, Quirk/Paranoid (Big Brother is Watching)

FILTVELT MILITARY ACADEMY

Characters may possess "questionable" Events in their past so long as the Event in question is not a conviction for a serious crime (such as espionage, premeditated murder, etc.; the GM should use his or her best judgment).

Traits: Promotion, Stigma/Graduate of a "Second-Rate" AFFC Academy

TIKONOV SCHOOL OF MILITARY DISCIPLINE

Characters may possess "questionable" Events in their past so long as the Event in question is not a conviction for a serious crime (such as espionage, premeditated murder, etc.; the GM should use his or her best judgment) or has an appropriate Contact that can cover up the Event.

Attribute Thresholds: WIL +2, SOC -2

Traits: Promotion, Toughness, Stigma/Tikonov separatist (whether or not it is true)

Skills: Streetwise/Tikonov Reaches +6, Bureaucracy/Federated Suns +4, Academic/Federated Suns History +2, Protocol/Federated Suns +2, Perception +2

STAGE 3: TRAINING BATTALION

Affiliation Federated Suns only (Lyran Alliance may attend prior to the secession of the Alliance in 3057). Cannot have Combat Paralysis or any of the following Traits above the second lowest possible level: Disabled, Lost Limb, Poor Vision, Poor Hearing. Additionally, characters may possess "questionable" Events in their past so long as the Event in question is not a conviction for a serious crime (such as espionage, premeditated murder, etc.; the GM should use his or her best judgment) or has an appropriate Contact that can cover up the Event.

With the formation of the Training Battalions, the ranks of the MechWarriors were no longer limited solely to the wealthy and politically connected. Literally anyone who could prove that he or she was capable could finally join the Federated Suns' elite cadre of modern-day knights. Opportunities for promotion and career advancement are often hard to come by for those hailing from a Training Battalion, as graduation from an academy is virtually required for anyone looking for a command position. Then again, most of these men and women are looking only to serve their nation or to fulfill a lifelong goal to become a MechWarrior.

Time: 1D6 + 3 Years

Attribute Minimums: DEX 3, RFL 4, EDG 5

Attribute Thresholds: WIL +1, SOC +1

Traits: Promotion to Sergeant (Rank 6), Stigma/Second-Rate MechWarrior (whether or not true)

Skills: +2 to all skills in the Basic Technician Field, Scrounge +1, +1 to any one learned skill for every year spent in the battalion (this bonus may not be combined together with itself)

Fields: Basic Training and MechWarrior

Previous Paths: Any

Next Path: Tour of Duty: Federated Suns (4)

Training Battalion Events

2 You blew up a friendly 'Mech in your first live-fire exercise. [Bad Reputation (2), Poverty (2), No more legitimate military Paths]

3 Some fool shot you and your 'Mech up in his first live-fire

exercise. [Addiction (2—Painkillers), Disabled (as effect of serious head wound, p. 125, *MW3*), Lost Limb (2)]
4 Fell in with the "wrong" crowd. [In for Life, Contact (2 – Mafia Boss)]
5 The Battalion was called to service early, and your best friend was killed. [–1 to all AIT skills, +3 to any two skills, Addiction, Quirk/Vow, Quirk/Temper]
6 Even a military life couldn't keep you away from the pleasures of youth. [–1 to all Military Field Skills, Escape Artist +1, Gambling +1, Seduction +1]
7 You hold the record for the most safety violations in any one year. [Bad Reputation (2)]
8 The sarge liked everyone but you. [Choose: WIL –1, Introvert or Timid]
9 Best of friends turn to bitter enemies over the smallest of things. [Enemy]
10 It was either helping out at the local soup kitchen or extra PT to make up for all those foul-ups. [Bad Reputation and either Streetwise +2 *or* Running +2]
11 You passed, and that's all that matters! [+1 to any one skill]
12 Could've been worse: you might've spent all that time mopping the floors instead of filing in triplicate! [Administration +3, Computers +3, Bureaucracy/Federated Suns +3]
13 Your skill with the hydrospanner almost equals your skill in the cockpit. [+2 to any three Technician skills]
14 A simple temporary assignment to the supply office uncovered some of your hidden talents. [Negotiation +4, Appraisal +3, Scrounge +3, Well-Equipped]
15 The other cadets looked up to you. [Leadership +3, Training +3, +2 in any three skills]
16 Officer thinking, cadet! [Leadership +2, may take OCS next, then return and continue this Path]
17 What are you doing *here*?! [Good Reputation, Contact; spend one year in this Path, gaining only a +2 to all Basic Training Field skills (no other benefits), but may attend a Military Academy as next Path, may take OCS next]
18 Your actions save the entire battalion from certain doom. [Good Reputation (2), Contact (3), Promotion (2), Custom Vehicle, and choose one of the following: Disabled, Lost Limb, Poor Hearing or Poor Vision, may take OCS next]
19 [Choose one event or roll twice and apply both events]
20 [Choose two events or roll three times and apply each event]

STAGE 4: TOUR OF DUTY: TRAINING CADRE

Available only to graduates of a Federated Suns Academy that fields a Training Cadre. This Tour may be taken only as the first Stage 4 Path, and only immediately after a character graduates from an Academy.

Time: 2 Years
Traits: Promotion, Contact
Skills: Protocol/Federated Suns +2, +1 to all Basic Training skills, +1 to half of Primary Field skills (round up)
Previous Paths: Any Federated Suns Military Academy with a Training Cadre
Events: Roll events on the Training Battalion Events Table
Next Path: Tour of Duty: Federated Suns or Inner Sphere (4)

STAGE 4: TOUR OF DUTY: FEDERATED SUNS

Available to characters who completed training at a Federated Suns Academy or Training Facility.

Time: 2 Years
Traits: Vehicle (2); if taken for a second time in a row, add Contact (2) and Stigma/either Victor or Katherine Loyalist (player choice)
Skills: Add +2 to any three Military Field skills, +1 to any three other skills
Previous Paths: Any which would allow entry into a regular Tour of Duty
Next Path: Covert Ops (4, Military Scientist, Scout or Special Forces-trained characters only), Ne'er-Do-Well (4), Police Academy (3), Tour of Duty: Federated Suns or Inner Sphere (4)

Tour of Duty: Federated Suns Events

2 You're having strange dreams about another life you never lead ... or did you?. [Amnesia, Quirk/Paranoid, GM should devise a "true" history for the character]
3 It was a "no-brainer," an assignment so simple you and your entire unit were taken prisoner. [–1 to all skills advanced this Path, Escape Artist +4, Survival +4, BOD –1, WIL –1 and add 1D6 years to the time this Path takes]
4 In the end, you took the coward's way out and watched your unit die, but you also took the hero's identity. [Acting +3, Alternate Identity, Combat Paralysis, Good Reputation (2), Quirk/Self-doubt]
5 You were there, and it was worse than anyone could've described. [+4 to any three Military Field Skills, and choose two of the following: Addiction (2), Combat Paralysis, Disabled (2), Glass Jaw, Lost Limb (3), Poor Hearing (2), Poor Vision (2) or Quirk/Phobia]
6 A bad bailout leaves you comatose for weeks. [Amnesia, Disabled]
7 Everyone else was on the take, so you had to at least take the money or become their patsy. [Player's choice: either In For Life, Stigma/On the Take and Wealth (4) *or* Enemy (3)]
8 You survived the fight that half your unit didn't walk away from, but not without some "problems." [+3 to any two Military Field skills, Interest/Astrology +4, Addiction/Astrology (must consult a seer before making any major decisions)]
9 Congratulations! You caught a rare disease in the Periphery and won an all-expenses paid vacation to the NAIS. [Bureaucracy/ Federated Suns +4, Contact (2), Well-Equipped, Allergy (2) *or* Disabled (player's choice), and add 1D6 ÷ 2 years (round up) to the time this Path takes]
10 Here comes the bride ... [Dependent]

11 Caught in the middle of the growing feud within the AFFC. [Bureaucracy/Federated Suns +2, Enemy]
12 The proverbial cake walk. [+1 to any one Military Field skill]
13 Action in the Chaos March leaves you wondering about your place in life. [+2 to any four skills and player's choice: WIL –2, Brave and Quirk/Gung-Ho *or* WIL +2, Timid and Quirk/Pacifist]
14 You know what assignment to the Outback means? Party time! [–1 to all skills advanced this Path, Gambling +3, Seduction +3, Streetwise +3,]
15 Saw action in St. Ives ... against your supposed "allies." [+3 to any two Military Field Skills, Stigma/Victor *or* Katherine Loyalist (player's choice)]
16 War is hell, and the Smoke Jaguars were its minions. [+3 to any three Military Field skills, Addiction, Combat Sense, Commission (Rank 1—if not an officer) *or* Promotion (if already an officer)]
17 Those men in the suits and sunglasses say you've got "potential." [Add +3 to any three skills, Contact (2), Character may take the Marine, Officer Training or Special Forces AIT path next]
18 You were the right man at the right place—awarded the Silver Sunburst for gallantry under fire. [+6 to any one Military Field skill, EDG +1, Good Reputation (2), Promotion, Vehicle (4), Custom Vehicle (2), Well-Equipped, Wealth (2); choose from *any* Stage 3 Academy Life Path next, applying a +2 to any resulting Event rolls]
19 [Choose one event or roll twice and apply both events]
20 [Choose two events or roll three times and apply each event]

INFILTRATOR MK. II BATTLE ARMOR

R&D Start Date: February 3056
Prototype Design and Production: January 3059
Standard Production: March 3060

Though the AFFC was the first Inner Sphere military to field a battle armor suit, and soon thereafter debuted two additional, highly specialized suits, the fact that it was also the largest military in the Inner Sphere prevented it from continuing its pioneering trend. Quite simply, the AFFC had not only to equip more line units than any other nation with battle armor formations but also to rebuild countless Regular Army regiments shattered or destroyed during the Clan Invasion. While the NAIS and other military think-tanks kept abreast of the latest developments pursued by their rivals throughout the Inner Sphere, R&D funding dropped significantly as priority was assigned toward recouping losses.

The Infiltrator battle armor suit, with its integral stealth capabilities, was a design almost inherently suited for special forces and intelligence-gathering missions. Purchases made by the elements of the Federated Commonwealth Intelligence Command (FCIC), both civilian and military, accounted for the majority of the Infiltrators built in the first few years of production. The Infiltrator's many flaws were immediately evident however, and the Intelligence Secretariat, the FCIC's command element, directed its Office of Science and Technology (OST) to coordinate with the NAIS to develop a replacement for the Infiltrator, one better suited for the FCIC's missions.

The actual breakthrough, though, came from a surprising source. In 3057 the Intelligence Secretariat found that the Gray Death Legion possessed several Mark XXI Nighthawk suits obtained from a Star League vault on Karbala. Grayson Carlyle, now Baron von Glengarry, had no choice but to comply with the Archon-Prince's request for the suits. By August they were in route to NAIS, where they were systematically dismantled and analyzed. Though they were destroyed in the process, the Nighthawks provided invaluable data.

The Mk. II suit surpasses the Mk. I in every way, though with a substantial increase in cost. It is faster, better-armored and far more agile than the "waddling toy robot," or "Waddle," as many operators euphemistically named the Mk. I suit. This Mk. II suit, unofficially known as the "Puma," includes an integral jump pack, as well as an optional para-foil attachment for high-altitude insertions, in addition to increased low-observability capabilities and electronic counter measures. Designed to take a more direct combat role than its predecessor, the Mk. II is fitted with two fully functional hands, though one of those hands is encased within the hardpoint when fitted with a specially-designed Gauss rifle, giving them a sniping weapon far superior to any other on the battlefield. As with most other battle armor designs, the Mk. II also has a mounting hardpoint for a secondary anti-personnel weapon.

BATTLETECH GAME RULES

The Infiltrator Mk. II is Level 2 equipment. It follows standard rules for Inner Sphere battle armor (p. 63, *BMR*) except as noted below.

Infiltrator Mk. II squads consist of four troopers each. Each member within a squad is equipped with the same weapon; the Magshot Gauss rifle. The Magshot Gauss rifle has the same range profile as an Inner Sphere medium laser and does 2 points of damage. Each Infiltrator Mk. II battlesuit has an Armor Value of 6 points, plus 1 additional point representing the trooper inside.

Additionally, the armor provides excellent stealth abilities. Attacks against Infiltrator Mk. II units suffer a +1 to-hit modifier at short range, a +3 to-hit modifier at medium range in place of the standard medium-range modifier and a +6 to-hit modifier at long range in place of the standard long-range modifier. Beagle active probes and their Clan equivalents cannot detect hidden Infiltrator Mk. II units.

Each Infiltrator squad has limited ECM capability. The effective radius is only the hex the squad occupies and functions as a Guardian ECM Suite (p. 126, *BMR*).

When using the *Dropping Troops* rule (p. 72, *BMR*),

Infiltrator squads land successfully if they roll a 3 or better. If the Infiltrator squad rolls a 2 on the 2D6 roll, it has failed its landing and each trooper in the unit suffers 1D6 damage.

The Infiltrator Mk. II cannot carry an SRM launcher.

MECHWARRIOR GAME RULES

The Infiltrator Mk. II follows the normal rules for Inner Sphere Battle Armor (pp. 142–143, *MW3*) except as noted below.

The Infiltrator mounts one primary weapon (the Magshot Gauss rifle, see below) and one secondary weapon, but no backpack SRM launcher.

The Infiltrator Mk. II is equipped with an integral IR and electronic signal suppression system, which can help the wearer avoid detection by a variety of electronic sensors. Any characters wearing a functional Infiltrator Mk. II suit gains +6 to any *Stealth* skill rolls (or, alternately, any character attempting to sight an Infiltrator Mk. II suit suffers a blanket +4 to all *Perception* TNs, in addition to any other modifiers, including range and conditions). This bonus (or penalty) applies only when the character could be potentially monitored by either an IR detector or an electronic sensor, such as a radar (including BattleMech sensor sweep and lock, but not including standard television cameras or optical sights).

Equipment Ratings: E/E/F
Cost: 514,000 C-Bills
Armor Values: 8/7/6/6
Coverage: Full
Attributes: STR +3, DEX –1, RFL –1
Movement: –1m to all movement rates

Federated-Barrett "Magshot" Gauss Rifle

When the FCIC released its requirements for the Infiltrator Mk. II, one of those requirements was for an extremely accurate long-range ballistic weapon with the capability to defeat light vehicle armor, such as mounted by many APCs and battle armor units. Though unwritten, they were also looking for a sniper weapon capable of taking down any "soft target" with a single shot at extremely long ranges. Well known within the Federated Suns for their quality specialized rifles, Federated-Barrett was the OST's first choice on a short list of contractors to produce this new weapon.

Federated Arms, the parent of Federated-Barret, already had much success with the Thunderstroke and Grand Mauler designs, giving the designers a long head start on developing this new weapon. As soon as an electronics suite was chosen for the Mk. II, the Magshot's designers incorporated a unique sight system into the rifle. Though the 45-kilogram rifle can be used independently from the Mk. II suit, its sights are designed to be linked with the suit's targeting system and HUD, providing a more accurate shot. Additionally, the suit's operator can shut down power to the Mk. II's motive system, immobilizing the suit and providing a more stable firing position than any bipod or tripod could provide.

The rifle is quite expensive when compared with other available weapons, though the cost is made up for by the weapon's capabilities. Outside the Infiltrator Mk. II units (where the weapon is standard-issue), the DMI's MI6 by far fields the majority of the few weapons in service, though a few score have been delivered to a number of elite Regular Army special forces units throughout the AFFC.

Game Notes

When used independently, treat the weapon's sight as a standard telescopic sight with IR and range finding capabilities. When tied to the Mk. II's targeting system (i.e., mounted on the Mk. II's hardpoint), the operator ignores any bonuses granted by the sight and instead applies a blanket –2 to any ranged attack TN. The bonus increases to –4 if the operator deactivates the suit's motive system, though only under the following conditions: the operator must be on terrain capable of supporting a ton of equipment, the operator is unable to make any rapid movements, and the attacker is treated as completely immobile. Characters must expend a Complex Action to deactivate the motive system and brace for the shot, and a Simple Action to reactivate it.

This weapon consumes ammunition like other ballistic weapons, and also requires a power pack like energy weapons. Each attack consumes one shot of ammo *and* 3 power points.

Skill: Support Weapons
Equipment Ratings: E/E/F
AP•Damage: 6•7D6
Type: B
Range S/M/L/E: 70/280/600/1250
Shots: 20, 3 power points per shot
Cost/Reload: 8,500 C-bills/10
Weight: 45 kg
Notes: Encumbering

CAVALIER BATTLE ARMOR

Prototype Design and Production: January 3058
Standard Production: June 3058

With the Infiltrator Mk. II battle armor already in the prototype stage, the AFFC High Command felt it was time to move away from the standard Inner Sphere battlesuit. Though the NAIS had pioneered the original suit, it was quickly becoming the norm in several Houses. Similar to the Combine's Raiden suit, whose battlefield performance is identical to the "standard" Inner Sphere battle armor but has a unique Japanese profile, the AFFC felt the same was appropriate for a new suit. This "redesign for Davion pride" project, which was originally spearheaded by Victor-Steiner Davion, led to the new suit, which upon completion was dubbed the Cavalier. As the suit had already been in service for years, only short months were needed to verify the new armor molding would not inhibit battlefield effectiveness.

BATTLETECH GAME RULES

The Cavalier follows the normal rules for Inner Sphere battle armor (p. 63, *BMR*).

MECHWARRIOR GAME RULES

The Cavalier follows the normal rules for Inner Sphere Battle Armor (pp. 142–143, *MW3*).

NEW EQUIPMENT

Ever since its formation, the NAIS has been at the cusp of advances in weapons research, both in unlocking the secrets lost in the fall of the Star League as well as in developing entirely new technologies. Though the majority of the university's R&D funding has been devoted to breaking the Clan technology barrier, the last several years have seen a resurgence of interest in developing new variations on the standard autocannon, as well as special munitions for use with standard autocannons.

All of these items are for use in Level 2 BattleTech and can be used in tournament play unless specifically prohibited under the rules of individual tournaments. Currently these items are only available to Federated Suns forces.

ROTARY AUTOCANNON

During the mid-3050's, the NAIS developed lighter-weight prototypes of several small-caliber autocannons, based partially on advances pioneered with the LB-X classes of autocannon. Unfortunately, the drop in mass also lead to a significant drop in accuracy at long ranges, and after months of lackluster combat trials, the Department of the Quartermaster killed the program when it passed on bids from two contractors to build these new weapons.

Undaunted, NAIS scientists continued to refine the design. The breakthrough came when a student revived interest in the long-theorized rotary autocannon. Autocannons, especially the LB-X class, have long used mulitple barrels. However, due to overheating problems and structural stress, multiple barrels did not appreciably increase the autocannons' rate of fire. The rotary autocannon was a multiple barrel concept, designed to increase rate of fire beyond the Ultra class autocannon. Deemed overly massive and unstable, the RAC was never even prototyped. Based on the advances made in the last few years, however, NAIS were able to quickly design and build several working models. Though the largest caliber weapons were still too massive and too unreliable, with several designs catastrophically failing even under lab conditions, two smaller-caliber RAC designs survived trials and were soon accepted by the DQ.

The principle of the RAC finally became viable with the development of the light autocannon. Taking the stock light autocannon, sustained rate of fire is increased by adding additional barrels, similar to the newer LB-X class autocannons being manufactured, and wedding those barrels with a more capable breech and loading mechanism—a design made possible after years of refining the "Ultra"-classes of autocannon. While the weapon generates additional excess heat, the stress on each individual barrel is reduced, leading to a longer service life. Unfortunately, the weapon is still prone to lock-ups and jams when firing at its highest rates of fire for lengthy periods.

Game Rules

A player firing a rotary autocannon must specify whether it is firing one, two, four or six shots. If firing one shot, all standard combat rules apply. If firing multiple shots, use the following special rules.

A rotary autocannon firing two shots generates twice as much heat and uses 2 shots of ammunition instead of the standard 1. A rotary autocannon firing four shots generates four times as much heat and uses 4 shots of ammunition. A rotary autocannon firing six shots generates six times as much heat and uses 6 shots.

When firing two shots and the standard to-hit roll is successful, the player rolls on the "2" column of the Missile Hits Table to determine how many shots struck the target. When firing four shots and the standard to-hit roll is successful, the player rolls on the "4" column of the Missile Hits Table to determine how many shots struck the target. When firing six shots and the standard to-hit roll is successful, the player rolls on the "6" column of the Missile Hits Table to determine how many shots struck the target.

When firing multiple shots, roll separately for a hit location for each attack; each hit inflicts the full amount of damage possible for an autocannon of the size used. All shots must be fired at the same target.

If a player is firing two shots and rolls a result of 2 on his to-hit roll, the autocannon's loading mechanism jams, making the weapon useless until it is successfully unjammed. If a player is firing four rounds and rolls a result of 2 or 3 on his to-hit roll, or is firing six rounds and rolls a result of 2, 3 or 4 on his to-hit roll, a jam ocurrs.

The player may attempt to unjam the weapon during the course of battle. The player must declare in the End Phase of any turn that he or she will attempt to unjam the weapon in the next turn. During the turn in which the player is attempting to unjam the weapon, the unit that suffered the jam must either stand still or expend walking/cruising MPs (the unit may not run/flank or jump). The unit must also make no weapon attacks (including TAG attacks or spotting for indirect or artillery attacks, though all other electronics, such as C3 and Guardian ECM, may operate normally). At the end of the Weapon Attack Phase, the player controlling the unit should make a Gunnery Skill Roll with a +3 modifier. A successful roll indicates the weapon is cleared and may be used in successive turns, while a failure means the jam was simply not cleared (though the player may attempt to unjam the weapon in a successive turn). The player may attempt to unjam a single weapon only once per turn, though he or she may attempt to unjam multiple RACs in the same turn (an "unjam" roll must be made for each jammed weapon, however). For purposes of a *MW3* game, consider the act of unjamming one or more RACs a Complex Action.

Targeting Computer: If the firing unit is using a targeting computer to aim at a specific hit location, and multiple shots hit, all the shots hit the targeted location.

Criticals: If a RAC suffers a critical hit while it is jammed, it causes an ammo explosion with a damage value equal to a single shot for the weapon.

AUTOCANNON SPECIAL MUNITIONS

Though many militaries in the Inner Sphere have scorned the venerable autocannon, replacing it with energy weapons or the sophisticated Ultra and LB-X autocannon types, the Federated Suns has traditionally had an almost romantic attachment to this much maligned weapon. With the NAIS devoting significant resources to fielding additional munitions that would improve the use of their beloved ACs, it was only a matter of time until such munitions would reach the AFFC in mass.

Unless noted otherwise, the following ammo types are only available in full-ton lots. None of the following munitions are available for the Ultra, LB-X or rotary autocannon. As with other types of special ammo, the player must announce which type he will use when declaring weapons attacks.

Armor-Piercing Ammunition

Standard autocannon ammunition consists of high-explosive armor-piercing (HEAP) rounds. Manufacturers have always experimented with ways to improve the armor-piercing qualities of standard AC rounds, but have made no significant advances in the last four centuries. However, researchers at the New Avalon Institute of Science (NAIS) have made a breakthrough that greatly increases the armor-piercing power of an AC round by using a combination of top-secret materials and explosives. Though the sheer weight of this type of round significantly reduces its muzzle velocity, cutting its accuracy and reducing the number of rounds per ton, its designers believe the increased armor penetration is worth the tradeoff.

Every hit with armor-piercing ammunition that strikes armor provides a chance for a critical hit, even if the internal structure was not damaged by the attack. After marking off the armor damage for the attack, roll once on the Determining Critical Hits Table. Apply a modifier to the die roll based on the type of autocannon used: –1 for AC/20; –2 for AC/10; –3 for AC/5; or –4 for AC/2. If the initial attack damages the internal structure, make the standard roll for possible critical hits. Armor-piercing ammunition has no additional effect for attacks damaging the internal structure.

The mass of armor-piercing ammunition means that a ton of armor-piercing ammo contains half as many shots as a ton of standard ammo (rounded down). In addition, armor-piercing rounds are more difficult to aim, adding a +1 modifier to the to-hit number at all ranges.

Flechette Ammunition

Designed to combat unarmored infantry, flechette rounds release a shower of metal slivers. These rounds are less effective against armored targets.

Autocannons firing flechette ammo inflict double the normal damage against unarmored infantry units. If such a unit struck by flechettes is in Clear terrain, the flechettes do four times normal damage. Flechette ammo inflicts normal damage against battle armor units. Against all other types of units the damage is halved (round fractions down).

Incendiary Ammunition

Coated with a thin film of magnesium and an inferno type warhead, these rounds ignite as soon as they leave the autocannon barrel and are almost impossible to extinguish, as they continue to burn until the magnesium material is consumed.

Autocannons firing incendiary ammo inflict 2 additional points of damage against unarmored infantry units.

Players can use incendiary ammunition to start fires in any suitable terrain. Rather than the usual 9+ for determining fires (p. 74, *BMR*) the Target Number to start fires is 5+. Incendiary ammunition otherwise operates like normal autocannon rounds.

When the autocannon is fired, the round illuminates the line of fire, making it much easier to adjust fire in darkness. Autocannons firing incendiary rounds reduce the to-hit modifier for night combat to +1.

A critical hit on an autocannon that has fired incendiary ammunition in the same Turn causes the incendiary coating to combust, resulting in an explosion with results identical to an ammunition explosion. Treat the result as an ammunition explosion in the location containing the autocannon, with a damage value equal to the damage of an autocannon being used.

Precision Ammunition

With an eye to creating a "smart" autocannon submunition, the NAIS spent long years in research that resulted in the precision ammunition. Incorporating advanced targeting circuitry and a modified, self-propelled gyrojet shell, the precision ammo can auto-correct towards a target in flight, using extremely limited microbursts delivered by the gyrojet shell. Because of the limited "intelligence" of the targeting circuitry, the ammunition only proved beneficial against moving targets.

When firing with precision ammunition, reduce the target movement modifier by 2 to a minimum of zero.

The weight of the targeting circuitry and micro thrusters means that a ton of precision ammo contains half as many shots as a ton of standard ammo (rounded down).

TARGETING COMPUTER

Always on the cutting edge, the scientists at the NAIS recently made what could possibly be one of the most important military break-throughs in recent memory. Having spent more then a decade analyzing numerous captured models, they

FEDERATED SUNS WEAPONS TABLE

Type	Heat	Dmg	Min.	Short	Med.	Long	Tons	Crit.	Ammo
Rotary AC/2	1	2	—	1–6	7–12	13–18	6	3	45
Rotary AC/5	1	5	—	1–5	6–10	11–15	8	6	20
Targeting Computer	—	—	—	—	—	—	*	*	—

* See special rules for this equipment.

AEROTECH 2 WEAPONS SUPPLEMENT

Type	Class	Heat	Attack Value	Max. Range	Tons	Ammo Per Ton
Rotary AC/2	Autocannon	6	8	Medium	8	45
Rotary AC/5	Autocannon	6	20	Medium	10	20

BATTLEFORCE 2 ROSTER SUPPLEMENT

Name	MP	Damage PB/M/L	Over-heat	Class	Armor/ Structure	Point Value	Specials
ENF-6T Enforcer III	5J	3/2/1	—	M	4/2	16	
GRM-01C Garm	5J	1/1/1	—	L	2/3	9	
JM7-F JagerMech	4	8/6/4	—	H	5/3	15	ecm
TLR1-O Templar	4	6/5/1	—	A	7/3	18	
TLR1-OA Templar	4J	6/4/3	—	A	7/3	20	ecm
TLR1-OB Templar	4	5/3/3	1	A	7/3	14	
Infiltrator Mk. II Squad	3J	1/1/—	—	I	3/—	1	car4, ecm

FEDERATED SUNS EQUIPMENT COSTS AND BATTLE VALUES

Submunitions Type	Cost Multiplier
Autocannon	
Armor-Piercing AC	4 x normal
Flechette AC	1.5 x normal
Incendiary AC	2 x normal
Precision AC	6 x normal

Type	Cost (unloaded)	Ammo Costs (per ton)	Battle Value
Rotary AC/2	175,000	3,000	118/15
Rotary AC/5	275,000	12,000	247/31
Targeting Computer	10,000 (per ton)	—	*
Infiltrator Mk. II BA Squad	2,600,000	—	121

*Same Battle Value as Clan Targeting Computer (see p. 144, *BMR*).

finally unveiled an Inner Sphere targeting computer. Although not as efficient as its Clan counterpart, it is every bit as deadly on the battle field.

Game Rules

Inner Sphere targeting computers follow the rules for Clan targeting computers on p. 134, *BMR*, with the following exceptions. The computer weighs 1 ton for every 4 tons (or fraction thereof) of direct-fire weapons the unit is carrying. The computer fills a number of critical slots equal to its tonnage.

BATTLEMECH VARIANTS

With the enforced peace with the Clans the AFFC is again able to flex its spending muscle on newly developed technologies. Two brand new chassis, along with several new variants on classic BattleMech designs, were introduced prior to the Star League offensive against the Clans. The DQ issued contracts for production runs of three more variants during the course of the campaign, with the first of this latest batch of new machines entering service just over a year ago.

Two of these three new variants make use of the recently developed Rotary Autocannon. Because of this new weapon system, the GRM-01C *Garm* is finally gaining acceptance within the AFFC mainstream. Likewise, units throughout the AFFC are clamoring for the upgraded JM7-F *JagerMech*. The ENF-6T *Enforcer*, a variant derived from a popular field refit, is the final of these 'Mechs purchased in any major quantity by the DQ.

TLR1-O TEMPLAR

Mass: 85
Chassis: Kallon Type XIX Endo Steel
Power Plant: KalTek 340 Extralight
Cruising Speed: 43 kph
Maximum Speed: 65 kph
Jump Jets: None
Jump Capacity: None
Armor: Starshield A with CASE
Armament: 40 tons of pod space available
Manufacturer: Kallon Industries
Primary Factory: Talon
Communications System: Newberg Telescan 17
Targeting and Tracking System: Spar 3C MegaTrak

OVERVIEW

With the Department of the Quartermaster concentrating mainly upon upgrading the AFFC's conventional BattleMech forces, many within the High Command voiced criticism toward this policy, arguing they needed to field a strong force of OmniMechs if the Federated Commonwealth were to maintain a technological parity with the rest of the Inner Sphere. Inevitably, the Prince, and later the Archon, quashed these arguments on economic reasons, maintaining the AFFC could not afford the costs to design and build a new OmniMech. Rather than let this continue, Field Marshal Duke George Hasek instead commissioned Kallon Industries to design and build an assault-class OmniMech, initially financing this project with his own family's fortune.

CAPABILITIES

Intended to take the place of some of the AFFC's most venerable assault-class 'Mechs, such as the *Awesome* (chassis it closely resembles) and the *Victor*, the *Templar* makes the most out of technologies that have just been introduced, coupling them with time-tested off-the-shelf equipment. Not only did this bring development time and costs down, but when considering its massive armor protection, the *Templar* is both a durable and easy to repair machine. The Primary configuration, with its powerful Gauss rifle and strong array of lasers, is designed to command a battlefield. With the addition of the NAIS' recently introduced advanced targeting system, this configuration is capable of defeating even the heaviest Clan OmniMech.

The most common variant pairs an Ultra 10-type autocannon with an ER PPC, again backed with the advanced targeting system. As the *Templar* was designed to showcase such new technologies, the inclusion of the expensive targeting computer on two of its three configurations was deemed appropriate. Added mobility is also provided, with the addition of jump jets.

A third variant, though not as common as the first two, is finding its way into the AFFC's Regular Army. Conceived to engage an opposing unit at the longest of ranges, the 'Mech mounts an extended-range particle cannon and three of the lightest-class RACs.

DEPLOYMENT

The first *Templar*s off of Kallon's lines were routed to the Syrtis Fusiliers, though machines from later runs have found their way to units like the Federated Suns Armored Cavalry and the Davion Assault Guards. Duke Hasek's people have been carefully parceling out these new OmniMechs, however, insuring that only select units receive the *Templar*. A majority remain stationed within the Capellan March, all of which have been assigned to units that show nothing but the greatest of loyalty to either Duke Hasek or Prince Victor. The remaining few have made their way to the Draconis March, in a show of solidarity between Duke Hasek and Duke Sandoval. To date, no unit claiming loyalty to Archon Katherine has been able to requisition one of these new *Templar*s.

Type: **Templar**
Technology Base: Inner Sphere OmniMech
Tonnage: 85
Battle Value: 1,770

Equipment		**Mass**
Internal Structure:	Endo Steel	4.5
Engine:	340 XL	13.5

Equipment		Mass
Walking MP:	4	
Running MP:	6	
Jumping MP:	0	
Heat Sinks:	13 [26]	3
Gyro:		4
Cockpit:		3
Armor Factor:	263	16.5
	Internal Structure	*Armor Value*
Head	3	9
Center Torso	27	44
Center Torso (rear)		10
R/L Torso	18	26
R/L Torso (rear)		10
R/L Arm	14	28
R/L Leg	18	36

Weight and Space Allocation

Location	*Spaces Fixed*	*Remaining*
Head	1 Endo Steel	0
Center Torso	None	2
Right Torso	3 Engine	7
	2 Endo Steel	
	CASE	
Left Torso	3 Engine	6
	2 Endo Steel	
Right Arm	1 Endo Steel	9
Left Arm	4 Endo Steel	6
Right Leg	2 Endo Steel	0
Left Leg	2 Endo Steel	0

Weapons And Ammo	Location	Critical	Tonnage
Primary Weapons Configuration			
Large Pulse Laser	LA	2	7
Medium Pulse Laser	LA	1	2
Double Heat Sink	LA	3	1
Gauss Rifle	RA	7	15
Targeting Computer	LT	7	7
Streak SRM 4	RT	1	3
Ammo (Streak) 25	RT	1	1
Ammo (Gauss) 16	RT	2	2
2 ER Medium Lasers	CT	2	2
Alternate Configuration A			
ER PPC	LA	3	7
Double Heat Sink	LA	1	3
Ultra AC/10	RA	7	13
Streak SRM 6	RA	2	4.5
Targeting Computer	LT	5	5
2 Jump Jets	LT	2	2
Ammo (Ultra) 30	RT	3	3
Ammo (Streak) 15	RT	1	1
2 Jump Jets	RT	2	2
Guardian ECM Suite	CT	2	1.5
Battle Value: 2,047			
Alternate Configuration B			
ER PPC	LA	3	7
Double Heat Sink	LA	1	3
3 RAC 2	RA	9	18
2 Double Heat Sinks	LT	6	2
Double Heat Sink	RT	3	1
Ammo (RAC) 135	RT	3	3
2 ER Medium Lasers	CT	2	2
Battle Value: 1,427			

AVALON-CLASS CRUISER

Design on the *Avalon* began over a decade ago, at the height of the Clan invasion. Saddled with the monstrous costs of rebuilding its shattered Regular Army, however, the High Command delayed construction of this new WarShip class for a number of years. Though decried by many within the AFFC's highest circles at the time, this decision actually saved billions in cost overruns as bureaucratic problems with ComStar delayed the shipment of components needed for completion of the first six *Foxes* for several years.

Rather than be forced to rely entirely upon ComStar to provide the components needed to build the massive interplanetary drives required for its WarShips, the High Command instead invested heavily in upgrading the naval production yards at Galax, Kathil and New Syrtis during these years of delay. By the time ComStar finally authorized delivery of the

delayed components, the Federated Suns' shipyards were capable of independent operation. Of course, soon afterward the Word of Blake seized Terra, an act that would have forced those shipyards into independence anyway. Yare Shipbuilders of Kathil finally laid the first *Avalon's* keel in early 3057.

Naturally, the *Avalon* underwent a number of design modifications during its years of dormancy. As advances in naval technology were debuted, design teams from the DQ and Yare continuously updated the vessel's modifications. Perhaps the most significant was its evolution into a missile cruiser. Originally mounting far more energy weaponry, representatives from the Admiralty lobbied for the inclusion of AR 10-class multi-purpose missile launchers as soon as they became available. Though the inclusion of so many launching hardpoints significantly reduced the ship's overall offensive capability, at least on paper, the ship was already overgunned, capable of using less than half its weaponry effectively in combat.

Designed to be at the heart of a naval task force, the *Avalon* is well armored and armed, and carries more than five times the fuel capacity of most other WarShips in service, giving it the capability to keep the entire task force operating for exceedingly lengthy periods of time. As AFFC naval doctrine requires each of the *Avalons* to be "escorted" by a dozen or more assault DropShips, the cruiser itself mounts only the most basic of point-defense weaponry, instead concentrating more on its anti-ship capabilities.

Undergoing final Contractors Trials at the outset of Operations Bulldog and Serpent, the first *Avalon* was delivered in time for operations in support of the defense of the St. Ives Compact. The *Melissa Davion* buttressed the heavy AFFC presence in the Kittery system, while the next three vessels were rushed through trials and immediately assigned to patrol routes throughout the Federated Suns. Both the *Lucien Davion* and the *Simon Davion* have subsequently suffered from a number of minor systems malfunctions, though representatives from each of the major contractors are on board to assist the crew in working the rest of the bugs out of their systems.

AVALON-CLASS CRUISER

Avalon (Cruiser)
Tech: Star League
Introduced: 3061
Mass: 770,000 Tons
Length: 812 meters
Sail Diameter: 1375 meters
Fuel: 10,000 tons (25000)
Tons/Burn Day: 39.52
Safe Thrust: 4
Maximum Thrust: 6
Sail Integrity: 4

KF Drive Integrity: 16
Heat Sinks: 937 (1,874)
Structural Integrity: 120
Armor:
Fore: 344
Fore-Sides: 310
Aft Sides: 302
Aft: 275
Cargo:
Bay 1: Fighters/Small Craft (6) 2 Doors
Bay 2: Fighters/Small Craft (6) 2 Doors
Bay 3: Cargo (14,220 tons) 4 Doors

DropShip Capacity: 6
Grav Deck: 2 (one 145 meter diameter and one 210 meter diameter)
Escape Pods: 35
Life Boats: 15
Crew: 24 Officers, 174 Enlisted/non-rated, 12 fighter/shuttle pilots, 64 Battle Armor Troopers (16 squads), 20 passengers

Weapons:		Capital Attack Values (Standard)				
Arc	Type	Short	Medium	Long	Extreme	Class
Nose (368 Heat)						
	2 NL 45	9	9	9	—	Capital Laser
	2 NAC/20	40	40	40	—	Capital AC
	2 AR10	—	—	—	—	AR10
	1 Screen Launcher	—	—	—	—	Screen
	4 Large Lasers	3 (32)	3 (32)	—	—	Laser
	2 Medium Pulse Lasers	1 (12)	—	—	—	Pulse Laser
	3 Anti-Missile Systems	—	—	—	—	AMS
	3 LRM 15 + Artemis	3 (36)	3 (36)	3 (36)	—	LRM
FL/FR (411 Heat)						
	2 NL 45	9	9	9	—	Capital Laser
	1 NAC/40	40	40	—	—	Capital AC
	1 Medium N-Gauss	25	25	25	25	Capital Gauss
	4 AR10	—	—	—	—	AR10
	2 Large Lasers	1 (16)	1 (16)	—	—	Laser
	2 Medium Pulse Lasers	1 (12)	—	—	—	Pulse Laser
	2 AMS	—	—	—	—	AMS
	3 LRM 15 + Artemis	3 (36)	3 (36)	3 (36)	—	LRM
LBS/RBS (140 Heat)						
	1 NAC/20	20	20	20	—	Capital AC
	2 Medium N-Gauss	50	50	50	50	Capital Gauss
	1 Screen Launcher	—	—	—	—	Screen
	2 Medium Pulse Lasers	1 (12)	—	—	—	Pulse Laser
	4 Anti-Missile Systems	—	—	—	—	AMS
	4 LB 10-X AC	2 (24)	2 (24)	—	—	LB-X AC
	4 LRM 15 + Artemis	4 (48)	4 (48)	4 (48)	—	LRM
AL/AR (155 Heat)						
	1 NAC/20	20	20	20	—	Capital AC
	1 Medium N-Gauss	25	25	25	25	Capital Gauss
	2 AR10	—	—	—	—	AR10
	2 Large Lasers	1 (16)	1 (16)	—	—	Laser
	2 Medium Pulse Lasers	1 (12)	—	—	—	Pulse Laser
	1 Anti-Missile System	—	—	—	—	AMS
	3 LRM 15 + Artemis	3 (36)	3 (36)	3 (36)	—	LRM
Aft (328 Heat)						
	2 NL 55	11	11	11	11	Capital Laser
	1 NAC 40	40	40	—	—	Capital AC
	2 Medium Pulse Lasers	1 (12)	—	—	—	Pulse Laser
	4 Anti-Missile Systems	—	—	—	—	AMS
	3 LRM 15 + Artemis	3 (36)	3 (36)	3 (36)	—	LRM

Ammunition: 1,200 rounds AMS ammunition (100 tons), 500 rounds LB 10-X ammunition (50 tons), 1,200 rounds LRM 15 ammunition (150 tons), 300 rounds NAC/20 ammunition (120 tons), 150 rounds NAC/40 ammunition (180 tons), 300 round Medium N-Gauss ammunition (120 tons), 150 Killer Whale Missiles (7500 tons), 150 Barracuda Missiles (4,500 tons), 25 rounds Screen Launchers (250 tons).

Notes: Equipped with Lithium-Fusion Battery and 1,801 tons of Lamellor Ferro-Carbide Armor

BATTLETECH®

BATTLEMECH RECORD SHEET

ARMOR DIAGRAM

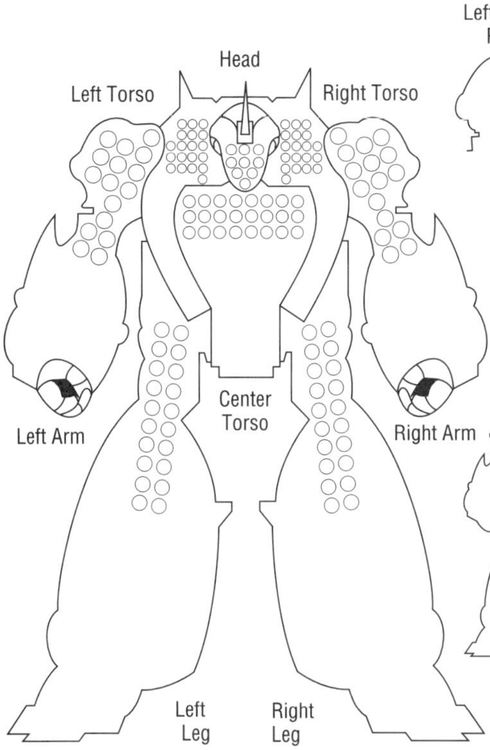

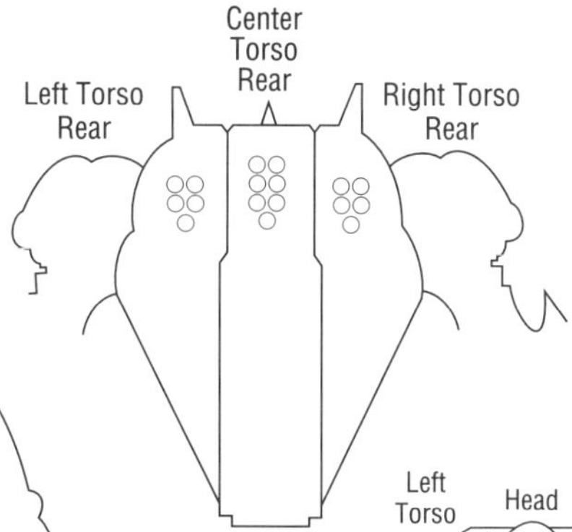

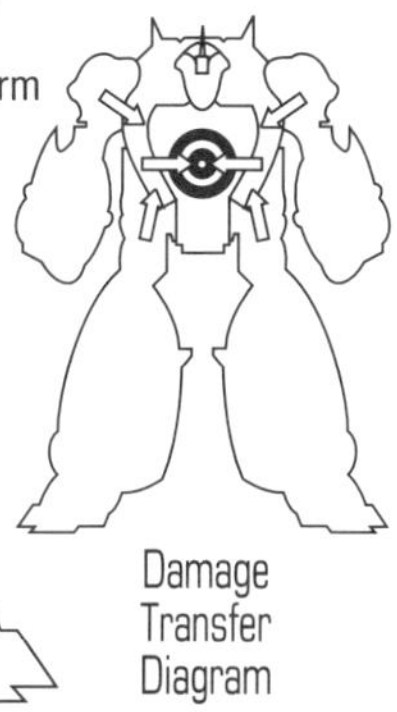

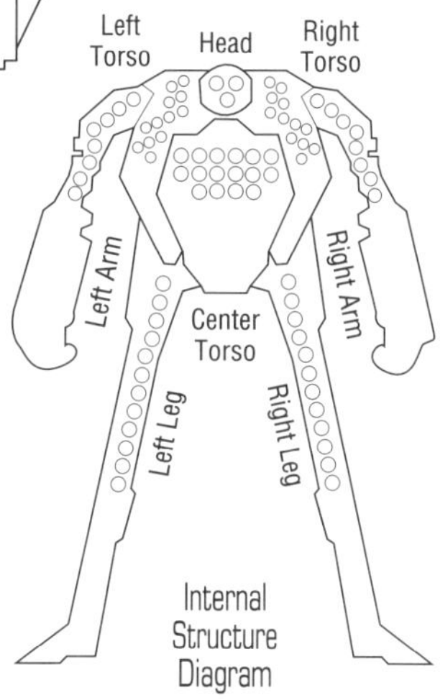

Mech Data

Type: ENF-6T Enforcer III — Tonnage: 50

Movement Points:
- Walking: 5
- Running: 8
- Jumping: 5

Technology Base: 3062 Inner Sphere

Weapons Inventory

Type	Location	Heat	Damage	Min.	Short	Med.	Long
1 Gauss Rifle	RA	1	15	2	7	15	22
1 ER Medium Laser	CT	5	5	–	4	8	12
Targeting Computer	RT	–	–	–	–	–	–

Ammo Type	Rounds
Gauss Rifle	16

Total Heat Sinks: 10 (20) Double

○○○○○○○○○○

Auto Eject

☐ Operational ☐ Disabled

Warrior Data

Name: ____________

Gunnery Skill: ______ Piloting Skill: ______

Hits Taken	1	2	3	4	5	6
Consciousness #	3	5	7	10	11	Dead

Critical Hit Table

Left Arm

1-3
1. Shoulder
2. Upper Arm Actuator
3. Lower Arm Actuator
4. Endo Steel
5. Endo Steel
6. Endo Steel

4-6
1. Endo Steel
2. Endo Steel
3. Endo Steel
4. Roll Again
5. Roll Again
6. Roll Again

Left Torso

1-3
1. XL Engine
2. XL Engine
3. XL Engine
4. Endo Steel
5. Endo Steel
6. Endo Steel

4-6
1. Endo Steel
2. Endo Steel
3. Endo Steel
4. Endo Steel
5. Endo Steel
6. Roll Again

Left Leg

1. Hip
2. Upper Leg Actuator
3. Lower Leg Actuator
4. Foot Actuator
5. Jump Jet
6. Jump Jet

Head

1. Life Support
2. Sensors
3. Cockpit
4. Roll Again
5. Sensors
6. Life Support

Center Torso

1-3
1. XL Engine
2. XL Engine
3. XL Engine
4. Gyro
5. Gyro
6. Gyro

4-6
1. Gyro
2. XL Engine
3. XL Engine
4. XL Engine
5. Jump Jet
6. ER Medium Laser

Engine Hits	○○○
Gyro Hits	○○
Sensor Hits	○○
Life Support	○

Cost: 8,525,000
Battle Value: 1,614

Right Arm

1-3
1. Shoulder
2. Upper Arm Actuator
3. Lower Arm Actuator
4. Gauss Rifle
5. Gauss Rifle
6. Gauss Rifle

4-6
1. Gauss Rifle
2. Gauss Rifle
3. Gauss Rifle
4. Gauss Rifle
5. Roll Again
6. Roll Again

Right Torso

1-3
1. XL Engine
2. XL Engine
3. XL Engine
4. Targeting Computer
5. Targeting Computer
6. Targeting Computer

4-6
1. Targeting Computer
2. Ammo (Gauss) 8
3. Ammo (Gauss) 8
4. Roll Again
5. Roll Again
6. Roll Again

Right Leg

1. Hip
2. Upper Leg Actuator
3. Lower Leg Actuator
4. Foot Actuator
5. Jump Jet
6. Jump Jet

Heat Scale

	Heat Overflow
30	Shutdown
29	
28	Ammo Explosion, avoid on 8+
27	
26	Shutdown, avoid on 10+
25	-5 Movement Points
24	+4 Modifier to Fire
23	Ammo Explosion, avoid on 6+
22	Shutdown, avoid on 8+
21	
20	-4 Movement Points
19	Ammo Explosion, avoid on 4+
18	Shutdown, avoid on 6+
17	+3 Modifier to Fire
16	
15	-3 Movement Points
14	Shutdown, avoid on 4+
13	+2 Modifier to fire
12	
11	
10	-2 Movement Points
09	
08	+1 Modifier to Fire
07	
06	
05	-1 Movement Points
04	
03	
02	
01	
00	

BATTLETECH is a Registered Trademark of FASA Corporation. Copyright 2000. Permission to photocopy for personal use.

BATTLETECH®
BATTLEMECH RECORD SHEET

ARMOR DIAGRAM

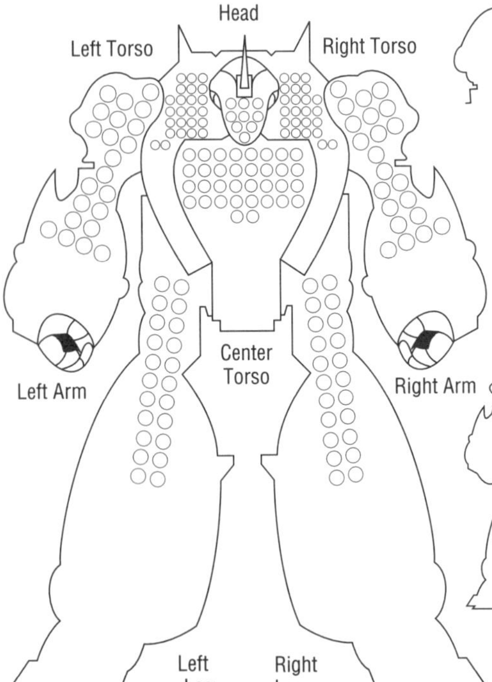

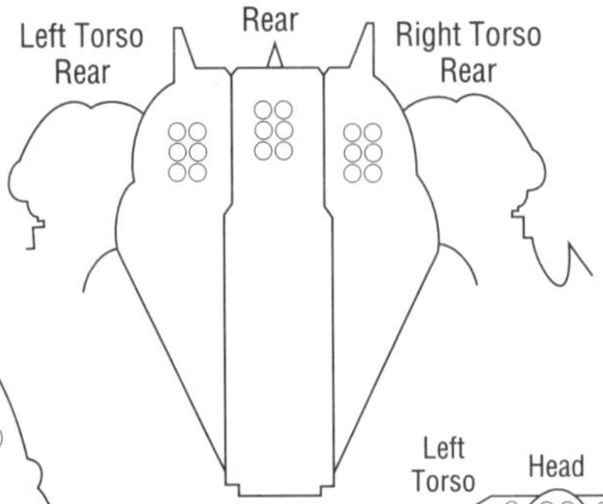

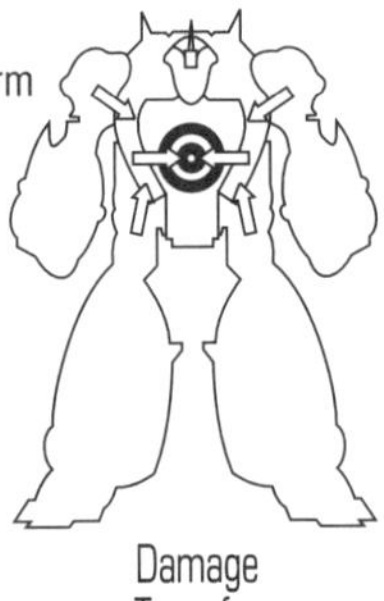

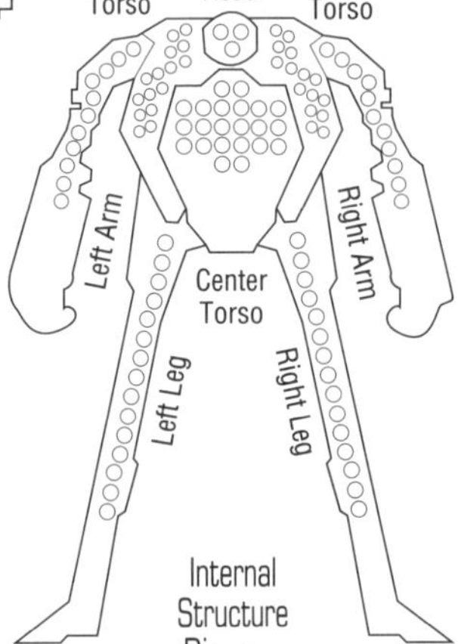

Mech Data

Type: JM7-F JagerMech Tonnage: 70

Movement Points:

Walking: 4

Running: 6

Jumping: 0

Technology Base: 3062 Inner Sphere

Weapons Inventory

Type	Location	Heat	Damage	Min.	Short	Med.	Long
Rotary AC/5	RA	1	5	–	5	10	15
Rotary AC/5	LA	1	5	–	5	10	15
2 Medum Pulse Lasers	CT	4	6	–	2	4	6
Targeting Computer	RT	–	–	–	–	–	–
Guardian ECM Suite	LT	0	–	–	–	–	6

Ammo Type	Rounds
Rotary AC/5	80

Total Heat Sinks: 12 (24) Double

OOOOOOOOOO OO

Auto Eject

☐ Operational ☐ Disabled

Warrior Data

Name: ______________________

Gunnery Skill: ________ Piloting Skill: ________

Hits Taken	1	2	3	4	5	6
Consciousness #	3	5	7	10	11	Dead

Heat Scale

(Overflow)	Heat Overflow
30	Shutdown
29	
28	Ammo Explosion, avoid on 8+
27	
26	Shutdown, avoid on 10+
25	-5 Movement Points
24	+4 Modifier to Fire
23	Ammo Explosion, avoid on 6+
22	Shutdown, avoid on 8+
21	
20	-4 Movement Points
19	Ammo Explosion, avoid on 4+
18	Shutdown, avoid on 6+
17	+3 Modifier to Fire
16	
15	-3 Movement Points
14	Shutdown, avoid on 4+
13	+2 Modifier to fire
12	
11	
10	-2 Movement Points
09	
08	+1 Modifier to Fire
07	
06	
05	-1 Movement Points
04	
03	
02	
01	
00	

Critical Hit Table

Left Arm

1. Shoulder
2. Upper Arm Actuator

1-3
3. Rotary AC/5
4. Rotary AC/5
5. Rotary AC/5
6. Rotary AC/5

4-6
1. Rotary AC/5
2. Rotary AC/5
3. Ferro-Fibrous
4. Ferro-Fibrous
5. Ferro-Fibrous
6. Ferro-Fibrous

Head

1. Life Support
2. Sensors
3. Cockpit
4. Roll Again
5. Sensors
6. Life Support

Right Arm

1. Shoulder
2. Upper Arm Actuator

1-3
3. Rotary AC/5
4. Rotary AC/5
5. Rotary AC/5
6. Rotary AC/5

4-6
1. Rotary AC/5
2. Rotary AC/5
3. Ferro-Fibrous
4. Ferro-Fibrous
5. Ferro-Fibrous
6. Ferro-Fibrous

Left Torso (Case)

1-3
1. XL Engine
2. XL Engine
3. XL Engine
4. Guardian ECM Suite
5. Guardian ECM Suite
6. Ammo (RAC) 20

4-6
1. Ammo (RAC) 20
2. Ammo (RAC) 20
3. Ammo (RAC) 20
4. Ferro-Fibrous
5. Ferro-Fibrous
6. CASE

Center Torso

1-3
1. XL Engine
2. XL Engine
3. XL Engine
4. Gyro
5. Gyro
6. Gyro

4-6
1. Gyro
2. XL Engine
3. XL Engine
4. XL Engine
5. Medium Pulse Laser
6. Medium Pulse Laser

Right Torso

1-3
1. XL Engine
2. XL Engine
3. XL Engine
4. Double Heat Sink
5. Double Heat Sink
6. Double Heat Sink

4-6
1. Targeting Computer
2. Targeting Computer
3. Targeting Computer
4. Targeting Computer
5. Targeting Computer
6. Targeting Computer

Left Leg

1. Hip
2. Upper Leg Actuator
3. Lower Leg Actuator
4. Foot Actuator
5. Ferro-Fibrous
6. Ferro-Fibrous

Engine Hits	OOO
Gyro Hits	OO
Sensor Hits	OO
Life Support	O

Cost: 13,715,034

Battle Value: 1,514

Right Leg

1. Hip
2. Upper Leg Actuator
3. Lower Leg Actuator
4. Foot Actuator
5. Ferro-Fibrous
6. Ferro-Fibrous

BATTLETECH is a Registered Trademark of FASA Corporation. Copyright 2000. Permission to photocopy for personal use.

BATTLETECH®

BATTLEMECH RECORD SHEET

ARMOR DIAGRAM

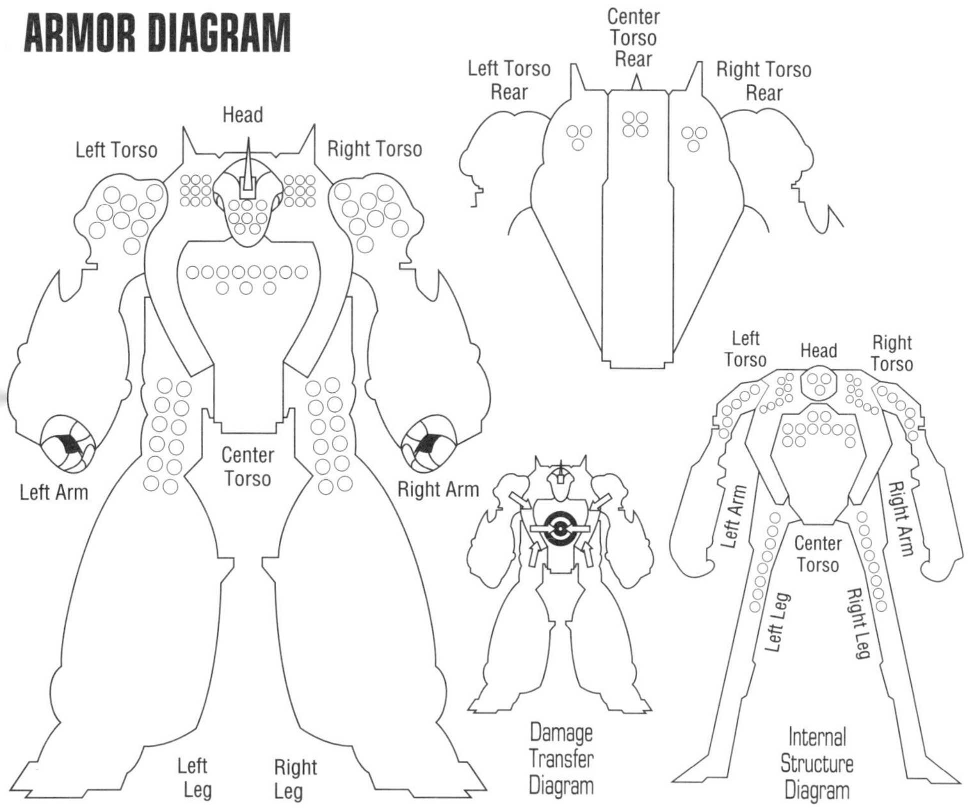

Critical Hit Table

Left Arm

1-3
1. Shoulder
2. Upper Arm Actuator
3. Lower Arm Actuator
4. Hand Actuator
5. Ferro-Fibrous
6. Ferro-Fibrous

4-6
1. Ferro-Fibrous
2. Ferro-Fibrous
3. Ferro-Fibrous
4. Ferro-Fibrous
5. Ferro-Fibrous
6. Ferro-Fibrous

Left Torso (Case)

1-3
1. Heat Sink
2. ER Medium Laser
3. Ammo (RAC) 20
4. Ammo (RAC) 20
5. Ferro-Fibrous
6. Ferro-Fibrous

4-6
1. Ferro-Fibrous
2. Ferro-Fibrous
3. Ferro-Fibrous
4. Ferro-Fibrous
5. CASE
6. Roll Again

Left Leg

1. Hip
2. Upper Leg Actuator
3. Lower Leg Actuator
4. Foot Actuator
5. Jump Jet
6. Jump Jet

Head

1. Life Support
2. Sensors
3. Cockpit
4. Endo Steel
5. Sensors
6. Life Support

Center Torso

1-3
1. XL Engine
2. XL Engine
3. XL Engine
4. Gyro
5. Gyro
6. Gyro

4-6
1. Gyro
2. XL Engine
3. XL Engine
4. XL Engine
5. Heat Sink
6. Jump Jet

Engine Hits	○○○
Gyro Hits	○○
Sensor Hits	○○
Life Support	○

Cost: 2,968,560
Battle Value: 934

Right Arm

1-3
1. Shoulder
2. Upper Arm Actuator
3. Lower Arm Actuator
4. Hand Actuator
5. Endo Steel
6. Endo Steel

4-6
1. Endo Steel
2. Endo Steel
3. Endo Steel
4. Endo Steel
5. Endo Steel
6. Endo Steel

Right Torso

1-3
1. Heat Sink
2. Rotary AC/5
3. Rotary AC/5
4. Rotary AC/5
5. Rotary AC/5
6. Rotary AC/5

4-6
1. Rotary AC/5
2. Endo Steel
3. Endo Steel
4. Endo Steel
5. Endo Steel
6. Endo Steel

Right Leg

1. Hip
2. Upper Leg Actuator
3. Lower Leg Actuator
4. Foot Actuator
5. Jump Jet
6. Jump Jet

Mech Data

Type: GRM-01C Garm — Tonnage: 35

Movement Points:
- Walking: 5
- Running: 8
- Jumping: 5

Technology Base: 3062 Inner Sphere

Weapons Inventory

Type	Location	Heat	Damage	Min.	Short	Med.	Long
1 Rotary AC/5	RT	1	5	–	5	10	15
1 ER Medium Laser	LT	5	5	–	4	8	12

Ammo Type	Rounds
Rotary AC/5	40

Total Heat Sinks: 10 — Single
○○○○○○○○○○

Auto Eject
☐ Operational ☐ Disabled

Warrior Data

Name: ____________________

Gunnery Skill: ________ Piloting Skill: ________

Hits Taken	1	2	3	4	5	6
Consciousness #	3	5	7	10	11	Dead

Heat Scale

	Heat Overflow
30	Shutdown
29	
28	Ammo Explosion, avoid on 8+
27	
26	Shutdown, avoid on 10+
25	-5 Movement Points
24	+4 Modifier to Fire
23	Ammo Explosion, avoid on 6+
22	Shutdown, avoid on 8+
21	
20	-4 Movement Points
19	Ammo Explosion, avoid on 4+
18	Shutdown, avoid on 6+
17	+3 Modifier to Fire
16	
15	-3 Movement Points
14	Shutdown, avoid on 4+
13	+2 Modifier to fire
12	
11	
10	-2 Movement Points
09	
08	+1 Modifier to Fire
07	
06	
05	-1 Movement Points
04	
03	
02	
01	
00	

BATTLETECH is a Registered Trademark of FASA Corporation. Copyright 2000. Permission to photocopy for personal use.

BATTLETECH®

BATTLEMECH RECORD SHEET

ARMOR DIAGRAM

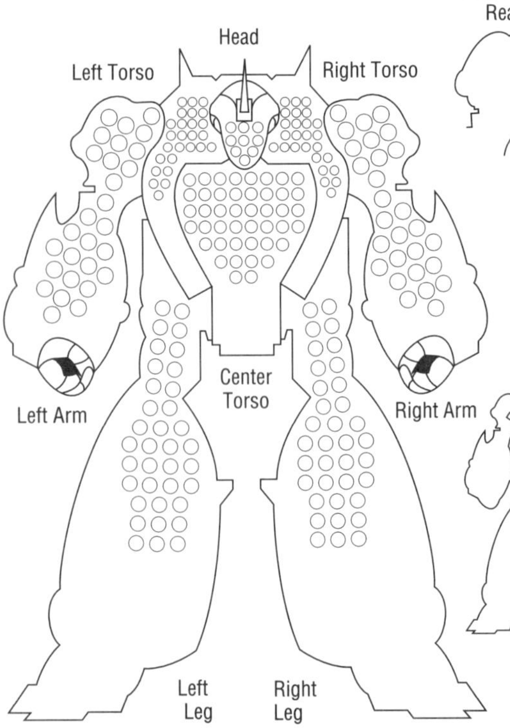

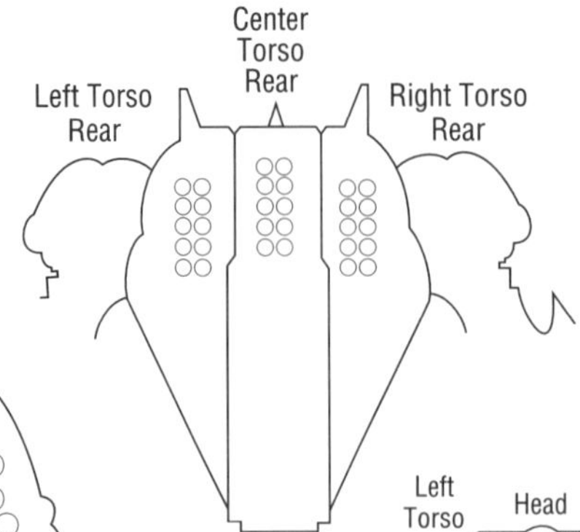

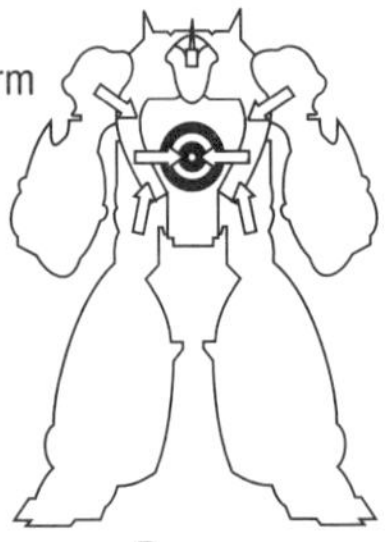

Damage Transfer Diagram

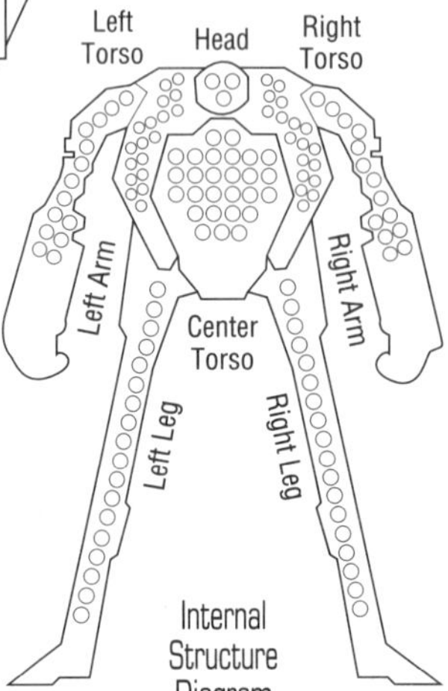

Internal Structure Diagram

Mech Data

Type: TLR1-O Templar Tonnage: 85

Movement Points:

Walking: 4

Running: 6

Jumping: 0

Technology Base: 3062 Inner Sphere

Weapons Inventory

Type	Location	Heat	Damage	Min.	Short	Med.	Long
1 Gauss Rifle	RA	1	15	2	7	15	22
1 Large Pulse Laser	LA	10	9	–	3	7	10
1 Medium Pulse Laser	LA	4	6	–	2	4	6
1 Streak SRM 4	RT	3	2	–	3	6	9
2 ER Medium Lasers	CT	5	5	–	4	8	12
Targeting Computer	LT	–	–	–	–	–	–

Ammo Type	Rounds
Gauss Rifle	16
Streak SRM 4	25

Total Heat Sinks: 14 (28) Double

○○○○○○○○○○ ○○○○

Auto Eject

☐ Operational ☐ Disabled

Critical Hit Table

Left Arm

1-3
1. Shoulder
2. Upper Arm Actuator
3. Double Heat Sink
4. Double Heat Sink
5. Double Heat Sink
6. Large Pulse Laser

4-6
1. Large Pulse Laser
2. Medium Pulse Laser
3. *Endo Steel*
4. *Endo Steel*
5. *Endo Steel*
6. *Endo Steel*

Head

1. Life Support
2. Sensors
3. Cockpit
4. *Endo Steel*
5. Sensors
6. Life Support

Right Arm

1-3
1. Shoulder
2. Upper Arm Actuator
3. Gauss Rifle
4. Gauss Rifle
5. Gauss Rifle
6. Gauss Rifle

4-6
1. Gauss Rifle
2. Gauss Rifle
3. Gauss Rifle
4. *Endo Steel*
5. *Roll Again*
6. *Roll Again*

Left Torso

1-3
1. XL Engine
2. XL Engine
3. XL Engine
4. Targeting Computer
5. Targeting Computer
6. Targeting Computer

4-6
1. Targeting Computer
2. Targeting Computer
3. Targeting Computer
4. Targeting Computer
5. *Endo Steel*
6. *Endo Steel*

Center Torso

1-3
1. XL Engine
2. XL Engine
3. XL Engine
4. Gyro
5. Gyro
6. Gyro

4-6
1. Gyro
2. XL Engine
3. XL Engine
4. XL Engine
5. ER Medium Laser
6. ER Medium Laser

Right Torso (Case)

1-3
1. XL Engine
2. XL Engine
3. XL Engine
4. Streak SRM 4
5. Ammo (Streak) 25
6. Ammo (Gauss) 8

4-6
1. Ammo (Gauss) 8
2. *Endo Steel*
3. *Endo Steel*
4. *CASE*
5. *Roll Again*
6. *Roll Again*

Left Leg

1. Hip
2. Upper Leg Actuator
3. Lower Leg Actuator
4. Foot Actuator
5. *Endo Steel*
6. *Endo Steel*

Engine Hits ○○○
Gyro Hits ○○
Sensor Hits ○○
Life Support ○

Cost: 25,338,449

Battle Value: 1,770

Right Leg

1. Hip
2. Upper Leg Actuator
3. Lower Leg Actuator
4. Foot Actuator
5. *Endo Steel*
6. *Endo Steel*

Warrior Data

Name: ____________________

Gunnery Skill: ________ Piloting Skill: ________

Hits Taken	1	2	3	4	5	6
Consciousness #	3	5	7	10	11	Dead

Heat Scale

	Heat Overflow
30	Shutdown
29	
28	Ammo Explosion, avoid on 8+
27	
26	Shutdown, avoid on 10+
25	-5 Movement Points
24	+4 Modifier to Fire
23	Ammo Explosion, avoid on 6+
22	Shutdown, avoid on 8+
21	
20	-4 Movement Points
19	Ammo Explosion, avoid on 4+
18	Shutdown, avoid on 6+
17	+3 Modifier to Fire
16	
15	-3 Movement Points
14	Shutdown, avoid on 4+
13	+2 Modifier to fire
12	
11	
10	-2 Movement Points
09	
08	+1 Modifier to Fire
07	
06	
05	-1 Movement Points
04	
03	
02	
01	
00	

BATTLETECH is a Registered Trademark of FASA Corporation. Copyright 2000. Permission to photocopy for personal use.

BATTLETECH®

BATTLEMECH RECORD SHEET

ARMOR DIAGRAM

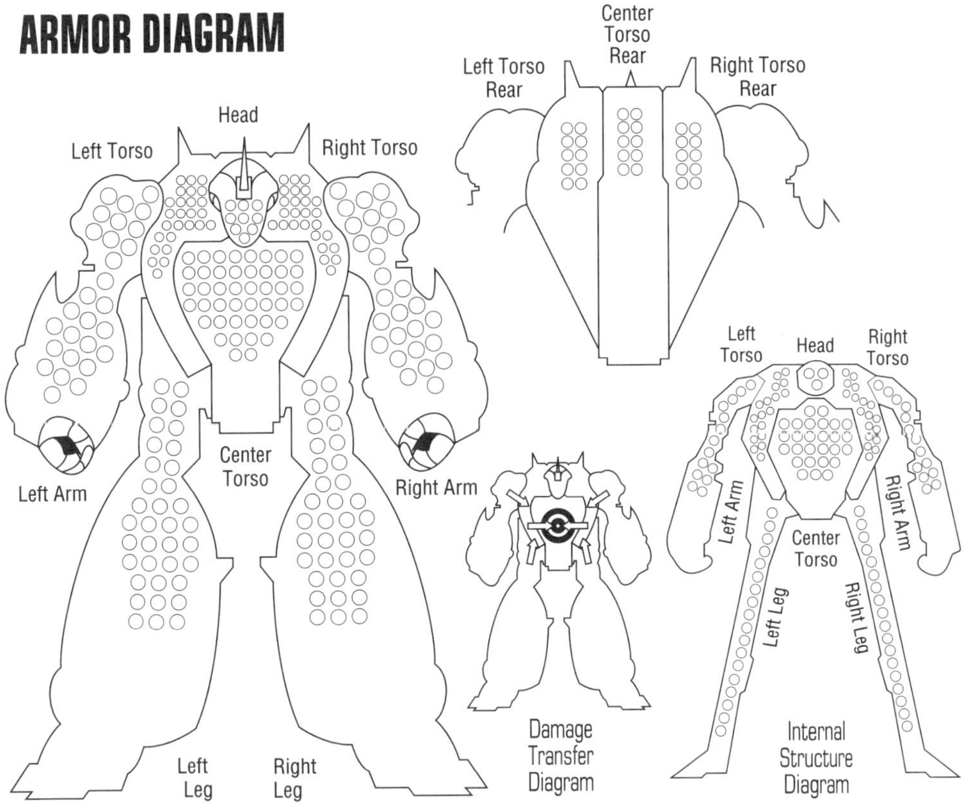

Mech Data

Type: TLR1-OA Templar — Tonnage: 85

Movement Points:
- Walking: 4
- Running: 6
- Jumping: 4

Technology Base: 3062 Inner Sphere

Weapons Inventory

Type	Location	Heat	Damage	Min.	Short	Med.	Long
1 Ultra AC/10	RA	4	10	–	6	12	18
1 Streak SRM 6	RA	4	2	–	3	6	9
1 ER PPC	LA	15	10	–	7	14	23
Guardian ECM Suite	CT	0	–	–	–	–	6
Targeting Computer	LT	–	–	–	–	–	–

Ammo Type	Rounds
Ultra AC/10	30
Streak SRM 6	15

Total Heat Sinks: 14 (28) — Double

○○○○○○○○○○ ○○○○

Auto Eject

☐ Operational ☐ Disabled

Warrior Data

Name: ____________________

Gunnery Skill: ________ Piloting Skill: ________

Hits Taken	1	2	3	4	5	6
Consciousness #	3	5	7	10	11	Dead

Critical Hit Table

Left Arm

1-3
1. Shoulder
2. Upper Arm Actuator
3. Double Heat Sink
4. Double Heat Sink
5. Double Heat Sink
6. ER PPC

4-6
1. ER PPC
2. ER PPC
3. *Endo Steel*
4. *Endo Steel*
5. *Endo Steel*
6. *Endo Steel*

Left Torso

1-3
1. XL Engine
2. XL Engine
3. XL Engine
4. Jump Jet
5. Jump Jet
6. Targeting Computer

4-6
1. Targeting Computer
2. Targeting Computer
3. Targeting Computer
4. Targeting Computer
5. *Endo Steel*
6. *Endo Steel*

Left Leg

1. Hip
2. Upper Leg Actuator
3. Lower Leg Actuator
4. Foot Actuator
5. *Endo Steel*
6. *Endo Steel*

Head

1. Life Support
2. Sensors
3. Cockpit
4. *Endo Steel*
5. Sensors
6. Life Support

Center Torso

1-3
1. XL Engine
2. XL Engine
3. XL Engine
4. Gyro
5. Gyro
6. Gyro

4-6
1. Gyro
2. XL Engine
3. XL Engine
4. XL Engine
5. Guardian ECM Suite
6. Guardian ECM Suite

Engine Hits	○○○
Gyro Hits	○○
Sensor Hits	○○
Life Support	○

Cost: 26,270,387
Battle Value: 2,047

Right Arm

1-3
1. Shoulder
2. Upper Arm Actuator
3. Ultra AC/10
4. Ultra AC/10
5. Ultra AC/10
6. Ultra AC/10

4-6
1. Ultra AC/10
2. Ultra AC/10
3. Ultra AC/10
4. Streak SRM 6
5. Streak SRM 6
6. *Endo Steel*

Right Torso (Case)

1-3
1. XL Engine
2. XL Engine
3. XL Engine
4. Jump Jet
5. Jump Jet
6. Ammo (Ultra) 10

4-6
1. Ammo (Ultra) 10
2. Ammo (Ultra) 10
3. Ammo (Streak) 15
4. *Endo Steel*
5. *Endo Steel*
6. *CASE*

Right Leg

1. Hip
2. Upper Leg Actuator
3. Lower Leg Actuator
4. Foot Actuator
5. *Endo Steel*
6. *Endo Steel*

Heat Scale

	Heat Overflow
30	Shutdown
29	
28	Ammo Explosion, avoid on 8+
27	
26	Shutdown, avoid on 10+
25	-5 Movement Points
24	+4 Modifier to Fire
23	Ammo Explosion, avoid on 6+
22	Shutdown, avoid on 8+
21	
20	-4 Movement Points
19	Ammo Explosion, avoid on 4+
18	Shutdown, avoid on 6+
17	+3 Modifier to Fire
16	
15	-3 Movement Points
14	Shutdown, avoid on 4+
13	+2 Modifier to fire
12	
11	
10	-2 Movement Points
09	
08	+1 Modifier to Fire
07	
06	
05	-1 Movement Points
04	
03	
02	
01	
00	

BATTLETECH is a Registered Trademark of FASA Corporation. Copyright 2000. Permission to photocopy for personal use.

BATTLETECH®

BATTLEMECH RECORD SHEET

ARMOR DIAGRAM

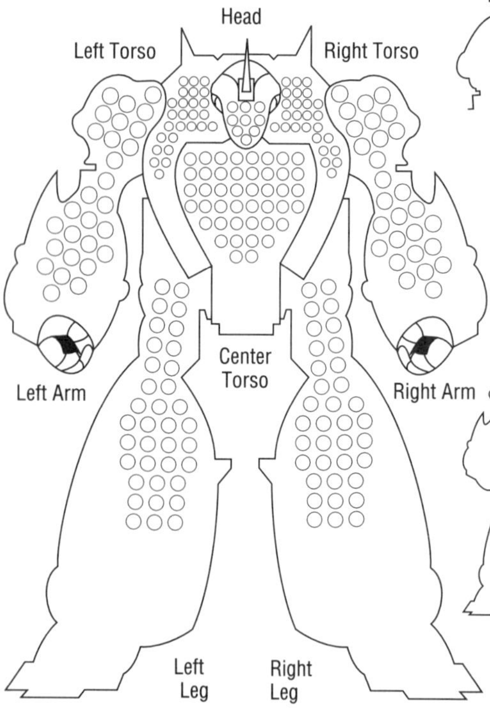

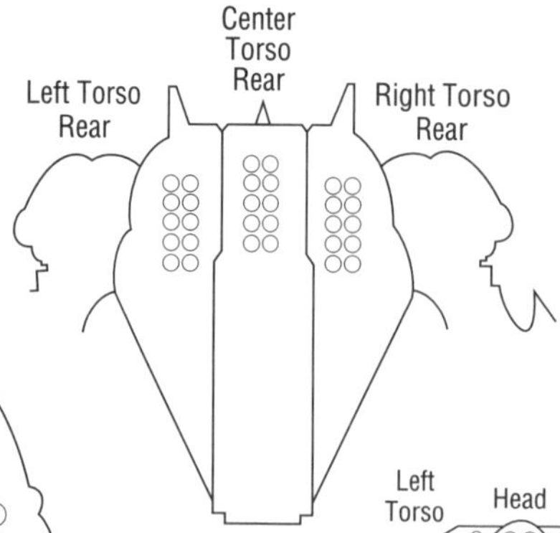

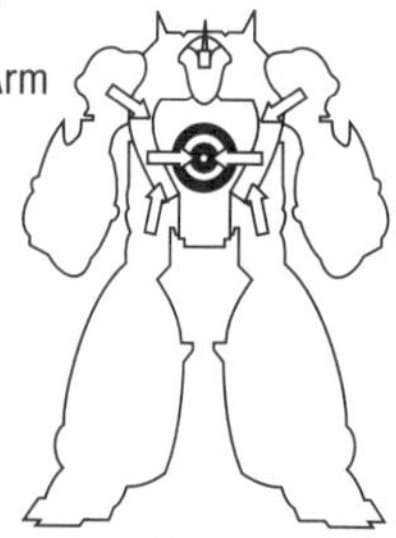

Damage Transfer Diagram

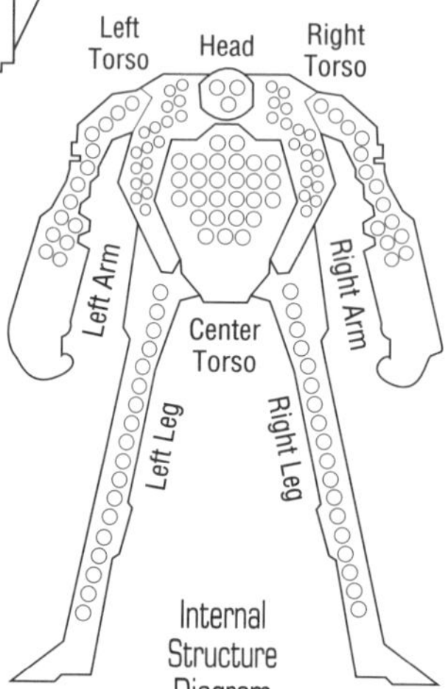

Internal Structure Diagram

Mech Data

Type: **TLR1-OB Templar** Tonnage: 85

Movement Points:
- Walking: **4**
- Running: **6**
- Jumping: **0**

Technology Base: 3062 Inner Sphere

Weapons Inventory

Type	Location	Heat	Damage	Min.	Short	Med.	Long
3 Rotary AC/2	RA	1	2	–	6	12	18
1 ER PPC	LA	15	10	–	7	14	23
2 ER Medium Lasers	CT	5	5	–	4	8	12

Ammo Type	Rounds
Rotary AC/2	135

Total Heat Sinks: 17 (34) Double

○○○○○○○○○○ ○○○○○○○

Auto Eject

☐ Operational ☐ Disabled

Warrior Data

Name: ______________________

Gunnery Skill: ________ Piloting Skill: ________

Hits Taken	1	2	3	4	5	6
Consciousness #	3	5	7	10	11	Dead

Critical Hit Table

Left Arm

1-3
1. Shoulder
2. Upper Arm Actuator
3. Double Heat Sink
4. Double Heat Sink
5. Double Heat Sink
6. ER PPC

4-6
1. ER PPC
2. ER PPC
3. *Endo Steel*
4. *Endo Steel*
5. *Endo Steel*
6. *Endo Steel*

Left Torso

1-3
1. XL Engine
2. XL Engine
3. XL Engine
4. Double Heat Sink
5. Double Heat Sink
6. Double Heat Sink

4-6
1. Double Heat Sink
2. Double Heat Sink
3. Double Heat Sink
4. *Endo Steel*
5. *Endo Steel*
6. *Roll Again*

Left Leg

1. Hip
2. Upper Leg Actuator
3. Lower Leg Actuator
4. Foot Actuator
5. *Endo Steel*
6. *Endo Steel*

Head

1. Life Support
2. Sensors
3. Cockpit
4. *Endo Steel*
5. Sensors
6. Life Support

Center Torso

1-3
1. XL Engine
2. XL Engine
3. XL Engine
4. Gyro
5. Gyro
6. Gyro

4-6
1. Gyro
2. XL Engine
3. XL Engine
4. XL Engine
5. ER Medium Laser
6. ER Medium Laser

Engine Hits ○○○
Gyro Hits ○○
Sensor Hits ○○
Life Support ○

Cost: 25,484,137
Battle Value: 1,427

Right Arm

1-3
1. Shoulder
2. Upper Arm Actuator
3. Rotary AC/2
4. Rotary AC/2
5. Rotary AC/2
6. Rotary AC/2

4-6
1. Rotary AC/2
2. Rotary AC/2
3. Rotary AC/2
4. Rotary AC/2
5. Rotary AC/2
6. *Endo Steel*

Right Torso (Case)

1-3
1. XL Engine
2. XL Engine
3. XL Engine
4. Double Heat Sink
5. Double Heat Sink
6. Double Heat Sink

4-6
1. Ammo (RAC) 45
2. Ammo (RAC) 45
3. Ammo (RAC) 45
4. *Endo Steel*
5. *Endo Steel*
6. *CASE*

Right Leg

1. Hip
2. Upper Leg Actuator
3. Lower Leg Actuator
4. Foot Actuator
5. *Endo Steel*
6. *Endo Steel*

Heat Scale

	Heat Overflow
30	Shutdown
29	
28	Ammo Explosion, avoid on 8+
27	
26	Shutdown, avoid on 10+
25	-5 Movement Points
24	+4 Modifier to Fire
23	Ammo Explosion, avoid on 6+
22	Shutdown, avoid on 8+
21	
20	-4 Movement Points
19	Ammo Explosion, avoid on 4+
18	Shutdown, avoid on 6+
17	+3 Modifier to Fire
16	
15	-3 Movement Points
14	Shutdown, avoid on 4+
13	+2 Modifier to fire
12	
11	
10	-2 Movement Points
09	
08	+1 Modifier to Fire
07	
06	
05	-1 Movement Points
04	
03	
02	
01	
00	

BATTLETECH is a Registered Trademark of FASA Corporation. Copyright 2000. Permission to photocopy for personal use.

BATTLETECH®

INFILTRATOR MK. II ARMOR RECORD FORM

MOVEMENT POINTS: 3 JUMP

PLATOON NUMBER_____

SQUAD NUMBER 1

1	06	05	04	03	02	01	00
2	06	05	04	03	02	01	00
3	06	05	04	03	02	01	00
4	06	05	04	03	02	01	00

MAGSHOT RIFLE: Dmg 2, Range 3/6/9

SQUAD NUMBER 2

1	06	05	04	03	02	01	00
2	06	05	04	03	02	01	00
3	06	05	04	03	02	01	00
4	06	05	04	03	02	01	00

MAGSHOT RIFLE: Dmg 2, Range 3/6/9

SQUAD NUMBER 3

1	06	05	04	03	02	01	00
2	06	05	04	03	02	01	00
3	06	05	04	03	02	01	00
4	06	05	04	03	02	01	00

MAGSHOT RIFLE: Dmg 2, Range 3/6/9

SQUAD NUMBER 4

1	06	05	04	03	02	01	00
2	06	05	04	03	02	01	00
3	06	05	04	03	02	01	00
4	06	05	04	03	02	01	00

MAGSHOT RIFLE: Dmg 2, Range 3/6/9

PLATOON NUMBER_____

SQUAD NUMBER 1

1	06	05	04	03	02	01	00
2	06	05	04	03	02	01	00
3	06	05	04	03	02	01	00
4	06	05	04	03	02	01	00

MAGSHOT RIFLE: Dmg 2, Range 3/6/9

SQUAD NUMBER 2

1	06	05	04	03	02	01	00
2	06	05	04	03	02	01	00
3	06	05	04	03	02	01	00
4	06	05	04	03	02	01	00

MAGSHOT RIFLE: Dmg 2, Range 3/6/9

SQUAD NUMBER 3

1	06	05	04	03	02	01	00
2	06	05	04	03	02	01	00
3	06	05	04	03	02	01	00
4	06	05	04	03	02	01	00

MAGSHOT RIFLE: Dmg 2, Range 3/6/9

SQUAD NUMBER 4

1	06	05	04	03	02	01	00
2	06	05	04	03	02	01	00
3	06	05	04	03	02	01	00
4	06	05	04	03	02	01	00

MAGSHOT RIFLE: Dmg 2, Range 3/6/9

PLATOON NUMBER_____

SQUAD NUMBER 1

1	06	05	04	03	02	01	00
2	06	05	04	03	02	01	00
3	06	05	04	03	02	01	00
4	06	05	04	03	02	01	00

MAGSHOT RIFLE: Dmg 2, Range 3/6/9

SQUAD NUMBER 2

1	06	05	04	03	02	01	00
2	06	05	04	03	02	01	00
3	06	05	04	03	02	01	00
4	06	05	04	03	02	01	00

MAGSHOT RIFLE: Dmg 2, Range 3/6/9

SQUAD NUMBER 3

1	06	05	04	03	02	01	00
2	06	05	04	03	02	01	00
3	06	05	04	03	02	01	00
4	06	05	04	03	02	01	00

MAGSHOT RIFLE: Dmg 2, Range 3/6/9

SQUAD NUMBER 4

1	06	05	04	03	02	01	00
2	06	05	04	03	02	01	00
3	06	05	04	03	02	01	00
4	06	05	04	03	02	01	00

MAGSHOT RIFLE: Dmg 2, Range 3/6/9

PLATOON NUMBER_____

SQUAD NUMBER 1

1	06	05	04	03	02	01	00
2	06	05	04	03	02	01	00
3	06	05	04	03	02	01	00
4	06	05	04	03	02	01	00

MAGSHOT RIFLE: Dmg 2, Range 3/6/9

SQUAD NUMBER 2

1	06	05	04	03	02	01	00
2	06	05	04	03	02	01	00
3	06	05	04	03	02	01	00
4	06	05	04	03	02	01	00

MAGSHOT RIFLE: Dmg 2, Range 3/6/9

SQUAD NUMBER 3

1	06	05	04	03	02	01	00
2	06	05	04	03	02	01	00
3	06	05	04	03	02	01	00
4	06	05	04	03	02	01	00

MAGSHOT RIFLE: Dmg 2, Range 3/6/9

SQUAD NUMBER 4

1	06	05	04	03	02	01	00
2	06	05	04	03	02	01	00
3	06	05	04	03	02	01	00
4	06	05	04	03	02	01	00

MAGSHOT RIFLE: Dmg 2, Range 3/6/9

BATTLE ARMOR ATTACK TABLE

Die Roll (2D6)	Squad Members Active			
	1	2	3	4
2	1	1	1	1
3	1	1	1	2
4	1	1	2	2
5	1	1	2	2
6	1	1	2	2
7	1	2	2	3
8	1	2	2	3
9	1	2	3	3
10	1	2	3	4
11	1	2	3	4
12	1	2	3	4

SWARM HIT LOCATION TABLE

Die Roll (2D6)	Location	Die Roll (2D6)	Location
2	Head	7	Front Center Torso
3	Rear Center Torso	8	Left Arm
4	Rear Right Torso	9	Front Left Torso
5	Front Right Torso	10	Rear Left Torso
6	Right Arm	11	Rear Center Torso
		12	Head

SWARM ATTACKS TABLE

Battle Armored Troopers Active	Base To-Hit Number
4	7
1–3	10
—	No attack possible

LEG ATTACKS TABLE

Battle Armored Troopers Active	Base To-Hit Number
4	4
3	7
2	10
1	12
—	No attack possible

Permission granted to photocopy for personal use. Copyright © 2000 FASA Corporation.

INDEX

INDEX

BATTLETECH®
NOVELS

STRIDE THE BATTLEFIELDS OF A THOUSAND WORLDS AND LEARN THE PRICE OF HONOR AND AMBITION.

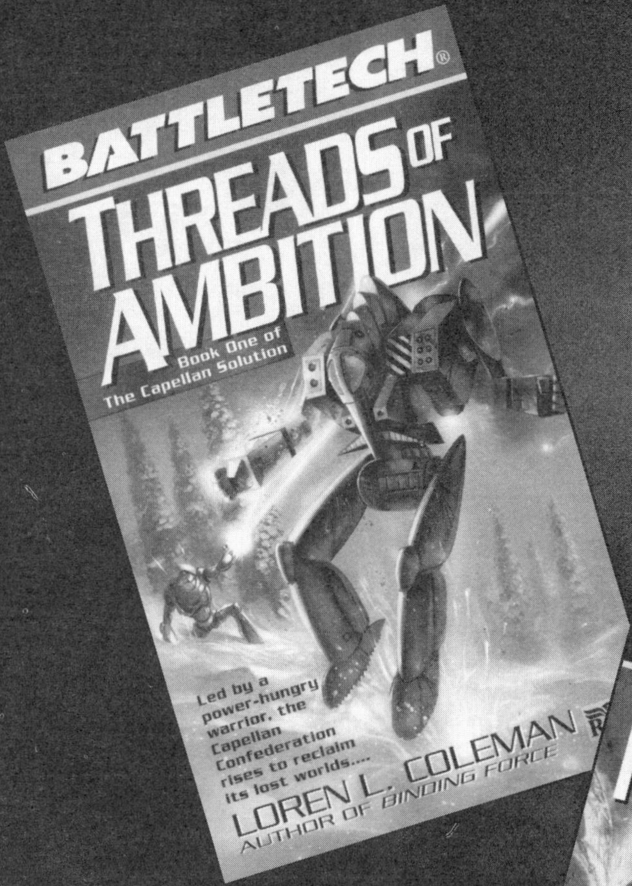

AVAILABLE FROM BETTER GAME, COMIC AND BOOK STORES. TO FIND THE STORE NEAREST YOU USE OUR RETAILER LOCATOR AT WWW.FASA.COM

BATTLETECH IS A REGISTERED TRADEMARK OF FASA CORP. COPYRIGHT 2000 FASA CORP. ALL RIGHTS RESERVED

THE STAR LEAGUE
WANTS YOU
FASA
CORPORATION
Do you know BattleTech® history better than the history of your own country?
Have you taught all your friends to play BattleTech?
Would you like to get the inside scoop on what's going on with the game?
Then we want you in the Marauders, FASA's official demo group.
If you're a fan of BattleTech and want to become part of this elite team, check out our website at www.fasa.com for complete details on how to enlist.
Battletech® is a registered trademark of FASA Corporation. Copyright © 2000 FASA Corporation.